SCIENTIFIC
AMERICAN

How
THINGS
WORK
TODAY

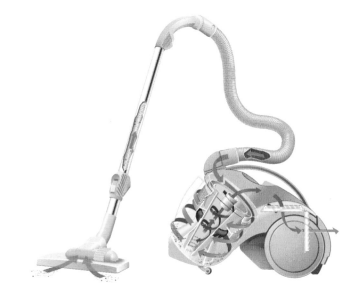

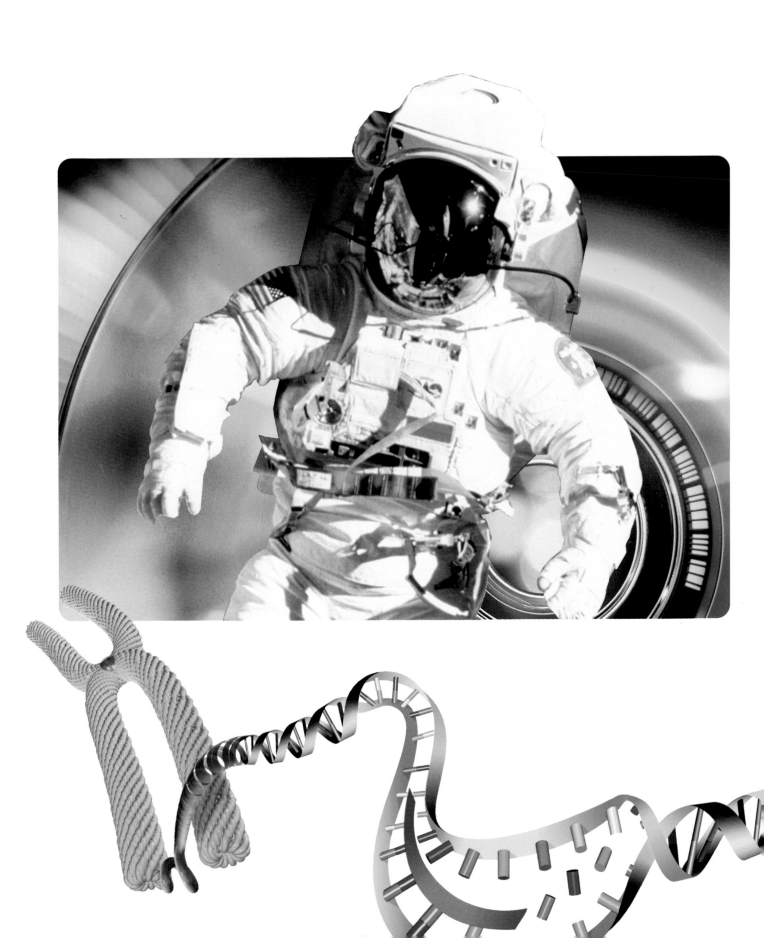

SCIENTIFIC
AMERICAN

How
THINGS
WORK
TODAY

Edited by Michael Wright and M. N. Patel

CROWN

Published by Crown Publishers, New York, New York. Member of the Crown Publishing Group.

Random House, Inc. New York, Toronto, London, Sydney, Auckland
www.randomhouse.com

CROWN is a trademark and the Crown colophon is a registered trademark of Random House, Inc.

Project Editor	Peter Adams	**Contributors**	David Baker
Editors	Robert Dinwiddie		Robert Dinwiddie
	Dan Green		Jolyon Goddard
	Giles Sparrow		Dan Green
Consultant Editors	Mukul Patel		Mukul Patel
	Michael Wright		Giles Sparrow
Editorial Researchers	Thea Hicks		Matt White
	Dan Green		Chris Woodford
			Michael Wright
Senior Art Editor	Thomas Keenes		
Art Editors	Iain Stuart	**Illustrators**	David Ashby
	Anne Fisher		Peter Bull
	Fehmi Cömert		Roy Flooks
Designer	Chuck Wunderman		Kuo Kang Chen
Design Assistant	Mike Grigoletti		Dr Rajeev Doshi
			Mike Fuller
Managing Editor	Paul Docherty		Alan Hancocks
Managing Art Editor	Patrick Carpenter		Tim Loughhead
Editorial Director	Ellen Dupont		Patrick Mulrey
Art Director	Dave Goodman		Ian Naylor
			Graham Parrish
Editorial Coordinator	Ros Highstead		Mike Richardson
Picture Research	Elaine Willis		Ed Stuart
	Zilda Tandy		Kevin Jones Associates
Editorial Assistant	Emily Salter		
Production	Nikki Ingram		
	Anna Pauletti		
	Chris Hawkins		

Library in Congress Cataloging-in-Publication Data
Scientific American: how things work today / edited by Michael Wright and M.N. Patel.
   p. cm.
   ISBN 0-375-41023-6
   1. Technology—Popular works. I. Title: How things work today. II. Wright, Michael, 1941- III. Patel, M.N. (Mukul N.)
   T47 .S43 2000
   600—dc21   00-059027

10 9 8 7 6 5 4 3 2

First Edition

Originated in Italy by **Articolor**.   Printed by **Printer Portuguesa**.

Contents

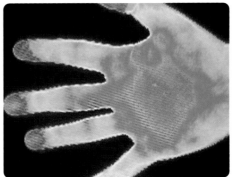
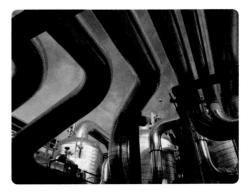

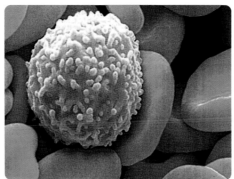
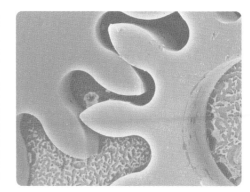

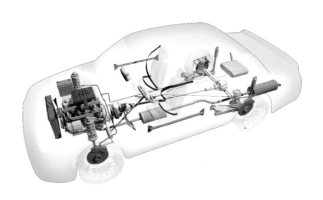

Introduction

All of us, from the cyber kid surfing the Internet, to the technophobe who dreads programming the VCR, are increasingly reliant on technology. Machines and systems pervade every aspect of our lives–the mechanical locks that secure our homes, the encryption software that safeguards financial transactions on computers, the microwave ovens that cook our meals, and the radio transmissions that convey cell phone calls. Many of the devices that we use incorporate layers of different technologies, making it difficult to understand just how these things work.

During the last century, people witnessed increasing mechanization, electrification, miniaturization, and mass production. Everyday lives were transformed by transport and communications technologies, such as the automobile, airliners, radio, telephones, communications satellites, and television. Equally revolutionary was the development of materials such as plastics and composites and the birth of digital computers. Perhaps the greatest single achievement was the landing of men on the Moon.

In contrast to the grand scale of many 20th-century advances, the technological frontier at the dawn of the 21st century is at the microscopic scale. Increasing miniaturization in the computer industry has led to the development of laptop computers that are thousands of times more powerful (and much less expensive) than machines that occupied whole rooms 50 years ago. The same technology that is used to imprint millions of transistors on a quarter-inch square of silicon is now being used to make micromachines, assemblages of tiny mechanical devices. Imminent breakthroughs in this and other applications of nanotechnology–engineering at the scale of several nanometers (billionths of a meter)–will enable faster computers, microscopically-engineered drugs, and miniature bionic implants.

The digital revolution is transforming education, commerce, entertainment, and medicine, and will have even greater impact as computers become more "intelligent." Yet for much of the time, this data and intelligence exists merely as ephemeral pulses of light or electric current. Equally intangible are the advances produced by the ongoing revolution in biotechnology. The new science of genetic engineering, which involves the manipulation of an organism's genes (units of heredity) to alter its physiological characteristics, is a case in point. Genetic engineering can improve the quality and quantity of agricultural output, although its greatest potential may lie in the growing range of applications in the prevention and treatment of disease. *The Way Things Work Today* demystifies these "hidden" developments using the most up-to-date information available, clear, accessible text, and a range of cross-sectional, exploded, and three-dimensional computer-generated illustrations.

Using the book

The book is arranged in thematic chapters, but extensive cross-referencing emphasizes the multiple applications of various technologies. For example, most TV sets are based on the cathode-ray tube, in which a beam of high-speed electrons is directed towards a target (the screen) and similar technology can be found in the giant particle accelerators used by research physicists. A Principles section provides an introduction to the basic science that underlies modern technology—Matter, Electricity and Magnetism, Light, Electronics, and Mechanics. Concise definitions of common scientific and technological terms are provided in a glossary. And at the end of the book, a list of World Wide Web addresses provides gateways to much more information.

The Urban and Domestic Environment

Technology surrounds us in our everyday lives. The modern city and the comfortable home are packed with devices that we all too often take for granted. How many of us know how a microwave oven works, or give a thought to the maze of service pipes and tunnels that riddle the ground beneath our feet? Yet without these devices, our lives would be far less comfortable and more difficult. This chapter explores how the infrastructure and essential services of cities have been reshaped by technological innovations and reveals the ingenuity that has created and improved the domestic appliances we all rely on.

Skyscrapers

The manufacture of stronger, lighter steel combined with the development of elevators that could carry people safely made possible the first high-rise buildings in the 1880s. By the year 2000, improvements in technology and design brought skyscrapers to almost 1500 ft (460 m) in height, with the promise of even taller megastructures using superframes in the near future.

A skyscraper needs to be strong, stable, and able to withstand natural forces, such as wind and earth tremors. The weight of the building is carried by the foundations, which are set firmly in the soil or rock bed. Types of foundations include spread footings, which spread the weight of the building over a wide area, and caisson piles, which transmit the load to stable bedrock.

Above ground, a skyscraper is constructed around a frame made of steel or concrete, which bears the load of walls, roof, and floors, and so supports the building. The design of skyscrapers has evolved to make them taller or lighter or more specifically geared to the conditions in the local environment. The steel frames of the early skyscrapers in Chicago and New York were followed in cities throughout the world by frames attached by cables to a central concrete core. Many modern skyscrapers, such as the Hong Kong Bank and the World Trade Center, are built around the steel-framed tube design, in which the load is borne by steel columns.

Transamerica Pyramid
The 215-ft (65-m) Transamerica Pyramid, built in San Francisco, is designed to withstand earthquakes. During the 1989 Loma Prieta earthquake, which measured 7.1 on the Richter scale, the building swung by more than one foot (30 cm) but sustained no damage. The two windowless abutments rising from the 29th floor house the elevators and stairwells.

Frames

Skyscraper frames provide the main structural support for the building and are composed of materials that are designed to withstand the force of the wind. Although there are many variations, most frames fall into two categories—skeleton and cable hung. Skeleton skyscrapers, which include steel-framed tube designs, use a structural frame to support the walls and floors. In a cable-hung skyscraper the main support is provided by a central concrete core, which acts as a wind brace and as a shaft for elevators and other utilities.

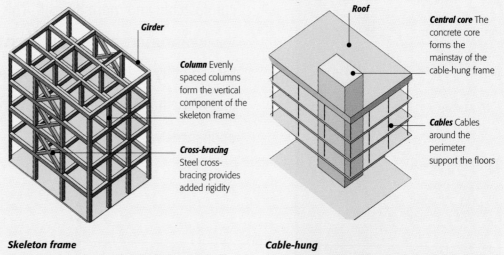

Girder

Column Evenly spaced columns form the vertical component of the skeleton frame

Cross-bracing Steel cross-bracing provides added rigidity

Roof

Central core The concrete core forms the mainstay of the cable-hung frame

Cables Cables around the perimeter support the floors

Skeleton frame
A compact lattice of columns and girders, often supported by cross-bracing, spreads the load of the overall structure throughout the building.

Cable-hung
The floors of each story are supported by the central core and by cables that are attached to the roof of the frame at the top of the structure.

The Hong Kong Bank

Eight masts—composed of four tubular steel columns—provide the support for this 44-story structure. The centerpiece of the bank is the atrium, comprising stories 3 through 12. The galleries on each of these levels is illuminated in the daytime by sunlight collected by a sunscoop fitted to the outside of the building.

Viewing gallery Panoramic views of Hong Kong can be seen from the gallery at the top of the building

Maintenance crane Making repairs to the bank's exterior and cleaning the windows are carried out from the building's maintenance cranes

Suspension truss The coat-hanger shaped trusses support the weight of the floors that are suspended from them and provide double-height floors

Mast Each mast has four columns that are braced by rectangular beams, called vierendeels. The foundations supporting each column rest upon bedrock 100 ft (30 m) underground

Hanger Each floor is attached to steel tubes called hangers that are suspended from the center and ends of the suspension trusses located above it

Cross-brace Adjoining masts are supported by cross-braces

Sunscoop The sunscoop has 480 glass mirrors which reflect sunlight onto the aluminum reflectors at the top of the atrium

Atrium The centerpiece of the building, the atrium is a wide, sunlit space in the center of the building surrounded by galleries that look over its glazed underbelly

Giant cross-brace This cross-brace is the focal point of the atrium

Main banking hall

Stairwell

Service module The services for each floor are in self-contained, prefabricated modules that are "bolted" to the sides of the building

Glazed underbelly of atrium

Escalator Two long-span escalators carry people from the plaza to the atrium, passing through its glazed underbelly

Plaza The plaza is a large, open space under the atrium. Retractable glazed screens offer protection from Hong Kong's typooons

Basement banking hall

SEE ALSO: BENEATH THE STREETS p.14 | ELEVATORS & ESCALATORS p.16 | STEEL & CONCRETE p.24 | LIGHTING p.28

Beneath the streets

From the steam heating pipes under Manhattan to the unmanned mail trains of London, the systems that keep cities working go largely unnoticed unless they fail. Beneath city streets are labyrinthine networks of pipes, cables, ducts, and tunnels that supply the city with water, electricity, and gas, transport its people, remove their waste, and connect them with the rest of the world.

Beneath the streets of any modern city is a tortuous maze of hidden service systems, connecting every building on every block. Such intricate networks are a relatively modern phenomenon. During the industrial revolution of the 19th century, large numbers of people flooding into urban areas put enormous strain on the service systems already in place. Substandard water supply and sewage systems gave rise to cholera epidemics that killed millions of people. In the interests of public health, treated drinking water was supplied to cities in dedicated pipes. Sewers were closed over and kept apart from clean water. During the 20th century, telecommunications, electricity, gas, and underground transportation networks have revolutionized modern city life and have increased the complexity of the networks below street level. These service systems are managed by dedicated bodies responsible for the maintenance and improvement of each system.

Population continues to grow in cities, demanding ever more reliable clean water supplies. This demand has led to ambitious engineering projects such as New York City's Water Tunnel No.3, a 6-mile (10-km) tunnel bored through bedrock at 250–800 ft (75–240 m) below ground, which was inaugurated in 1998.

Skyscraper foundations
Caisson piers sunk down to bedrock support the weight of skyscrapers

Water distribution hub
Clean water is fed to water mains pipes

Water uptake shaft
Vertical shafts draw water from deep water pipes to water distribution hubs

200 FEET (60 METERS)

400 FEET (120 METERS)

Pigs

The term "pig" once referred to a bundle of rags or leather sent down a pipe to clean it. Pigs are now sophisticated pipeline tools, specifically adapted for particular tasks, and used by service companies to maintain the soundness of their pipelines. Using a range of onboard technologies, pigs can detect leaks, scan for defects in a pipe, make 3-D maps of a pipeline, or simply clean the pipe and clear obstructions.

Scoutscan
This pig enables companies to make accurate 3-D maps of their pipelines. It uses onboard accelerometers and a gyroscope to measure its position. This data is correlated with control units outside the pipe that measure time and position using GPS satellites.

Deep water pipe Concrete pipes, typically 25 ft (7.5 m) in diameter, carry clean water to the heart of the city

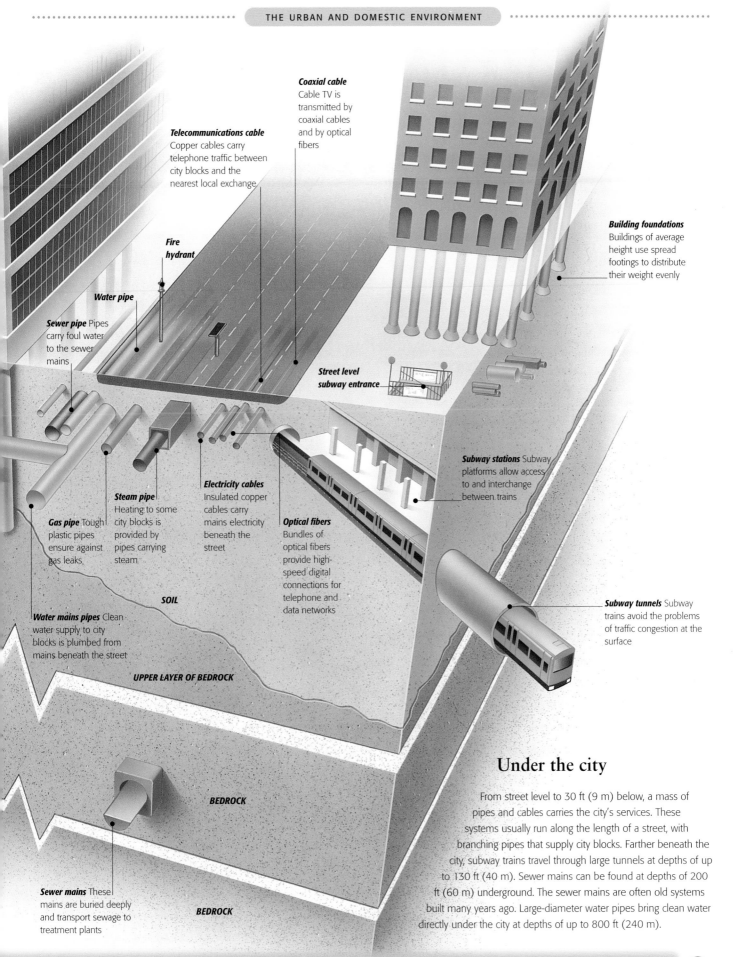

Coaxial cable
Cable TV is transmitted by coaxial cables and by optical fibers

Telecommunications cable
Copper cables carry telephone traffic between city blocks and the nearest local exchange

Fire hydrant

Water pipe

Sewer pipe Pipes carry foul water to the sewer mains

Building foundations
Buildings of average height use spread footings to distribute their weight evenly

Street level subway entrance

Gas pipe Tough plastic pipes ensure against gas leaks

Steam pipe
Heating to some city blocks is provided by pipes carrying steam

Electricity cables
Insulated copper cables carry mains electricity beneath the street

Optical fibers
Bundles of optical fibers provide high-speed digital connections for telephone and data networks

Subway stations Subway platforms allow access to and interchange between trains

Subway tunnels Subway trains avoid the problems of traffic congestion at the surface

SOIL

Water mains pipes Clean water supply to city blocks is plumbed from mains beneath the street

UPPER LAYER OF BEDROCK

BEDROCK

Sewer mains These mains are buried deeply and transport sewage to treatment plants

BEDROCK

Under the city

From street level to 30 ft (9 m) below, a mass of pipes and cables carries the city's services. These systems usually run along the length of a street, with branching pipes that supply city blocks. Farther beneath the city, subway trains travel through large tunnels at depths of up to 130 ft (40 m). Sewer mains can be found at depths of 200 ft (60 m) underground. The sewer mains are often old systems built many years ago. Large-diameter water pipes bring clean water directly under the city at depths of up to 800 ft (240 m).

SEE ALSO: SKYSCRAPERS *p.12* | SUBWAY STATIONS *p.18* | THE WATER SYSTEM *p.30* | CABLE TECHNOLOGY *p.46*

Elevators and escalators

The first passenger elevator was installed in the Haughwout Department Store in New York in 1857. It was designed by Elisha G. Otis and was driven by steam power. Improvements in elevator technology gave architects the opportunity to design increasingly tall buildings, contributing to the rise of the skyscraper. Escalators, also developed during the late 19th century, allowed better access to airports, subway stations, and department stores.

Over two million elevators carry people and freight up and down buildings worldwide, a fifth of them are located in North America alone where 350 million people ride on elevators each day.

Nearly all elevators derive power from two kinds of electric traction. Gearless traction elevators can reach speeds of up to 2,000 ft (600 m) per minute and are generally used in offices of more than 10 stories and in apartment buildings of more than 30 stories. Geared traction elevators, equipped with a reduction gear that slows the sheave, or pulley, move at a quarter that speed. Elevators that are hydraulically driven travel at about 100 ft (30 m) per minute and are used as industrial freight elevators or in low-rise office or apartment buildings of six stories or less.

An escalator is a people conveyor in which drive chains, powered by a high-speed electric motor, carry a series of metal steps around in an inclined loop. The direction in which these staircases move can be reversed to carry people upward or downward. The first escalators were installed in New York City and Paris, France, in 1900. They resembled what we would now call a travelator—a smooth incline with no folding steps. Handrails were added later, and the modern escalator appeared in 1921. Escalators typically run at a speed of 150 ft (45 m) per minute and can transport as many as 10,000 people an hour.

How an escalator works

Escalators have three major structural components: the landings, the truss, and the tracks. The landings are flush with the floor at the top and bottom of an escalator. Hinged plates allow access to the machinery that operates the escalator, such as the drive machinery under the top landing. Comb plates mesh with the grooves on the tread of each step. The truss (not shown) is the structure that bridges the gap between the floors and holds the tracks. The tracks guide the step chain as it moves in its loop. Hidden rollers under each step run on these tracks, which are positioned so that the steps fold out as they travel up or down the front of the escalator, and fold flat when they travel underneath it.

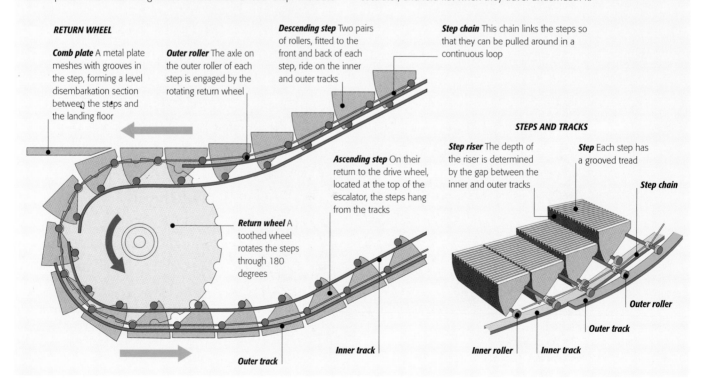

RETURN WHEEL

Comb plate A metal plate meshes with grooves in the step, forming a level disembarkation section between the steps and the landing floor

Outer roller The axle on the outer roller of each step is engaged by the rotating return wheel

Descending step Two pairs of rollers, fitted to the front and back of each step, ride on the inner and outer tracks

Step chain This chain links the steps so that they can be pulled around in a continuous loop

Ascending step On their return to the drive wheel, located at the top of the escalator, the steps hang from the tracks

Return wheel A toothed wheel rotates the steps through 180 degrees

Outer track

Inner track

STEPS AND TRACKS

Step riser The depth of the riser is determined by the gap between the inner and outer tracks

Step Each step has a grooved tread

Step chain

Outer roller

Outer track

Inner roller   **Inner track**

Elevator

Elevators transport passengers vertically in fireproof cars. Before the elevator can move, both sets of doors, on the car and on the floor, must properly close. The drive sheave on the hoist machine then engages and raises or lowers the car to the desired floor. A concrete counterbalance, roughly equal in weight to the empty car plus 40 percent of the car's maximum load, counterbalances the weight of the car, effectively reducing the load that the motor has to lift or lower. The illustration shows a gearless traction elevator with the drive sheave attached directly to the hoist cables. This system is used to propel express elevators and double-deck elevators in high-rise buildings such as the Empire State Building in New York.

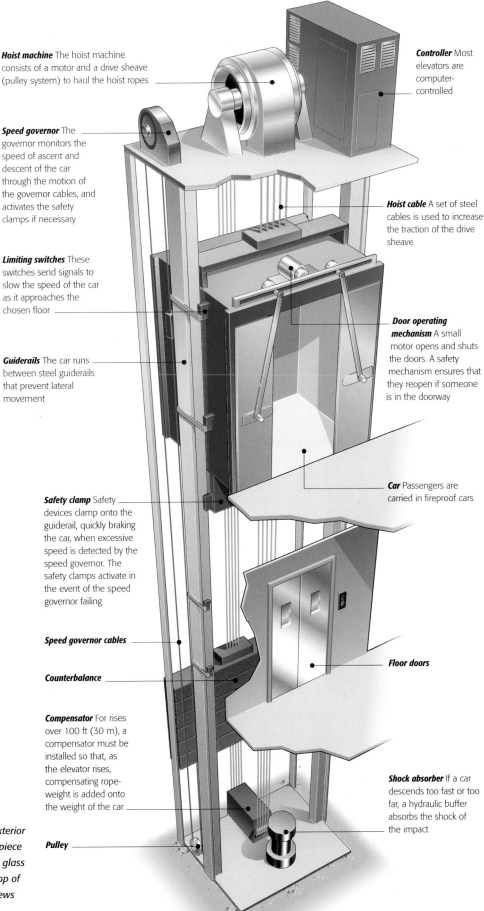

Hoist machine The hoist machine consists of a motor and a drive sheave (pulley system) to haul the hoist ropes

Speed governor The governor monitors the speed of ascent and descent of the car through the motion of the governor cables, and activates the safety clamps if necessary

Limiting switches These switches send signals to slow the speed of the car as it approaches the chosen floor

Guiderails The car runs between steel guiderails that prevent lateral movement

Safety clamp Safety devices clamp onto the guiderail, quickly braking the car, when excessive speed is detected by the speed governor. The safety clamps activate in the event of the speed governor failing

Speed governor cables

Counterbalance

Compensator For rises over 100 ft (30 m), a compensator must be installed so that, as the elevator rises, compensating rope-weight is added onto the weight of the car

Pulley

Controller Most elevators are computer-controlled

Hoist cable A set of steel cables is used to increase the traction of the drive sheave

Door operating mechanism A small motor opens and shuts the doors. A safety mechanism ensures that they reopen if someone is in the doorway

Car Passengers are carried in fireproof cars

Floor doors

Shock absorber If a car descends too fast or too far, a hydraulic buffer absorbs the shock of the impact

La Grande Arche, Paris
This innovative elevator is attached to the exterior of La Grande Arche, the architectural centerpiece of Paris's La Défense. Passengers travel in a glass capsule that takes seconds to travel to the top of La Grande Arche and affords spectacular views over the whole of Paris.

Subway stations

Subways were merely engineering novelties at the start of the 20th century, but today these compact railroads that run underground through congested city centers are the most efficient urban transportation systems. The New York City subway, which opened with 28 stations in 1904, now has 468 stations and carries more than a billion passengers each year. The world's busiest subway network, in Moscow, carries more than three times as many people.

Subways and the stations that serve them present major challenges for architects and engineers since many of the stations and much of the track must be constructed underground without disrupting life in the city streets above. This is typically done by sinking deep vertical shafts into the ground and excavating horizontally beneath existing buildings. Shallow tunnels are constructed by the "cut-and-cover" method, which involves digging an open trench, reinforcing the sides, then adding a roof on top.

To speed up transit times through large subway stations, automatic ticket dispensers and barriers are installed on many networks, and in deep stations heavy-duty escalators, sometimes stacked in several flights, ferry people between the trains and the street.

Subway station

This illustration of a typical deep-level subway station incorporates many of the key features that allow passengers to move swiftly and safely between the trains and the streets above. The ticket hall incorporates an automated vending system and electronic barriers to control access to the platforms. The layout of the station allows disembarking passengers to be routed up one set of escalators, leaving the other set free for embarking passengers.

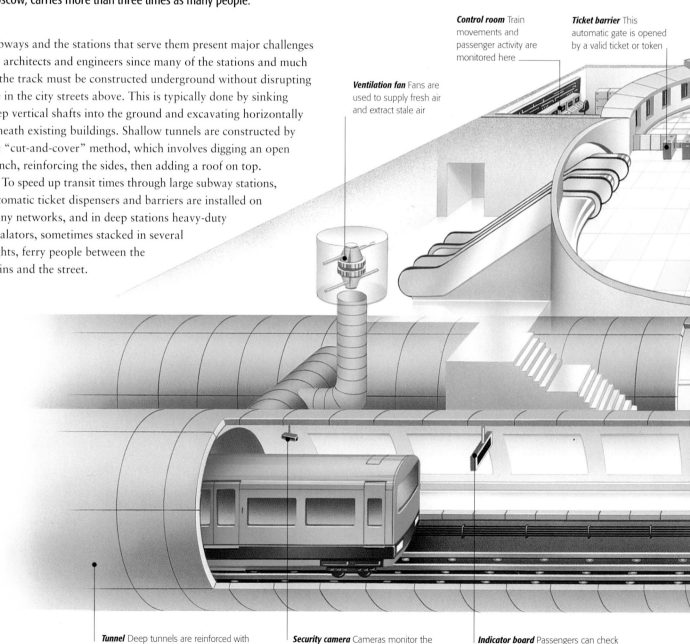

Control room Train movements and passenger activity are monitored here

Ticket barrier This automatic gate is opened by a valid ticket or token

Ventilation fan Fans are used to supply fresh air and extract stale air

Tunnel Deep tunnels are reinforced with segments of cast iron

Security camera Cameras monitor the activity of passengers on the platform

Indicator board Passengers can check train times on electronic display boards

Platform-edge doors

Some modern subway stations have a toughened glass safety barrier between the platform edge and the track that opens only when a train stops at the station. A signal from an automatic control system inside the train triggers a sensor on the platform, causing the platform-edge doors to open at the same time as the doors of the train. In addition to increasing safety, platform-edge doors reduce noise, vibration, and dust.

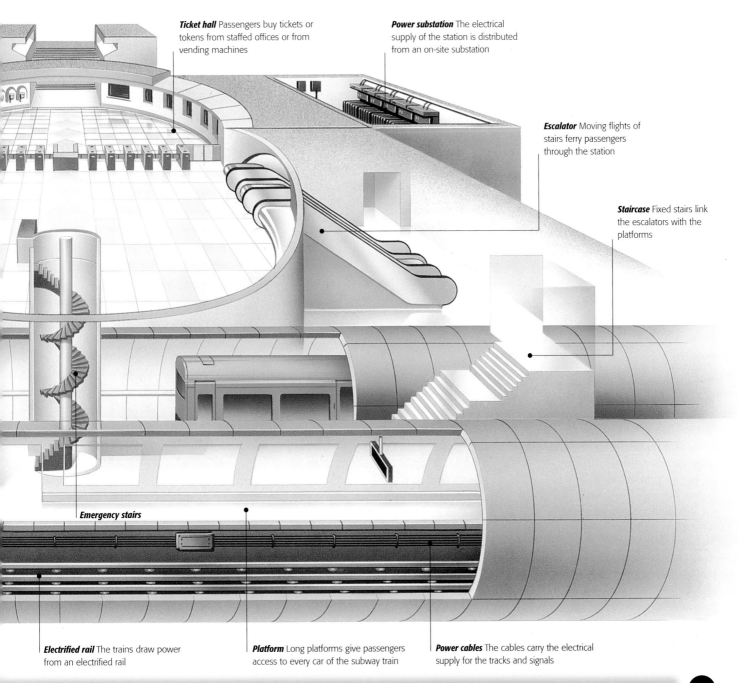

Ticket hall Passengers buy tickets or tokens from staffed offices or from vending machines

Power substation The electrical supply of the station is distributed from an on-site substation

Escalator Moving flights of stairs ferry passengers through the station

Staircase Fixed stairs link the escalators with the platforms

Emergency stairs

Electrified rail The trains draw power from an electrified rail

Platform Long platforms give passengers access to every car of the subway train

Power cables The cables carry the electrical supply for the tracks and signals

SEE ALSO: BENEATH THE STREET *p.14* | ELEVATORS & ESCALATORS *p.16* | TUNNELS *p.20*

Tunnels

Tunnels provide direct underground routes that bypass obstacles encountered on the surface. Many cities rely on an elaborate network of tunnels to deliver water and dispose of sewage and for railroad and road links underneath congested streets. Tunnels also provide underground passages through mountains and under bodies of water such as Chesapeake Bay.

There are two main methods of tunneling, depending on the depth of tunnel. Shallow tunnels are dug by the "cut-and-cover" method, which involves excavating a trench and adding either reinforced sides, a roof, and a floor or a prefabricated pipe, then covering the whole structure over with the original excavated material. Deep tunnels in hard rock must be bored by drilling, then blasting out the rock with explosives. Boring carries several hazards: the tunnel may collapse before the roof has been reinforced; water from aquifers (natural underground water reservoirs) may cause flooding; and there are other risks caused by debris and trapped gas when using explosives underground. Today, gigantic tunnel-boring machines (TBMs), or moles, are used to dig through soil and soft rock, making tunneling both quicker and safer. These machines were extensively used during the construction of the Channel Tunnel, which provides a railroad link between France and England.

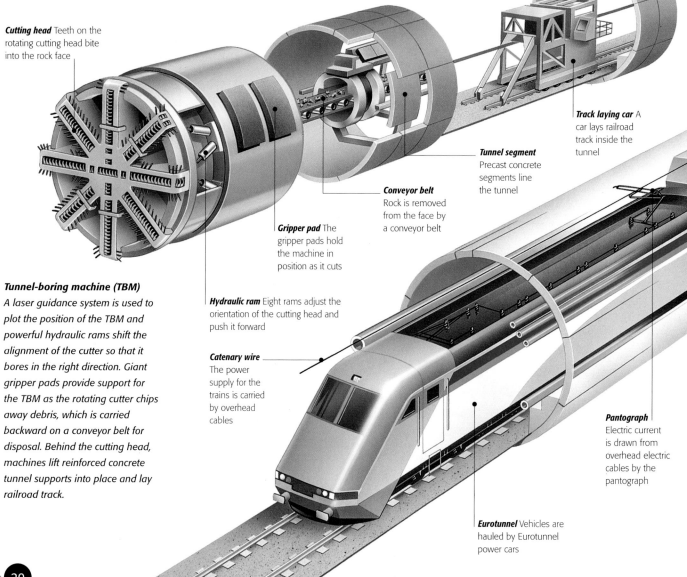

Cutting head Teeth on the rotating cutting head bite into the rock face

Tunnel-boring machine (TBM)
A laser guidance system is used to plot the position of the TBM and powerful hydraulic rams shift the alignment of the cutter so that it bores in the right direction. Giant gripper pads provide support for the TBM as the rotating cutter chips away debris, which is carried backward on a conveyor belt for disposal. Behind the cutting head, machines lift reinforced concrete tunnel supports into place and lay railroad track.

Gripper pad The gripper pads hold the machine in position as it cuts

Hydraulic ram Eight rams adjust the orientation of the cutting head and push it forward

Catenary wire The power supply for the trains is carried by overhead cables

Conveyor belt Rock is removed from the face by a conveyor belt

Tunnel segment Precast concrete segments line the tunnel

Track laying car A car lays railroad track inside the tunnel

Pantograph Electric current is drawn from overhead electric cables by the pantograph

Eurotunnel Vehicles are hauled by Eurotunnel power cars

Constructing the Channel Tunnel
A total of 11 tunnel-boring machines and nearly 13,000 workers were employed in the construction of the Channel Tunnel. The cutting heads of the TBMs used to bore the railroad tunnels were nearly 30 ft (9 m) in diameter. Rotating two to three times a minute, they advanced at rates of up to 245 ft (75 m) per day.

Ventilation shaft More than 2,470 cubic feet (70 cubic meters) of air are pumped from the surface into the service tunnel each second via two ventilation shafts located onshore

Piston relief duct These ducts, which are fitted with fireproof doors, relieve pressure waves from fast-moving trains

Cross passage Air is circulated and access provided by cross passages

Car transporter Automobiles are transported in two-deck cars

Laser source Computerized laser guidance is used to keep the TBM on the right course

Control and communications cables

Drainage pipe

Freight vehicle A separate freight service transports trucks using Eurotunnel power cars

Service tunnel This tunnel provides maintenance access and an emergency exit for the railroad tunnels

Main lighting cable

Eurostar The passenger service uses electric traction Eurostar trains

Service vehicle These vehicles are self-steering and follow wires buried in the floor

Water pipe Water is pumped along large pipes to cool the tunnel

The Channel Tunnel

The 30-mile (50-km) long Channel Tunnel that links England and France runs at a depth of approximately 150 ft (45 m) below the sea floor. It consists of two parallel railroad tunnels and a smaller service tunnel, linked by cross tunnels. Construction began in December 1987 and tunnel-boring machines, which worked simultaneously outward from Folkestone in England and Calais in France, completed the service tunnel in December 1990 and the last of the railroad tunnels in June 1991.

SEE ALSO: BENEATH THE STREETS *p.14* | STEEL & CONCRETE *p.24* | HIGH-SPEED TRAINS *p.118* | LASERS *p.220*

Bridges

From an engineering viewpoint, bridges are designed not only to span a physical obstacle, but also to distribute loads evenly through a few key supports using tension (pulling forces) and compression (pushing forces). By employing beams, arches, and cables in specific ways, engineers seek to channel loads to abutments (supports at either end of a bridge) or piers (supports in the middle).

The design of a bridge depends on its length, what it will carry, the nature of the obstacle it will cross, and the firmness of the ground beneath it. The load that a bridge supports includes the weight of the bridge itself and of traffic, and forces caused by wind, water currents, and expansion and contraction of the bridge due to temperature changes. Today, most bridges are constructed from a combination of steel and concrete. Steel can withstand high tension, while concrete, which is relatively inexpensive, can withstand high compression. The technique of prestressing concrete by casting it around steel wires held under high tension has enabled longer bridge decks to be used, the longest of which are supported by steel cables in suspension bridges. One of the first modern suspension bridges, New York's Brooklyn Bridge, built in 1883, was considered an engineering masterpiece for its span of 1,595 ft (485 m). The world's longest suspension bridge, the Akashi Kaikyo in Japan, completed in 1998, is nearly 2½ miles (4 km) long.

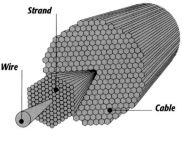

Strand

Wire

Cable

Tower The main towers have diagonal bracing to reduce the effect of strong winds

Detail of suspension cable
The two main cables of the Akashi Kaikyo Bridge are composed of 290 strands, each containing 127 separate wires made of high-tensile steel.

Cable

Hanger

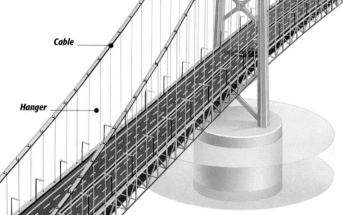

Akashi Kaikyo Bridge

The 100,000-ton (91,000-tonne) weight of this bridge is supported by two 985-ft (300-m) high towers and 186,500 miles (300,000 km) of cable. It is constructed to withstand winds of 180 mph (290 km/h) and earthquakes up to 8.5 on the Richter scale.

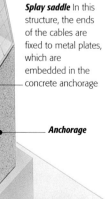

Splay saddle In this structure, the ends of the cables are fixed to metal plates, which are embedded in the concrete anchorage

Anchorage

Positioning a caisson
The caissons, which are underwater foundations sunk into solid bedrock, were prefabricated in a factory and towed into position in the Akashi Strait by tugs. Here they were submerged to a depth of 200 ft (60 m) and filled with concrete.

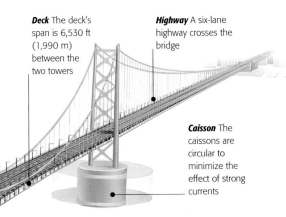

Deck The deck's span is 6,530 ft (1,990 m) between the two towers

Highway A six-lane highway crosses the bridge

Caisson The caissons are circular to minimize the effect of strong currents

Cable-stayed bridge
The central section of the Sunshine Skyway, which runs for about 15 miles (25 km) across Tampa Bay in Florida, is spanned by a cable-stayed bridge. Supporting cables run diagonally from the towers to the deck of the bridge and therefore do not need to be anchored onshore.

Types of bridges

Different bridges distribute loads in a variety of ways and are suited to particular environments. Simple beam bridges can span only limited distances without extra support. Arch bridges are stronger than beam bridges and are used when supporting piers cannot be constructed easily; the arch may be built either above or below the bridge deck. Cantilever bridges consist of beams or metal frames (cantilevers) that are balanced on either side of supporting piers and are often linked by a central beam. The piers are built in shallow water, but the central beam may span a deep-water channel. Suspension bridges and cable-stayed bridges are more visually attractive than cantilevers and can span much longer distances. Other types of bridge include the permanently floating pontoon bridge, which can carry light traffic over calm water.

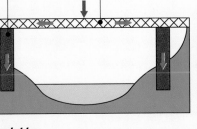

Beam bridge
Because the load on a beam bridge is carried by the beam and transferred to piers, the beams are prone to cracking if the span is too long.

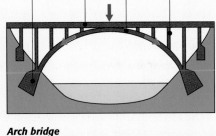

Arch bridge
Resting on columns, the deck of an arch bridge is supported by the arch, which transfers the load downward toward the abutments.

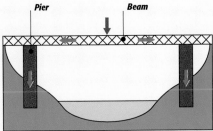

Cantilever bridge
A framework of metal trusses supports the deck of a cantilever bridge and helps to transfer the load to firmly anchored piers.

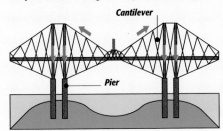

Suspension bridge
The deck is attached by hangers to thick cables anchored to the bank. The cables transfer the load to supporting towers that hold them aloft.

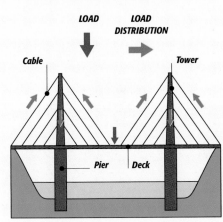

Cable-stayed bridge
The cables on a cable-stayed bridge support the deck of the bridge directly and transfer the load on the deck to the towers and the piers.

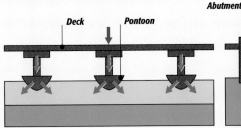

Pontoon bridge
The load on a pontoon bridge pushes down on floating pontoons, which provide less support than piers with a solid foundation.

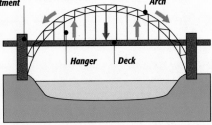

Bowstring arch bridge
The arch of a bowstring arch bridge is attached to the deck by hangers and distributes the load diagonally to abutments on either bank.

SEE ALSO: STEEL & CONCRETE *p.24* | COMPOSITE MATERIALS *p.144* | MECHANICS *p.266*

Steel and concrete

Steel and concrete are the most widespread structural materials in the world today, and both can be traced back through thousands of years of history. Although steel has been mass-produced only since the middle of the 19th century, iron, its major constituent, has been found in tools dating back to 3000 BCE. Concrete was discovered in the floor of a Stone Age Central European village built about 5600 BCE, and was widely used in Roman bridges and aqueducts from 200 BCE onward. But the secret of concrete was later forgotten and only rediscovered in the 18th century.

What is steel?

Although steel's main constituent is iron, it is an alloy containing other elements that give it its unique strength. Most steels are general-purpose carbon steel alloys, which contain up to two percent carbon. The other major types of steel are: extremely tough tool steels (containing tungsten and molybdenum for added strength), stainless steels (containing around 12 percent chromium to prevent rusting and to give them a shiny decorative appearance), and specialized alloy steels (with other elements added to give them properties for specific jobs). All steels are fundamentally a mixture of ferrite (ordinary, soft iron) with the alloys cementite (iron carbide, a hard and brittle iron-carbon alloy), and pearlite (a strong iron-carbon alloy with properties somewhere between ferrite and cementite). Steel can be made harder by increasing the amount of pearlite and cementite relative to ferrite (which increases the carbon content), but this also makes the steel more brittle. Steel can be made much harder by heating it to about

Cement crystal
Cement hardens when its crystals grow and interlock with one another in the presence of water. This electron-microscope photograph shows an area about a thousandth of an inch (25 micrometers) wide.

14,500°F (8,000°C) and then quenching it (cooling it rapidly in water), to produce a hard and brittle steel called martensite, which is then toughened by tempering (raising and holding the temperature for an extended period).

What is concrete?

Concrete is a type of artificial stone made by mixing a dry bulk material called an aggregate (usually sand and gravel) with

---

Reinforced concrete

Concrete is good at bearing compressive stress (squeezing forces) but not tensile stress (stretching forces), which may cause it to crack. To overcome this problem, concrete can be reinforced by adding materials with high tensile strength, such as steel or glass fibers, to the mix or by casting it around a steel bar. Compressing the concrete inside a metal or timber frame as it sets strengthens it further. "Prestressed" concrete, which is even stonger, is made by molding wet concrete around stretched steel cables held under tension. As the concrete hardens, the tension in the cables is released, and they strengthen the concrete by both compressing and reinforcing it. Prestressed concrete requires less steel than ordinary reinforced concrete and so is lighter.

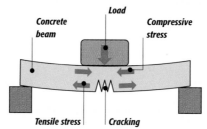

Ordinary concrete
When a load is placed on an ordinary concrete beam that is supported at the ends, it causes the beam to bend. This results in compressive forces on the upper part of the beam, and tensile forces on the lower part. Concrete is weak under tension, and the bottom of the beam cracks.

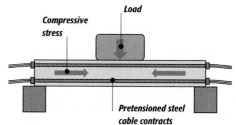

Prestressed concrete
A similar load does not bend or crack a prestressed concrete beam because the pretensioned steel cables set inside the concrete contract, compressing the concrete beam. This squeezing force counteracts the tensile stress and strengthens the beam.

cement (a substance made by firing lime-stone, clay, and gypsum). When water is added to this mixture large crystals grow, interlocking together and binding the aggregate into a solid mass.

Different types of concrete can be made by varying the basic ingredients—stronger concrete can be made by increasing the cement content and reducing the aggregate and water. Concrete also gets stronger as it gets older—it takes several days to set properly (the longer the drying period, the stronger the result), and continues to gain strength for at least five years after that. The properties of concrete can be varied in other ways. It can be made waterproof or porous, and smoothed off or textured to look like wood paneling.

Geodesic dome
Invented by architect Richard Buckminster Fuller (1895–1983), the geodesic dome does away with bulky beams and girders by distributing the weight of a building through an external steel skeleton. The impressive geodesic dome shown here is the Scince World exhibition building in Vancouver, Canada.

Steel and concrete structures

Steel is one of the most versatile materials, used in everything from jet engines to surgical instruments and from table knives to machine tools. Concrete is less versatile, but is still used widely in the construction industry. Today's modern cityscape is built large-ly from reinforced concrete—concrete strengthened by steel bars that are stronger than either material alone.

Steel and concrete have been partners in construction since at least the middle of the 19th century. The great advan-tage of both materials is that they allow mass production of identical parts. Skyscrapers are put together from thousands of identical steel "trees," and concrete roofs are assembled from large, precisely shaped precast shells. Concrete also has the enormous advantage that it can be transported in semi-liquid form, and then cast into the desired shape on the building site—for example, to create deep-foundation piles or concrete floors.

Concrete architecture was pioneered by French engineer François Hennebique (1842–1921), and French architects such as Le Corbusier were among the first to appreciate its remarkable qualities. Concrete allowed the construction of huge curving arches and other graceful sweeping shapes. Finnish architect Eero Saarinen (1910–1961) designed New York's first reinforced con-crete skyscraper, the CBS building, in 1965, and used the material in the famous swooping roof of the Trans World Airlines (TWA) building at New York's John F. Kennedy Airport.

Concrete-shell roofing
With suitable support or reinforcement, concrete can be used to make a wide range of imaginative structures. This elegant concrete roof at New York's JFK airport is internally reinforced with steel rods, and so it can carry its own weight, transmitting it to the ground at just a few points.

SEE ALSO: SKYSCRAPERS *p.12* | TUNNELS *p.20* | BRIDGES *p.22* | MINING & EXTRACTING METALS *p.186*

Traffic guidance

As more and more cars squeeze onto the world's road systems, journey times increase. The old cure of building more roads is no longer considered acceptable because of the environmental impact. Instead, the proposed solution is smarter traffic control and guidance systems. Effective guidance systems will shorten journey times, reduce fuel consumption, lower pollution, and make driving safer.

U.S. government experts have estimated that, apart from the environmental advantages, hundreds of billions of dollars could be saved in the U.S. each year by keeping vehicles out of traffic jams. Automated freeway systems—a lane or set of lanes in which specially equipped autos, trucks, and buses could travel together under computer control in small "platoons"— are envisaged to be an important component in achieving this, and prototype platooning vehicles have already been built. A typical freeway lane can handle around 2,000 vehicles an hour, but a lane equipped to guide traffic automatically should be able to carry about 6,000, depending on the spacing of entrances and exits. Extension of existing public traffic information systems, as well as private subscriber systems and technologies such as smart traffic lights and automatic tolls, will also play a part.

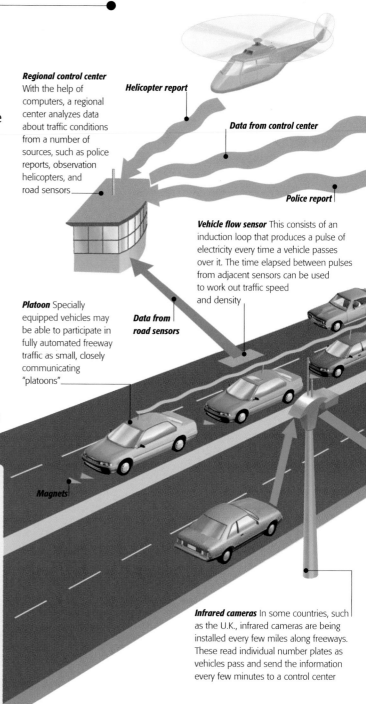

Regional control center With the help of computers, a regional center analyzes data about traffic conditions from a number of sources, such as police reports, observation helicopters, and road sensors

Helicopter report

Data from control center

Police report

Vehicle flow sensor This consists of an induction loop that produces a pulse of electricity every time a vehicle passes over it. The time elapsed between pulses from adjacent sensors can be used to work out traffic speed and density

Platoon Specially equipped vehicles may be able to participate in fully automated freeway traffic as small, closely communicating "platoons"

Data from road sensors

Magnets

Infrared cameras In some countries, such as the U.K., infrared cameras are being installed every few miles along freeways. These read individual number plates as vehicles pass and send the information every few minutes to a control center

Platooning vehicle

Radio antenna

Computer-controlled steering

Display panel

Radar unit

Rear magnetometer

Computer-controlled braking

Onboard computer

Computer-controlled throttle

Platooning vehicles for use on automated freeways require precise systems for determining their road position. This can be achieved by means of magnetometers under the bumpers, which interact with magnets embedded in the road, and a radar unit for monitoring the distance to the next vehicle or obstruction. The positioning data is fed to an onboard computer which automatically controls the throttle, brakes, and steering. In addition, a radio system is required for constant communication with the other vehicles in the platoon.

A range of guidance systems

The systems shown here range from the established to the experimental. Many urban areas have regional centers that provide traffic data to all road users via radio broadcasts. Speed traps and automatic tolls are also widespread. In some countries, additional data about freeway traffic is available to individual subscribers, based on a network of infrared cameras. Smart traffic lights and platooning systems are more experimental.

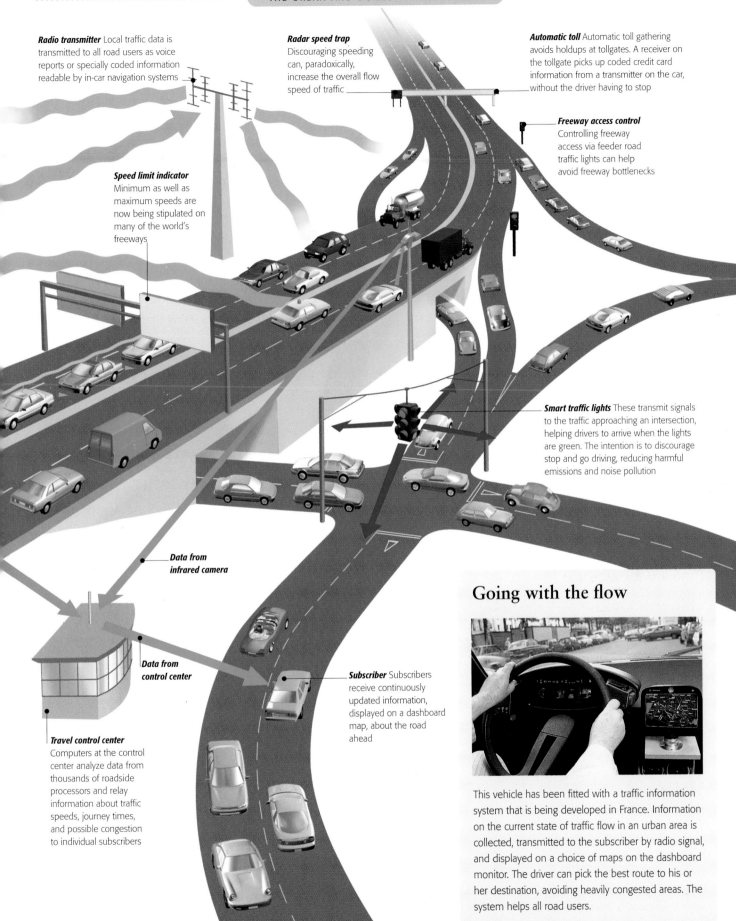

Radio transmitter Local traffic data is transmitted to all road users as voice reports or specially coded information readable by in-car navigation systems

Radar speed trap Discouraging speeding can, paradoxically, increase the overall flow speed of traffic

Automatic toll Automatic toll gathering avoids holdups at tollgates. A receiver on the tollgate picks up coded credit card information from a transmitter on the car, without the driver having to stop

Freeway access control Controlling freeway access via feeder road traffic lights can help avoid freeway bottlenecks

Speed limit indicator Minimum as well as maximum speeds are now being stipulated on many of the world's freeways

Smart traffic lights These transmit signals to the traffic approaching an intersection, helping drivers to arrive when the lights are green. The intention is to discourage stop and go driving, reducing harmful emissions and noise pollution

Data from infrared camera

Data from control center

Travel control center Computers at the control center analyze data from thousands of roadside processors and relay information about traffic speeds, journey times, and possible congestion to individual subscribers

Subscriber Subscribers receive continuously updated information, displayed on a dashboard map, about the road ahead

Going with the flow

This vehicle has been fitted with a traffic information system that is being developed in France. Information on the current state of traffic flow in an urban area is collected, transmitted to the subscriber by radio signal, and displayed on a choice of maps on the dashboard monitor. The driver can pick the best route to his or her destination, avoiding heavily congested areas. The system helps all road users.

SEE ALSO: RADIO *p.48* | VIDEO TECHNOLOGY *p.104* | CARS *p.108* | GLOBAL POSITIONING SYSTEM *p.136*

Lighting

Until the development of practical electric lighting in the late 19th century, artificial light sources were limited to candles and oil or gas lamps. Modern incandescent lamps work on the same basic principle as Edison's bulb of 1879. Other types of lamps, such as discharge lamps are more energy efficient and emit light of various colors. Increasing concern about environmental issues within buildings and in the wider world has led to innovations in the use of natural light. Devices called heliostats are used to collect sunlight, which can be piped around a building using hollow light guides.

An incandescent lamp consists of a long, coiled filament of the metal tungsten sealed in a glass bulb filled with an inert (unreactive) gas such as argon. When an electric current is passed through the filament, the electrical energy heats the filament. The filament may get as hot as 4,550°F (2,500°C), at which temperature it radiates significant amounts of light. Other types of lamps, such as discharge lamps and LEDs (light-emitting diodes), work on different principles and produce less waste heat. A discharge lamp has no filament; instead, it contains a gas at low pressure that glows when a current is passed through it. Neon signs, sodium-vapor street lamps, and fluorescent lamps are all discharge lamps. LEDs are semiconductor devices that emit specific colors of light and are used in electronic equipment.

Neon lighting

Neon lamps are a type of electric discharge lamp, in which a current is passed through low-pressure neon gas sealed in a glass tube. The neon atoms become energized, and release this energy by emitting a red-orange light. Different colors can be produced by using a colored glass tube, coating the tube with phosphor, or adding various gases to the tube.

Halogen lamps

In a conventional incandescent lamp, the hot tungsten filament gradually evaporates, making it thinner and more likely to break. Tungsten atoms form a deposit on the inside of the bulb, reducing its brightness. Halogen lamps solve these problems by containing a small amount of a halogen gas, such as iodine. Chemical reactions between the gas and the tungsten atoms return evaporated atoms to the filament, increasing filament life and maintaining brightness by preventing tungsten deposits on the bulb. The bulb is made of heat-resistant quartz. This allows a higher filament temperature, which produces brighter, whiter light.

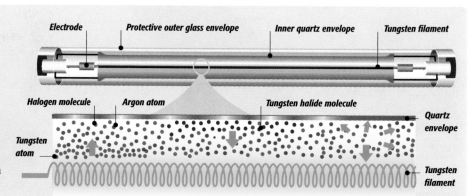

Electrode · Protective outer glass envelope · Inner quartz envelope · Tungsten filament

Halogen molecule · Argon atom · Tungsten halide molecule · Quartz envelope

Tungsten atom · Tungsten filament

1 Evaporation
The tungsten filament becomes very hot when an electric current passes through it, causing atoms to evaporate from its surface. Tungsten atoms mix with argon and halogen gases in the tube.

2 Combination
In the cooler part of the lamp near the bulb wall, tungsten atoms react chemically with molecules of halogen gas to form tungsten-halide molecules, which eventually drift back toward the filament.

3 Decomposition
Tungsten halide is unstable at high temperatures, so those molecules near the hot filament decompose, depositing tungsten back onto the filament, leaving behind molecules of halogen gas.

Light guides and heliostats

Light guides are used to transport light around buildings. They provide consistent illumination and reduce air-conditioning costs because they do not transport the waste heat generated by lamps. Light guides using special prismatic optical film are a new, less expensive, and simpler alternative to light guides based on lenses or optical fibers. Devices called extractors within the prismatic light guide scatter light for illumination where required. Light guides may be combined with heliostats (devices that collect sunlight).

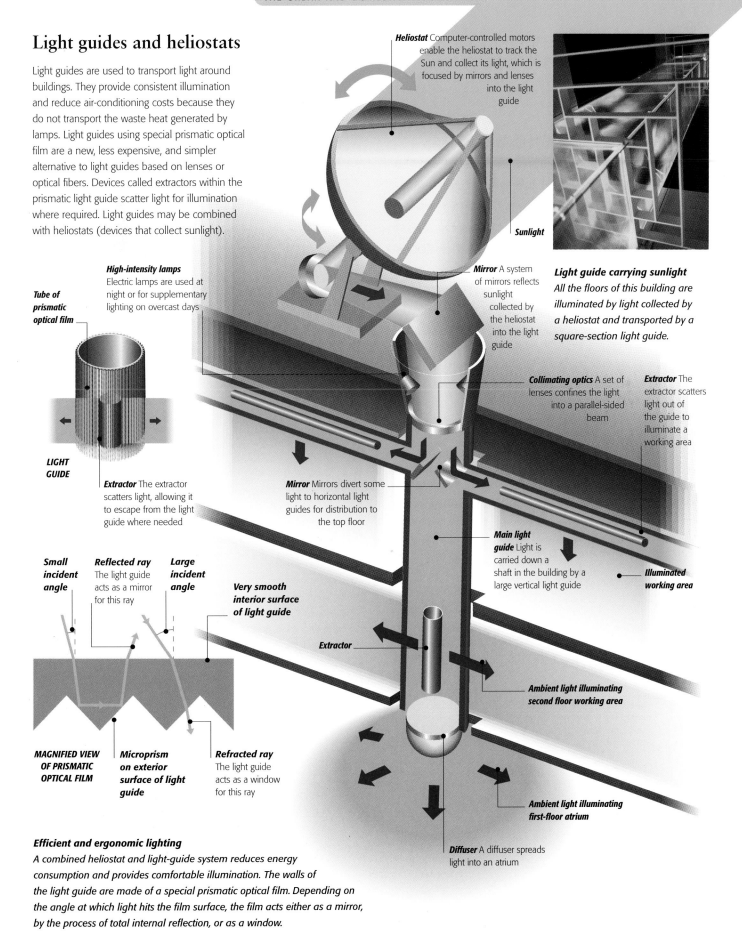

Heliostat Computer-controlled motors enable the heliostat to track the Sun and collect its light, which is focused by mirrors and lenses into the light guide

Sunlight

Mirror A system of mirrors reflects sunlight collected by the heliostat into the light guide

Light guide carrying sunlight
All the floors of this building are illuminated by light collected by a heliostat and transported by a square-section light guide.

High-intensity lamps
Electric lamps are used at night or for supplementary lighting on overcast days

Tube of prismatic optical film

LIGHT GUIDE

Extractor The extractor scatters light, allowing it to escape from the light guide where needed

Collimating optics A set of lenses confines the light into a parallel-sided beam

Extractor The extractor scatters light out of the guide to illuminate a working area

Mirror Mirrors divert some light to horizontal light guides for distribution to the top floor

Main light guide Light is carried down a shaft in the building by a large vertical light guide

Illuminated working area

Small incident angle

Reflected ray The light guide acts as a mirror for this ray

Large incident angle

Very smooth interior surface of light guide

Extractor

Ambient light illuminating second floor working area

MAGNIFIED VIEW OF PRISMATIC OPTICAL FILM

Microprism on exterior surface of light guide

Refracted ray The light guide acts as a window for this ray

Ambient light illuminating first-floor atrium

Diffuser A diffuser spreads light into an atrium

Efficient and ergonomic lighting
A combined heliostat and light-guide system reduces energy consumption and provides comfortable illumination. The walls of the light guide are made of a special prismatic optical film. Depending on the angle at which light hits the film surface, the film acts either as a mirror, by the process of total internal reflection, or as a window.

SEE ALSO: SKYSCRAPERS *p.12* | CABLE TECHNOLOGY *p.46* | SOLAR ENERGY *p.174* | LIGHT *p.262*

The water system

Water is essential for life and, with population density rapidly increasing, is set to become our most valuable natural resource in the 21st century. Waterborne bacteria and parasites are a major cause of disease and mortality, so water must be treated to make it safe for human consumption. Waste water and sewage must also be treated to ensure that it does not contaminate fresh water supplies.

Fresh water drawn from lakes, reservoirs, rivers, springs, and underground aquifers must be purified to remove microorganisms and harmful chemicals and unpalatable tastes and odors before it is fit for consumption. In regions where there is little fresh water, desalination plants are used to remove salt from seawater or mineral-rich groundwater prior to purification. Purified water is then piped to areas of demand. Pressure in the distribution system is maintained by using the weight of water stored in elevated tanks, or by pumps. Wastewater from industry and homes, and runoff from storm drains is piped into sewage treatment plants. Here, it is filtered and disinfected before it is discharged into rivers or the sea, or pumped back to purification plants. Earth's growing population increases demands on water systems. To maintain this vital resource requires more efficient collection and distribution, fewer leaks, less water overuse, and better sewage management.

Dams and reservoirs
Dams are usually built in mountainous areas and collect rainfall in reservoirs behind them. Reservoir water becomes purified through long-term storage: suspended particles in the water gradually settle, and sunlight kills some of the bacteria present. But reservoirs must be carefully managed to prevent contamination and to control the growth of algae. Reservoir outflow must be controlled to provide water for agriculture and prevent flooding of areas downstream. Large reservoirs are often used to power hydroelectric plants.

Making water fit to drink

Making fresh water fit to drink involves several processes. Impurities are removed in order of decreasing size by screening, coagulation, flocculation, and sedimentation. The water is then filtered and disinfected, usually with chlorine, to kill bacteria. It may also be aerated (sprayed into air to remove tastes and odors). Finally, fluoride may be added to help prevent tooth decay.

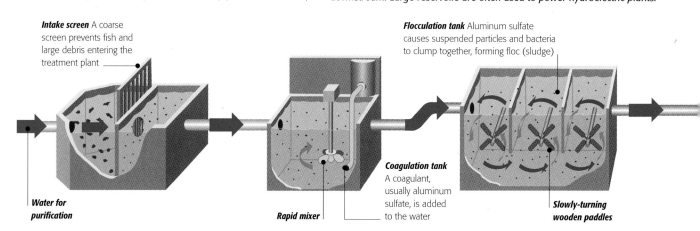

Intake screen A coarse screen prevents fish and large debris entering the treatment plant

Flocculation tank Aluminum sulfate causes suspended particles and bacteria to clump together, forming floc (sludge)

Water for purification

Rapid mixer

Coagulation tank A coagulant, usually aluminum sulfate, is added to the water

Slowly-turning wooden paddles

SCREENING

COAGULATION AND FLOCCULATION

Wastewater treatment

There are at least two stages to wastewater treatment. Primary treatment removes most of the floating and suspended particles from the sewage. Secondary treatment involves the removal of dissolved impurities and bacteria. Advanced treatment is necessary if water is to be reused or discharged near a freshwater source. Although expensive, advanced treatment removes 99 percent of all impurities, including ammonia, nitrates, and phosphates. Impurities removed from the sewage form sludge, which must be treated before it can be safely disposed of or reused as fertilizer.

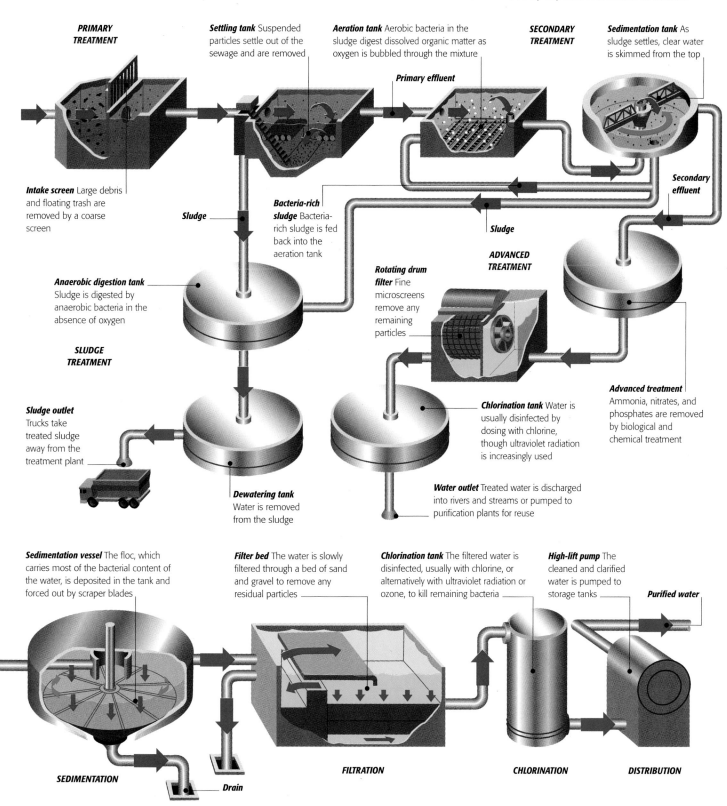

PRIMARY TREATMENT

Settling tank Suspended particles settle out of the sewage and are removed

Aeration tank Aerobic bacteria in the sludge digest dissolved organic matter as oxygen is bubbled through the mixture

SECONDARY TREATMENT

Sedimentation tank As sludge settles, clear water is skimmed from the top

Primary effluent

Intake screen Large debris and floating trash are removed by a coarse screen

Sludge

Bacteria-rich sludge Bacteria-rich sludge is fed back into the aeration tank

Sludge

Secondary effluent

Anaerobic digestion tank Sludge is digested by anaerobic bacteria in the absence of oxygen

Rotating drum filter Fine microscreens remove any remaining particles

ADVANCED TREATMENT

SLUDGE TREATMENT

Sludge outlet Trucks take treated sludge away from the treatment plant

Chlorination tank Water is usually disinfected by dosing with chlorine, though ultraviolet radiation is increasingly used

Advanced treatment Ammonia, nitrates, and phosphates are removed by biological and chemical treatment

Dewatering tank Water is removed from the sludge

Water outlet Treated water is discharged into rivers and streams or pumped to purification plants for reuse

Sedimentation vessel The floc, which carries most of the bacterial content of the water, is deposited in the tank and forced out by scraper blades

Filter bed The water is slowly filtered through a bed of sand and gravel to remove any residual particles

Chlorination tank The filtered water is disinfected, usually with chlorine, or alternatively with ultraviolet radiation or ozone, to kill remaining bacteria

High-lift pump The cleaned and clarified water is pumped to storage tanks

Purified water

SEDIMENTATION

Drain

FILTRATION

CHLORINATION

DISTRIBUTION

SEE ALSO: BENEATH THE STREETS p.14 | POWER STATIONS p.170 | AGRICULTURAL TECHNOLOGY p.188

Washing machines

The labor of washing clothes by hand was part of daily life all over the world until the mid-19th century. At that time, inventors began to create primitive washing machines in which the clothes were agitated when the user turned a crank by hand. Today's washing machines are motor driven and have electronic timers that control the wash, rinse, and drying cycles automatically.

Washing machines for domestic use are available in top-loading or front-loading designs. Front-loaders are becoming fashionable because they use less water. In top-loaders the clothes are agitated by spinning paddles to ensure maximum penetration of water and detergent. In front-loaders, sufficient agitation is provided by the rotating action of a ribbed drum. A control unit containing a timer determines when the water-inlet valves open. It also turns the motor on and off, regulates the speed of the drum, activates the pump, and determines the temperature of the wash. Microchip technology allows several sequences of operation—programs—to be stored in the control unit's memory, so that the machine can deal with a range of fabrics. Today's

washing machines work in combination with modern detergents, which began to be developed in the mid-20th century. These synthetic chemicals lift dirt and grease from fabrics with ease. They often contain additives such as fluorescent agents that make whites and colors appear brighter.

Machine manufacturers also devised ways of drying clothes more quickly. Spin dryers, in which water is expelled from the clothes through a rapidly-spinning, perforated drum, had appeared by the 1920s. In the 1950s, tumble dryers were introduced. These dry the clothes in a slowly rotating drum through which heated air is passed. The hot air absorbs moisture and is then vented through a duct. The latest washing machines include spin-drying and tumble-drying functions.

Front-loading washer-dryer

When the timer switches on, inlet valves are opened, and water and detergent are sprayed into the drum. When full, the drum rotates, washing the clothes. The pump removes dirty water. The valves then admit clean water, rinsing the clothes. Rapid spinning of the drum forces water out of the clothes. Finally, hot air dries the clothes.

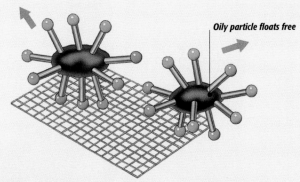

Detergent

Some dirt dissolves in water and rinses out of clothes, but greasy or oily stains do not, so additional cleaning agents are necessary to remove them. Modern detergents are specifically designed to perform this task and, unlike soap, do not leave scum in hard water. One end of the detergent molecule is acidic and is attracted to water molecules; the other, a hydrocarbon chain, is repelled by water molecules and is attracted to oil molecules.

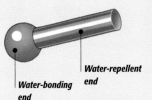

Water-bonding end

Water-repellent end

Detergent molecule
The acidic end of a detergent molecule bonds to water; the hydrocarbon end is repelled by water.

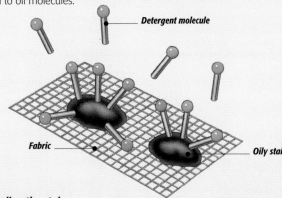

Detergent molecule

Fabric

Oily stain

Oily particle floats free

Surrounding the stains
The water-repellent ends of the detergent molecules are attracted to oily stains. Gradually, the detergent molecules build up around the stain until the dirt is surrounded by detergent.

Dissolving oil and grease
The detergent-surrounded particles can no longer adhere to the fabric and so float free. The water-bonding ends of the detergent molecules dissolve in water, allowing the oily particles to be rinsed away.

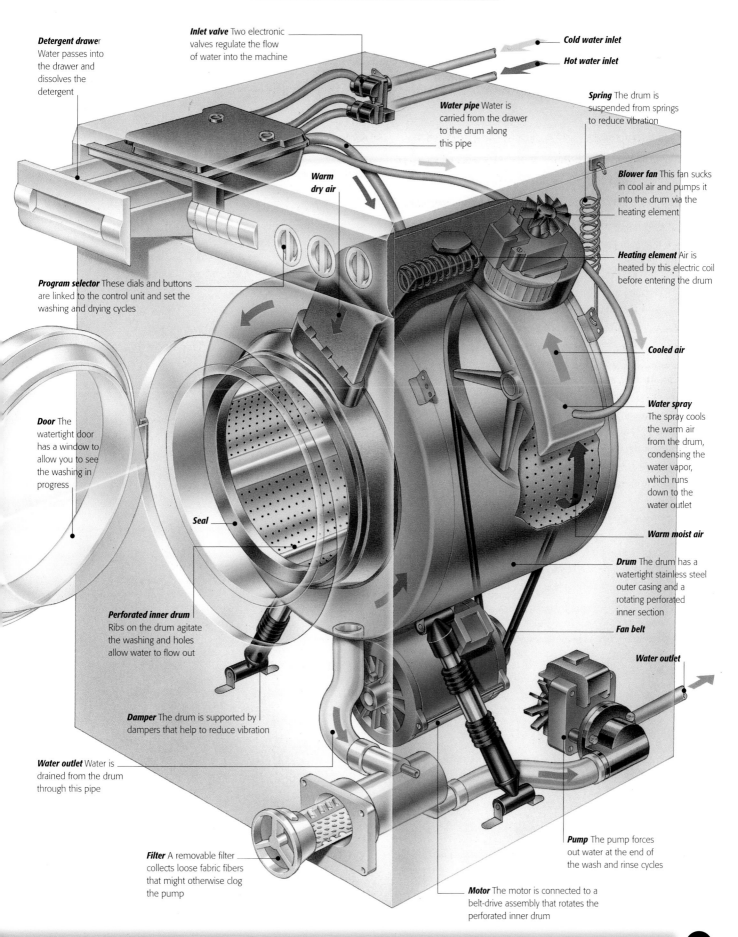

Detergent drawer Water passes into the drawer and dissolves the detergent

Inlet valve Two electronic valves regulate the flow of water into the machine

Cold water inlet

Hot water inlet

Water pipe Water is carried from the drawer to the drum along this pipe

Spring The drum is suspended from springs to reduce vibration

Blower fan This fan sucks in cool air and pumps it into the drum via the heating element

Warm dry air

Heating element Air is heated by this electric coil before entering the drum

Program selector These dials and buttons are linked to the control unit and set the washing and drying cycles

Cooled air

Water spray The spray cools the warm air from the drum, condensing the water vapor, which runs down to the water outlet

Door The watertight door has a window to allow you to see the washing in progress

Seal

Warm moist air

Perforated inner drum Ribs on the drum agitate the washing and holes allow water to flow out

Drum The drum has a watertight stainless steel outer casing and a rotating perforated inner section

Fan belt

Water outlet

Damper The drum is supported by dampers that help to reduce vibration

Water outlet Water is drained from the drum through this pipe

Pump The pump forces out water at the end of the wash and rinse cycles

Filter A removable filter collects loose fabric fibers that might otherwise clog the pump

Motor The motor is connected to a belt-drive assembly that rotates the perforated inner drum

SEE ALSO: THE WATER SYSTEM *p.30* | VACUUM CLEANERS *p.38* | MATTER *p.258*

Stoves and microwaves

Since humans learned how to create fire, heat has been used to process food. In conventional stoves, heat is generated either by a controllable gas flame or by an electric heating element and is transferred to the food by conduction. As over half the heat energy is lost to the environment by conventional appliances, stoves have been developed that transfer energy more directly to the food.

The most revolutionary of these new cooking technologies is the microwave oven, which uses a device called a magnetron to generate a type of electromagnetic radiation called microwaves. The magnetron was first developed during World War II for use in radar systems. It was soon discovered that microwaves penetrate food where they cause water molecules to spin at high speed, generating friction, and thus heat, which cooks the food very rapidly. At microwave frequencies (around 2450 MHz), electromagnetic waves are reflected by metal, so they can be trapped within a metal box. But microwaves pass through most glass and plastics, so these materials can be used to hold the food. The first microwave oven was patented in 1953, but early models were too cumbersome for use in the home. Smaller and more efficient microwave ovens were developed in the 1970s, and since then they have become increasingly popular both in homes and restaurants.

Rotating paddle This fan scatters the microwaves so that they reach all parts of the oven

Microwaves The metal casing of the oven reflects microwaves

Protective grille A metal screen in the glass door reflects microwaves back into the oven

Container microwaves pass through most paper, plastic, and glass containers

Induction stoves

Like microwave ovens, induction stoves use varying electromagnetic fields to produce heat. An alternating current passed through a coil in the heating element on the stovetop (the hob) produces a varying magnetic field—an effect known as electromagnetic induction. If the pan on the hob is made from a conductive material, the changing magnetic field in turn induces an alternating current in the material of the pan. The electrical resistance of the pan's material to this current causes the pan to heat up, cooking the food. Since there is no direct transfer of heat from the hob to the pan, there is also none of the heat loss that thermal electric and gas stoves produce. As well as being energy efficient, induction stoves are easy to control.

Microwave oven

In a microwave oven an electric current is used to generate microwaves in a magnetron. These microwaves are radiated by an antenna through a metal wave guide into the oven, where they are scattered by a rotating paddle. Some of the microwaves penetrate the food directly, while others are reflected off the walls into the food, ensuring that it cooks evenly. An electronic timer is used to control the cooking time.

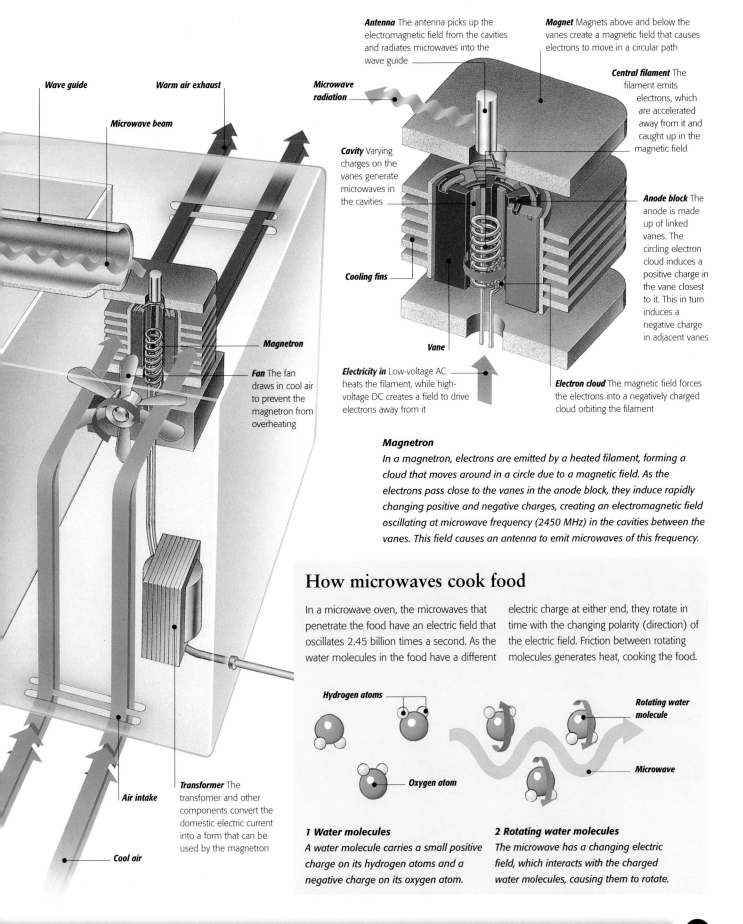

Antenna The antenna picks up the electromagnetic field from the cavities and radiates microwaves into the wave guide

Magnet Magnets above and below the vanes create a magnetic field that causes electrons to move in a circular path

Central filament The filament emits electrons, which are accelerated away from it and caught up in the magnetic field

Wave guide

Warm air exhaust

Microwave beam

Microwave radiation

Cavity Varying charges on the vanes generate microwaves in the cavities

Anode block The anode is made up of linked vanes. The circling electron cloud induces a positive charge in the vane closest to it. This in turn induces a negative charge in adjacent vanes

Cooling fins

Magnetron

Vane

Fan The fan draws in cool air to prevent the magnetron from overheating

Electricity in Low-voltage AC heats the filament, while high-voltage DC creates a field to drive electrons away from it

Electron cloud The magnetic field forces the electrons into a negatively charged cloud orbiting the filament

Magnetron

In a magnetron, electrons are emitted by a heated filament, forming a cloud that moves around in a circle due to a magnetic field. As the electrons pass close to the vanes in the anode block, they induce rapidly changing positive and negative charges, creating an electromagnetic field oscillating at microwave frequency (2450 MHz) in the cavities between the vanes. This field causes an antenna to emit microwaves of this frequency.

Transformer The transfomer and other components convert the domestic electric current into a form that can be used by the magnetron

Air intake

Cool air

How microwaves cook food

In a microwave oven, the microwaves that penetrate the food have an electric field that oscillates 2.45 billion times a second. As the water molecules in the food have a different electric charge at either end, they rotate in time with the changing polarity (direction) of the electric field. Friction between rotating molecules generates heat, cooking the food.

Hydrogen atoms

Rotating water molecule

Oxygen atom

Microwave

1 Water molecules
A water molecule carries a small positive charge on its hydrogen atoms and a negative charge on its oxygen atom.

2 Rotating water molecules
The microwave has a changing electric field, which interacts with the charged water molecules, causing them to rotate.

SEE ALSO: NAVIGATION *p.132* | SPACE TELESCOPES *p.254* | ELECTRICITY & MAGNETISM *p.260* | LIGHT *p.262*

Fridges and freezers

Refrigerators (fridges) and freezers are heat pumps—they move heat energy from cold (low temperature) regions to hot (high temperature) ones, against its natural flow, further cooling the cold regions. They work by compressing a substance called a refrigerant and then letting it expand. Early fridges used ammonia as a refrigerant. Domestic fridges today use different refrigerants, but operate in the same way.

Refrigerants are liquids with low boiling points, able to change between liquid and gaseous states with ease at room temperature. Ammonia, an otherwise eminently suitable refrigerant, is toxic, so less toxic chlorofluorocarbons (CFCs) were developed in the 1930s. Since the 1990s, CFCs have been replaced by hydro-chlorofluorocarbons (HCFCs) and hydrofluorocarbons (HFCs), which are thought to be less damaging to the Earth's ozone layer. Today's refrigerators store food at temperatures between 37°F (3°C) and 41°F (5°C). At these temperatures, the activity of microorganisms that cause food spoilage is slowed, but not halted. Freezers maintain a temperature of about 0°F (−18°C). Under these conditions, microorganisms cease reproduction and nearly all other activity, so spoilage of food is effectively halted.

Vacuum flask

A vacuum flask maintains the temperature of hot and cold liquids by thermally insulating them from their surroundings within a double-walled glass bottle. A vacuum between the double walls of the bottle minimizes heat transfer since neither convection nor conduction can occur within a vacuum. The glass has a silvered surface to reflect infrared radiation (heat energy) reducing heat transfer by radiation.

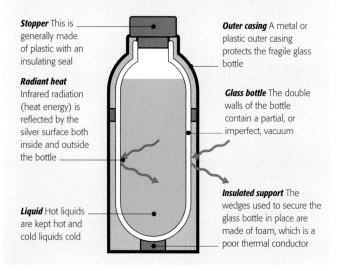

Stopper This is generally made of plastic with an insulating seal

Radiant heat Infrared radiation (heat energy) is reflected by the silver surface both inside and outside the bottle

Liquid Hot liquids are kept hot and cold liquids cold

Outer casing A metal or plastic outer casing protects the fragile glass bottle

Glass bottle The double walls of the bottle contain a partial, or imperfect, vacuum

Insulated support The wedges used to secure the glass bottle in place are made of foam, which is a poor thermal conductor

Household refrigerator

A refrigerator is a heat-transfer machine, which takes heat from inside the refrigerator and dumps it outside. This is accomplished through repeated evaporation and condensation of the refrigerant. Evaporation—changing state from a liquid to a gas—requires heat to take place, which is absorbed from the contents of the refrigerator. The reverse change—condensation—gives out heat, which is released from the refrigerant to the outside of the refrigerator.

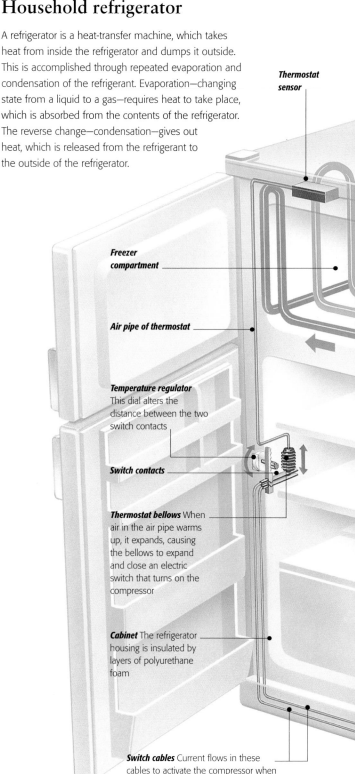

Thermostat sensor

Freezer compartment

Air pipe of thermostat

Temperature regulator This dial alters the distance between the two switch contacts

Switch contacts

Thermostat bellows When air in the air pipe warms up, it expands, causing the bellows to expand and close an electric switch that turns on the compressor

Cabinet The refrigerator housing is insulated by layers of polyurethane foam

Switch cables Current flows in these cables to activate the compressor when the thermostat bellows close the switch

Expansion valve

The expansion valve is a narrow opening that allows refrigerant to flow from the narrow condenser pipe into the broader evaporator pipe while maintaining pressure differences between the two pipes.

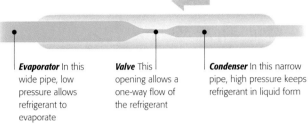

Evaporator In this wide pipe, low pressure allows refrigerant to evaporate

Valve This opening allows a one-way flow of the refrigerant

Condenser In this narrow pipe, high pressure keeps refrigerant in liquid form

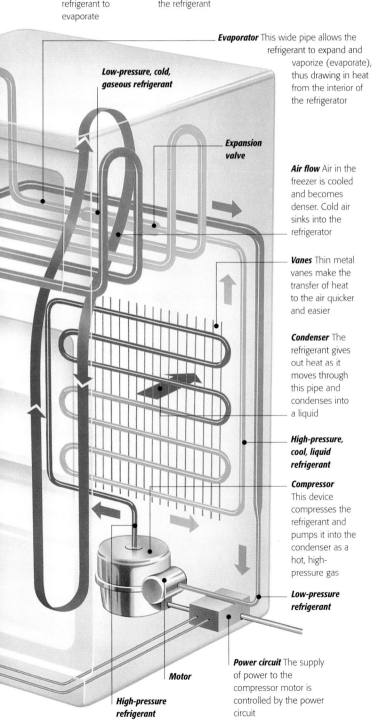

Low-pressure, cold, gaseous refrigerant

Evaporator This wide pipe allows the refrigerant to expand and vaporize (evaporate), thus drawing in heat from the interior of the refrigerator

Expansion valve

Air flow Air in the freezer is cooled and becomes denser. Cold air sinks into the refrigerator

Vanes Thin metal vanes make the transfer of heat to the air quicker and easier

Condenser The refrigerant gives out heat as it moves through this pipe and condenses into a liquid

High-pressure, cool, liquid refrigerant

Compressor This device compresses the refrigerant and pumps it into the condenser as a hot, high-pressure gas

Low-pressure refrigerant

Motor

High-pressure refrigerant

Power circuit The supply of power to the compressor motor is controlled by the power circuit

Ice-cube maker

Ice-cube makers are standard equipment in bars and other catering establishments. They are also included in some refrigerators. An icemaking machine requires connection to a drinking-water supply as well as to a power supply.

1 Water into ice
A measured volume of water flows into the ice-cube mold when the shutoff arm is down. A thermostatically controlled cooling mechanism that freezes the water is then activated.

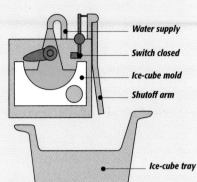

Water supply

Switch closed

Ice-cube mold

Shutoff arm

Ice-cube tray

2 Releasing the ice
When the thermostat senses that the water has frozen, a heating element warms the mold sufficiently to loosen the ice and a motor-driven ejector blade scoops the ice cube from the mold.

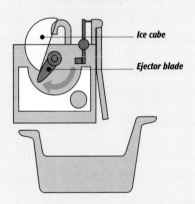

Ice cube

Ejector blade

3 Ejecting ice cubes
The ejector blade continues to rotate and pushes the ice cube up to the shutoff arm which opens, allowing the ice cube to drop down into the ice-cube tray. The refrigerant is compressed and pumped

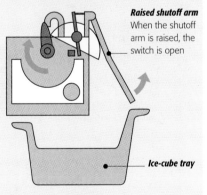

Raised shutoff arm When the shutoff arm is raised, the switch is open

Ice-cube tray

4 Full ice-cube tray
This sequence continues until the quantity of ice cubes in the tray prevents the shutoff arm from closing. The process is then suspended until some ice cubes are taken from the tray.

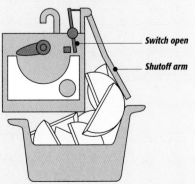

Switch open

Shutoff arm

SEE ALSO: SHIPS *p.140* | PRINCIPLES *p.258*

Vacuum cleaners

A new bagless machine transformed domestic vacuum cleaning in the early 1990s. British inventor James Dyson made one of the biggest breakthroughs in home cleaning with his dual cyclone system since the first portable vacuum cleaner to be powered by an electric motor was introduced in 1907. The Dyson cleaner replaces the traditional porous bag with two cyclone chambers that spin the air to extract dust from it.

Conventional upright and canister vacuum cleaners use a fan to create a partial vacuum and suck dirty air from carpets, rugs, upholstery, and bare floors through a paper or cloth bag. Air passes through the microscopic pores of the bag but dust is left behind in the bag, which can be disposed of later. Dyson's cyclone uses a spinning column of air instead of a bag to extract the dust. It spins dirty air rapidly inside a drum, forcing dust outward to the wall, just as water is forced outward in a spin dryer. The dust, separated from the air, collects in a plastic container.

As the bag in a conventional vacuum cleaner fills with dust, its pores become clogged so air can no longer flow through so easily. This clogging reduces the power of suction. The dual cyclone cleaner has no bag to clog, and retains its cleaning effectiveness.

Upright vacuum cleaner

The electric motor of an upright vacuum cleaner spins a fan at up to 300 revolutions per second, expelling air from the back of the machine. This creates a partial vacuum inside the machine, which causes dirt and air to be sucked up through the bottom of the machine into a dust bag. Most of the dust becomes trapped in the bag, while the clean air passes through pores in the bag and is drawn through a filter system.

Dirty air The Dyson sucks in as much as 5½ gallons (25 liters) of air every second at a speed of 70 mph (112 km/h)

Dust bag Most of the dust becomes trapped in the bag, while air escapes through pores in the bag

Filtering system Residual dust is removed from the air by the filtering system

Expelled air

Wand control The length of the wand is adjusted with this switch

Rotating brush The rotating brush loosens dirt from carpets and rugs

Fan The rotating fan expels clean air, creating a partial vacuum inside the dust bag

Head control Brushes on the underside of the head are adjusted with this control

The Dyson vacuum cleaner

The dual cyclone system of the Dyson spins dirty air at high speed inside two cyclones, forcing dust particles outward, to be collected in the collection container. The outer cyclone separates out large particles of dust while the inner cyclone, which rotates three times as fast and tapers downward, separates out smaller particles.

Dust particles

Filtering systems in vacuum cleaners

Household dust particles that cause allergy sufferers the most discomfort are less than 3 micrometers (0.003 mm) in diameter. They include molds, spores, pet dander, smoke particles, and dust mite feces, all of which cause lung irritation and may cause an allergic reaction. In the past, most vacuum cleaners made the household environment worse, not better, for allergy sufferers. Fine particles were sucked out of the carpet, then expelled into the air as they were too minute to be trapped by the cleaner's dust bag. The best modern machines have filters that extract particles as small as 0.1 micrometer (0.0001 mm) from expelled air.

Scanning electron micrograph of a dust mite
Millions of dust mites, such as the one shown here magnified 370 times, live in fabric in the average home. They feed off the dead scales of human skin and their bodies and excrement can cause allergic reactions.

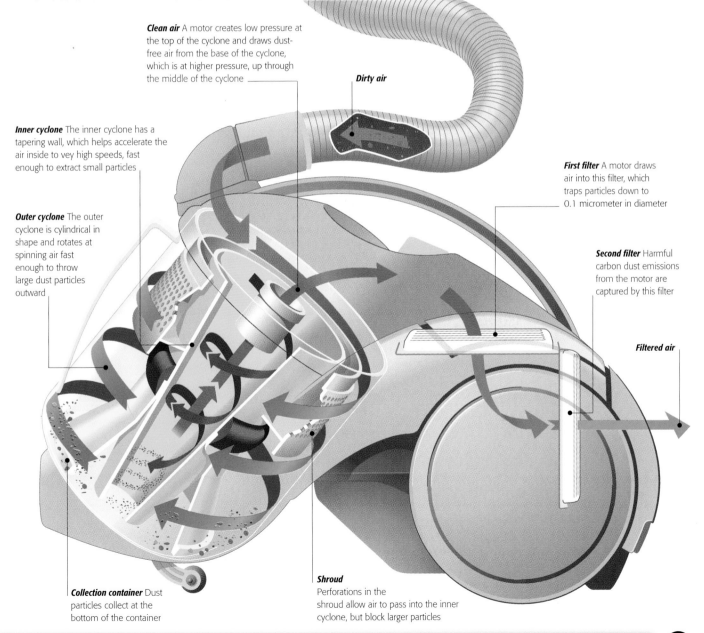

Clean air A motor creates low pressure at the top of the cyclone and draws dust-free air from the base of the cyclone, which is at higher pressure, up through the middle of the cyclone

Dirty air

Inner cyclone The inner cyclone has a tapering wall, which helps accelerate the air inside to vey high speeds, fast enough to extract small particles

First filter A motor draws air into this filter, which traps particles down to 0.1 micrometer in diameter

Outer cyclone The outer cyclone is cylindrical in shape and rotates at spinning air fast enough to throw large dust particles outward

Second filter Harmful carbon dust emissions from the motor are captured by this filter

Filtered air

Collection container Dust particles collect at the bottom of the container

Shroud Perforations in the shroud allow air to pass into the inner cyclone, but block larger particles

SEE ALSO: WASHING MACHINES *p.32* | ELECTRON MICROSCOPES *p.226* | MECHANICS *p.270*

Clocks and watches

For centuries, people mostly measured time by the motion of the Sun or by large, mechanical pendulum clocks. The need for accurate and portable time measurement arose only in the 17th and 18th centuries, when long-distance sea navigation necessitated a method of calculating longitude. Until recently, all clocks and watches were driven mechanically, but electronic timekeeping is now widespread.

The first accurate clocks and watches were built in the mid-1700s for navigation, when long sea voyages were becoming common. Although sailors could work out their latitude from the stars, the only practical way to calculate longitude was to know the time at the home port when it was noon on the ship—midday is four minutes earlier or later for each degree of longitude east or west.

Mechanical clocks and watches are powered by springs or moving weights. The best spring-driven mechanical clocks of the early 20th century gained or lost up to one tenth of a second per day. Quartz clocks and watches are battery powered and are accurate to just one-thousandth of a second per day. This accuracy is possible because they use stable electronic oscillations instead of springs or a pendulum, and so are not subject to physical disruption from heat and friction. Some new quartz watches have no batteries—instead, they generate electricity using a swinging weight powered by the wearer's motion. The most accurate clocks are atomic clocks. They are used to set a universal standard of time, which is essential for industry and telecommunications.

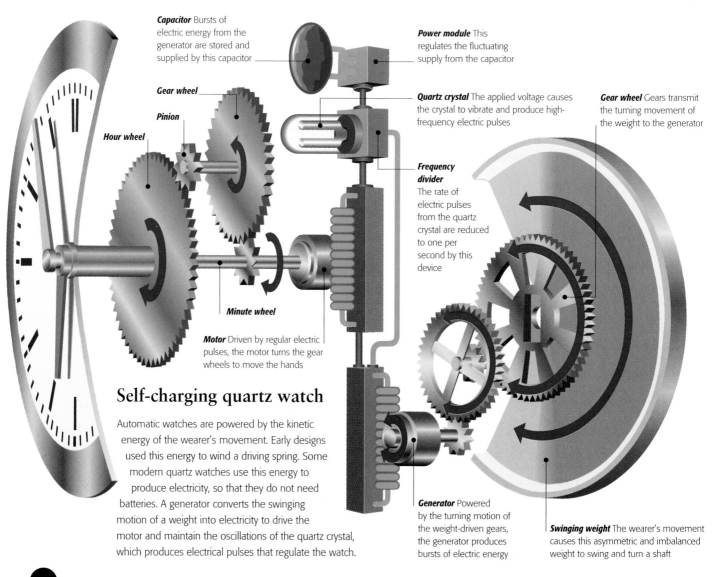

Capacitor Bursts of electric energy from the generator are stored and supplied by this capacitor

Power module This regulates the fluctuating supply from the capacitor

Gear wheel

Pinion

Hour wheel

Quartz crystal The applied voltage causes the crystal to vibrate and produce high-frequency electric pulses

Gear wheel Gears transmit the turning movement of the weight to the generator

Frequency divider The rate of electric pulses from the quartz crystal are reduced to one per second by this device

Minute wheel

Motor Driven by regular electric pulses, the motor turns the gear wheels to move the hands

Self-charging quartz watch

Automatic watches are powered by the kinetic energy of the wearer's movement. Early designs used this energy to wind a driving spring. Some modern quartz watches use this energy to produce electricity, so that they do not need batteries. A generator converts the swinging motion of a weight into electricity to drive the motor and maintain the oscillations of the quartz crystal, which produces electrical pulses that regulate the watch.

Generator Powered by the turning motion of the weight-driven gears, the generator produces bursts of electric energy

Swinging weight The wearer's movement causes this asymmetric and imbalanced weight to swing and turn a shaft

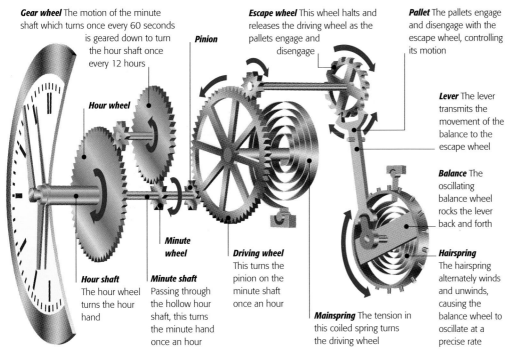

Gear wheel The motion of the minute shaft which turns once every 60 seconds is geared down to turn the hour shaft once every 12 hours

Pinion

Hour wheel

Escape wheel This wheel halts and releases the driving wheel as the pallets engage and disengage

Pallet The pallets engage and disengage with the escape wheel, controlling its motion

Lever The lever transmits the movement of the balance to the escape wheel

Balance The oscillating balance wheel rocks the lever back and forth

Hour shaft The hour wheel turns the hour hand

Minute shaft Passing through the hollow hour shaft, this turns the minute hand once an hour

Minute wheel

Driving wheel This turns the pinion on the minute shaft once an hour

Hairspring The hairspring alternately winds and unwinds, causing the balance wheel to oscillate at a precise rate

Mainspring The tension in this coiled spring turns the driving wheel

Mechanical watch

Mechanical watches are powered by the gradual unwinding of a mainspring and regulated by a system called a lever escapement, which periodically releases force from the mainspring to turn a driving wheel. An oscillating hairspring and balance wheel cause a lever to rock to and fro regularly. Pallets on the end of the lever alternately engage and disengage teeth on an escape wheel, which turns in steps, unwinding the mainspring.

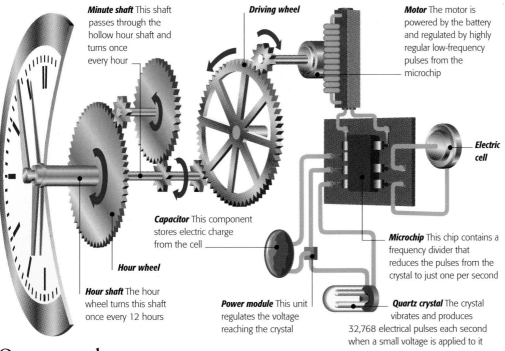

Minute shaft This shaft passes through the hollow hour shaft and turns once every hour

Driving wheel

Motor The motor is powered by the battery and regulated by highly regular low-frequency pulses from the microchip

Capacitor This component stores electric charge from the cell

Electric cell

Hour wheel

Microchip This chip contains a frequency divider that reduces the pulses from the crystal to just one per second

Hour shaft The hour wheel turns this shaft once every 12 hours

Power module This unit regulates the voltage reaching the crystal

Quartz crystal The crystal vibrates and produces 32,768 electrical pulses each second when a small voltage is applied to it

Quartz watch

The gear mechanism for turning the hour and minute wheels and the hands of a quartz watch is the same as in a mechanical watch. The difference is in the power supply, which is a cell (battery), and in the regulating mechanism, which is a quartz crystal that vibrates and produces electric pulses at a fixed rate. A microchip converts these high-frequency pulses into slower pulses that regulate the turning of an electric motor.

Pendulums

A pendulum swings at a constant frequency (rate), which depends only on the pendulum's length. In a pendulum clock, the swinging of the pendulum controls the rotation of the escape wheel, which drives the clock's hands.

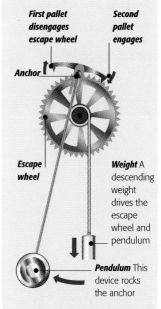

First pallet disengages escape wheel

Second pallet engages

Anchor

Escape wheel

Weight A descending weight drives the escape wheel and pendulum

Pendulum This device rocks the anchor

1 The pendulum swings left, disengaging the first pallet. The wheel turns slightly before the second pallet engages.

Second pallet

Escape wheel The constant period of the pendulum allows the wheel to turn a precise amount before the second pallet engages

2 The pendulum swings right, the second pallet disengages, and the wheel is released until the first pallet reengages.

SEE ALSO: NAVIGATION p.132 | BATTERIES p.178 | ATOMIC CLOCKS p.224 | MECHANICS p.266

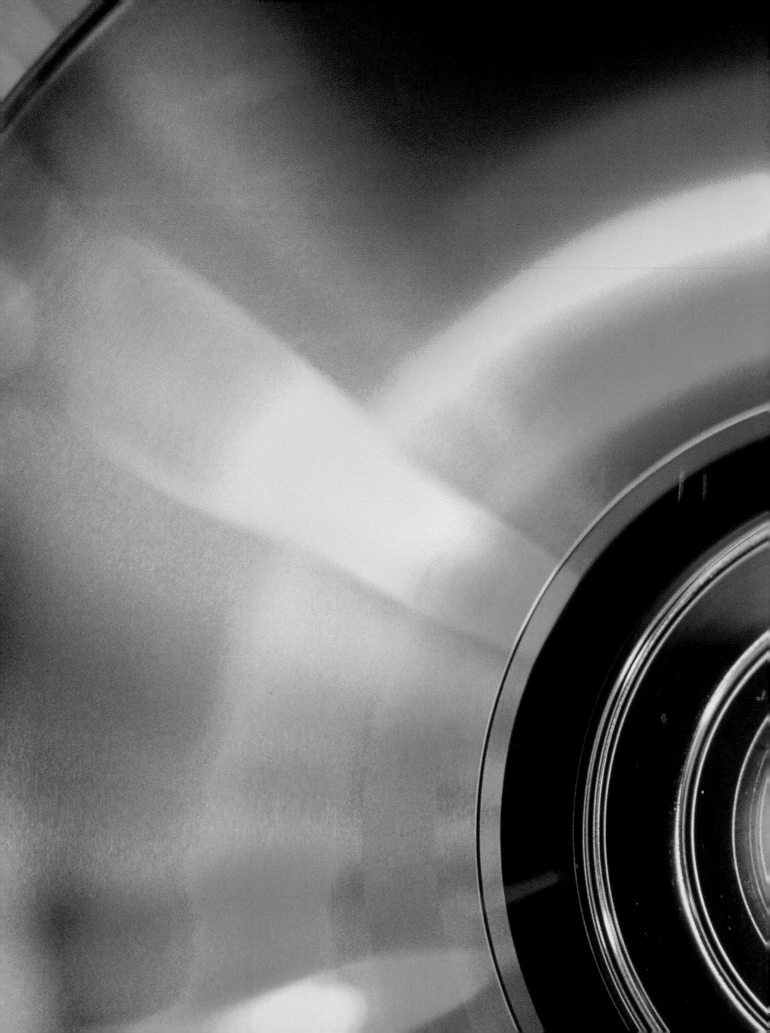

Communications and Leisure

The world is currently in the middle of a communications revolution as dramatic as the industrial revolution that created our modern society two centuries ago. The growth of telecommunications systems, the arrival of the Internet, and the proliferation of computers in every aspect of our lives are transforming both industrial and knowledge-based economies around the world. Higher disposable incomes and increased leisure time are also fueling demand for luxury electronic goods and new forms of entertainment. This chapter examines many of the cutting-edge technologies at the heart of the communications revolution.

The telephone network

Since the British inventor Alexander Graham Bell patented the telephone in 1877, telephony has become the most important form of distance communication. The telephone network is now truly global, with submarine cables and communications satellites linking every continent. This globe-spanning network handles phone calls, fax transmissions, and Internet traffic.

The long-distance "backbones" of the telephone network are high-capacity optical-fiber cables. Lower-capacity copper cables connect individual phones to the network. The analog signal from an ordinary phone is sampled 4,000 times per second and converted into an 8-bit digital signal. Several conversations can then be transmitted simultaneously down the same cable, using a technique known as multiplexing, which increases the capacity of the network. Routing calls through the telephone network is done automatically by telephone exchanges. In the near future, the telephone network may also carry video and music channels, interactive television, and videophone calls.

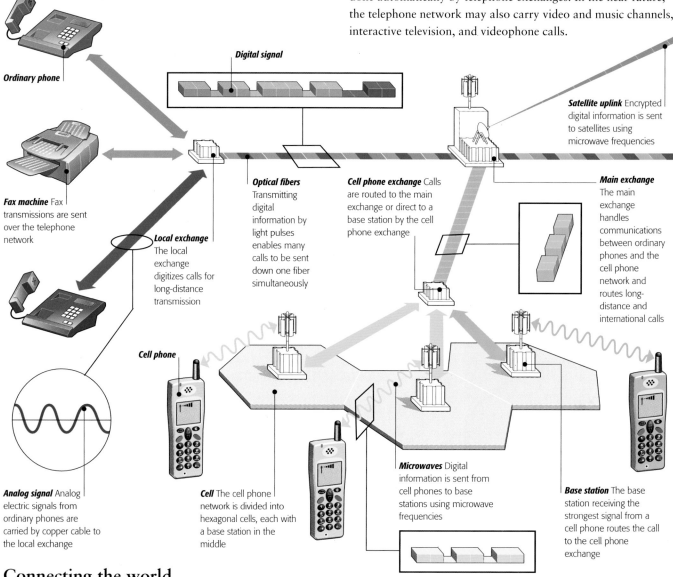

Ordinary phone

Digital signal

Satellite uplink Encrypted digital information is sent to satellites using microwave frequencies

Fax machine Fax transmissions are sent over the telephone network

Optical fibers Transmitting digital information by light pulses enables many calls to be sent down one fiber simultaneously

Cell phone exchange Calls are routed to the main exchange or direct to a base station by the cell phone exchange

Main exchange The main exchange handles communications between ordinary phones and the cell phone network and routes long-distance and international calls

Local exchange The local exchange digitizes calls for long-distance transmission

Cell phone

Microwaves Digital information is sent from cell phones to base stations using microwave frequencies

Base station The base station receiving the strongest signal from a cell phone routes the call to the cell phone exchange

Analog signal Analog electric signals from ordinary phones are carried by copper cable to the local exchange

Cell The cell phone network is divided into hexagonal cells, each with a base station in the middle

Connecting the world

This diagram shows the main components of the global telephone network. Local exchanges route calls to and from ordinary telephones. At the local exchange the signals are digitized and sent to a main exchange. The main exchange can communicate with cell phone exchanges and send calls long distance, or internationally, to other main exchanges. Telephone calls can be sent down copper and optical-fiber cables or via microwave links.

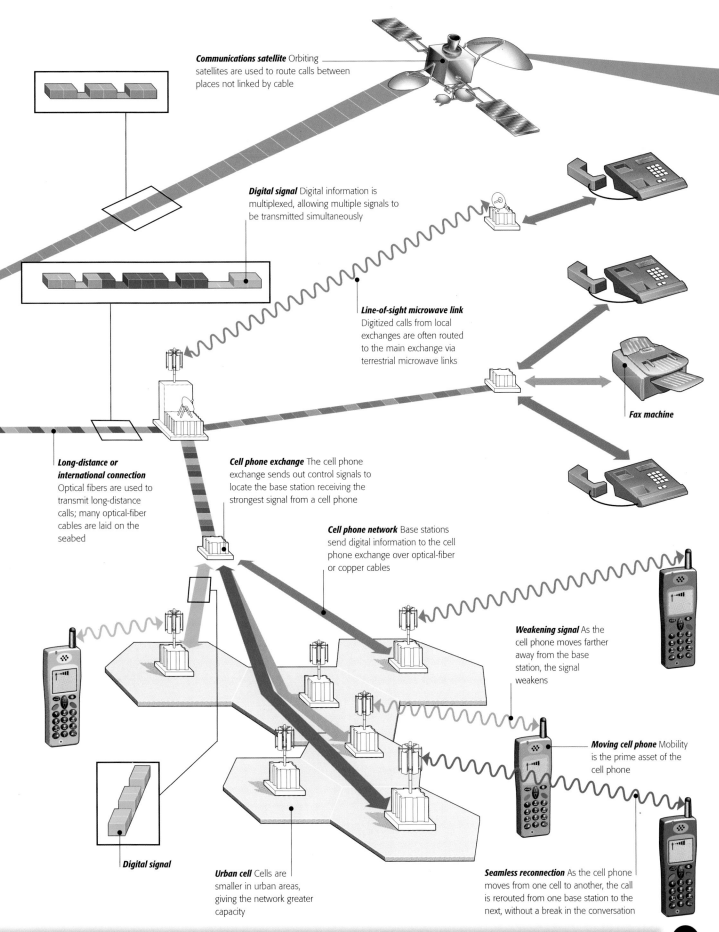

Communications satellite Orbiting satellites are used to route calls between places not linked by cable

Digital signal Digital information is multiplexed, allowing multiple signals to be transmitted simultaneously

Line-of-sight microwave link Digitized calls from local exchanges are often routed to the main exchange via terrestrial microwave links

Fax machine

Long-distance or international connection Optical fibers are used to transmit long-distance calls; many optical-fiber cables are laid on the seabed

Cell phone exchange The cell phone exchange sends out control signals to locate the base station receiving the strongest signal from a cell phone

Cell phone network Base stations send digital information to the cell phone exchange over optical-fiber or copper cables

Weakening signal As the cell phone moves farther away from the base station, the signal weakens

Moving cell phone Mobility is the prime asset of the cell phone

Digital signal

Urban cell Cells are smaller in urban areas, giving the network greater capacity

Seamless reconnection As the cell phone moves from one cell to another, the call is rerouted from one base station to the next, without a break in the conversation

Cable technology

Modern telecommunication relies on the accurate and rapid transmission of large volumes of information. Cables, the original information conduit, remain central to telecommunications networks, despite developments in satellite technology. Advances in cable technology have led to the development of telephone trunk cables that can transmit tens of thousands of calls simultaneously.

Communication over long distances is achieved by the transmission of data as waves or streams of pulses. An electric current alternating (switching direction) at radio wave frequency is modulated (shaped) by an analog or digital data signal. The modulated wave is sent through coaxial cables, or broadcast via groundstations and satellites as radio waves. Digital information may also be sent as streams of light pulses (or, less commonly, of electric pulses)—a pulse representing a "1," and the absence of a pulse representing a "0." Cables act as highways for these data streams, and must be designed to minimize interference between different data streams, which leads to loss of data. The bandwidth of a cable is a measure of the amount of data that it can transmit. Optical fibers have a greater bandwidth than coaxial cables, and can transmit signals farther and faster since light is less prone to attenuation (power loss) than electric currents, and propagates faster. The bandwidth of a cable system can be increased by installing more cables, but also by employing multiplexing techniques, which squeeze more data down each cable.

Coaxial cable

Coaxial cables have a central copper core running through the center of a cylindrical copper braid. The two conducting channels are separated by a layer of insulation. The inner channel carries information as modulated radio-frequency alternating current while the outer channel gives protection against electromagnetic interference. Coaxial cables are used in telephone, computer, and cable TV networks, and to link TV antennae to receivers.

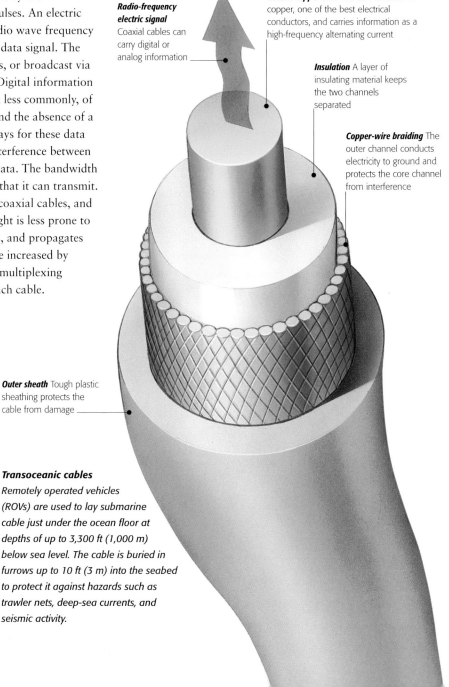

Radio-frequency electric signal Coaxial cables can carry digital or analog information

Central copper core The core is made of copper, one of the best electrical conductors, and carries information as a high-frequency alternating current

Insulation A layer of insulating material keeps the two channels separated

Copper-wire braiding The outer channel conducts electricity to ground and protects the core channel from interference

Outer sheath Tough plastic sheathing protects the cable from damage

Transoceanic cables
Remotely operated vehicles (ROVs) are used to lay submarine cable just under the ocean floor at depths of up to 3,300 ft (1,000 m) below sea level. The cable is buried in furrows up to 10 ft (3 m) into the seabed to protect it against hazards such as trawler nets, deep-sea currents, and seismic activity.

Optical-fiber cable

Optical fibers are hair-thin strands composed of two layers of very pure glass or plastic. An outer cladding layer confines light signals to the inner core by a process called total internal reflection. A plastic sheath protects the fiber from damage. Digital information is transmitted as pulses of light. However, transmission is not flawless. Each pulse may travel along several paths, or modes, through the fiber and tends to gradually spread out, so a clear input pulse becomes blurred and loses intensity as it travels. This pulse dispersion limits the rate and range of data transmission.

INPUT SIGNAL **OUTPUT SIGNAL**

Signal

Core

Cladding

Protective sheath

Light pulse Light rays travel along a single path along the axis of the core

Graded-index multimode fiber

Because of the variations in density of the fiber core from the center to the edges, light traveling along the shorter path down the center travels more slowly than light at the edges. Thus the light rays that make up each pulse reach the receiver at almost the same time irrespective of their path, resulting in relatively little pulse dispersion.

Single-mode fiber

Single-mode fiber has a narrow core which confines light to traveling along its axis in a single path or mode, so there is very little pulse dispersion. Light of a coherent wavelength must be used, so a laser is used as the light source. Single-mode fiber is used for long-distance telecommunications.

Strengthening steel core

Graded-index fiber These fibers are used to transmit information over local-area networks

Light pulse A light-emitting diode (LED) is used as the light source

INPUT SIGNAL **OUTPUT SIGNAL**

INPUT SIGNAL **OUTPUT SIGNAL**

Optical fiber

Light pulse Light rays zigzag down the core by bouncing off the cladding

Plastic sheath An opaque outer sheath prevents stray light from entering the fiber

Protective layer for trunk cable

Trunk cable

Step-index multimode fiber

Like graded-index fiber, this type of fiber has a wide core that allows multiple light paths. These paths vary in length and cause pulse dispersion. Step-index fiber is therefore used where only short lengths of fiber are required, for example, in endoscope tubes. Multimode fibers may be used with light-emitting diodes (LEDs) as sources, rather than expensive lasers.

Multiplexing

Multiplexing increases the capacity of cable systems by increasing the bandwidth of each cable. Multiple data streams are sent down a single cable by combining them to form one complex signal. Streams are then separated at the output end of the cable. Time-division multiplexing (TDM) interleaves streams of digital data, allowing many data streams to be sent in a fraction of the time that would be needed to send them separately. Frequency-division multiplexing (FDM) sends thousands of streams at different frequencies. TDM and FDM can be used together to maximize cable bandwidth.

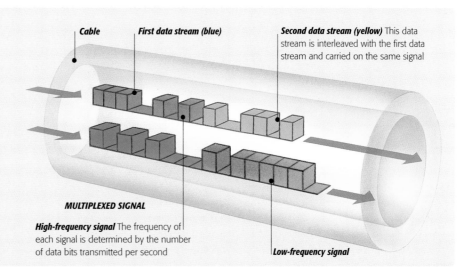

Cable

First data stream (blue)

Second data stream (yellow) This data stream is interleaved with the first data stream and carried on the same signal

MULTIPLEXED SIGNAL

High-frequency signal The frequency of each signal is determined by the number of data bits transmitted per second

Low-frequency signal

SEE ALSO: THE TELEPHONE NETWORK *p.44* | RADIO *p.48* | INTERNET & E-MAIL *p.70* | LASERS *p.220*

Radio

Radio waves are a type of electromagnetic radiation that is all around us, generated naturally in electrical storms and constantly flooding the skies from outer space. The discovery that they could be used to transmit information revolutionized the early 20th-century world. Today, radio waves are used for traffic guidance, cell phones, satellite communications, and radar, as well as TV and radio broadcasts.

Radio waves are electromagnetic radiation with wavelengths between about 0.04 in (1 mm) and 6 miles (10 km) or more. They are often described by their frequency, measured in hertz (cycles per second), which ranges from 30 kHz to 300 GHz. Guglielmo Marconi sent the first Morse code messages using simple pulses of radio waves in 1895, and methods of modulating them—altering the waveform by superimposing a sound or other signal—to carry complex information/data were invented shortly afterward. Radio waves may be either amplitude modulated (AM) or frequency modulated (FM). The signals of FM radio broadcasts have larger bandwidths than those of AM broadcasts—that is, they occupy a wider range of frequencies. Electrical blips and random radio "noise" from the atmosphere diminish the amplitude of radio signals, but do not alter their frequencies. This means FM broadcasts are less prone to interference than AM broadcasts. To avoid radio broadcasts interfering with each other, the frequencies used by radio stations and other transmitters are regulated by international agreement.

Radio broadcast

A radio transmitter works by generating an electric "carrier wave" at the same frequency as the required radio signal. Data is superimposed onto the carrier in a process known as modulation. The transmitter feeds the modulated signal to a mast, where the fluctuating current generates a radio signal that reflects the shape of the modulated carrier wave. Radio waves are usually multidirectional, and a signal's range depends on its frequency. A receiver picks up all frequencies, but uses a tuning circuit to extract one specific signal that is then demodulated to extract the original data.

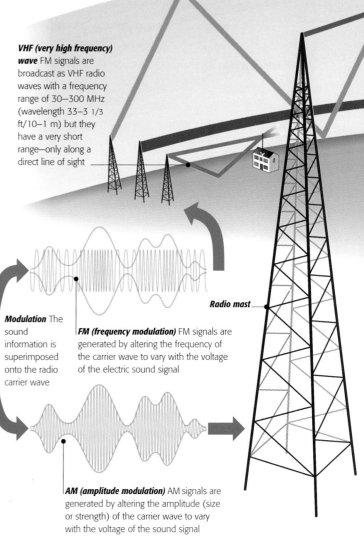

Sound wave Speech or music consists of pressure waves with varying frequency and amplitude

Amplifier This device boosts the tiny electric signal produced by the microphone

Electric sound signal Sound information is represented by the varying voltage of this signal

Microphone Sound waves are converted into electric signals by microphones

Oscillator An oscillator generates a highly regular, radio-frequency electric carrier wave

Electric Carrier wave

Modulation The sound information is superimposed onto the radio carrier wave

VHF (very high frequency) wave FM signals are broadcast as VHF radio waves with a frequency range of 30–300 MHz (wavelength 33–3 1/3 ft/10–1 m) but they have a very short range—only along a direct line of sight

Radio mast

FM (frequency modulation) FM signals are generated by altering the frequency of the carrier wave to vary with the voltage of the electric sound signal

AM (amplitude modulation) AM signals are generated by altering the amplitude (size or strength) of the carrier wave to vary with the voltage of the sound signal

Signal modulation
Radio waves that carry sound or picture data must be modulated (shaped) by that data. To carry sound data, a sound wave is first converted into an electric signal of varying voltage by a microphone. The signal is then superimposed on a radio-frequency carrier wave. Modulation adjusts either the frequency (FM) or the amplitude (AM) of the carrier wave.

Receiving loud and clear

A nondirectional radio antenna transmits radio waves in all directions and is used for broadcasting. Radio waves can be reflected from surfaces such as buildings and hills, so a receiver may simultaneously receive transmissions in the same bandwidth that have all traveled different distances. This causes a type of interference known as multipath interference, distorting the sound. Analog AM and FM signals are particularly prone to this type of interference in built-up areas. Digital radio transmissions avoid this problem by coding the signal so that the receiver can reconstruct it even if there is interference.

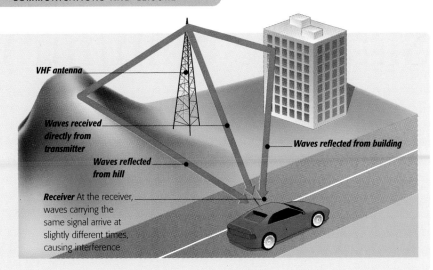

VHF antenna

Waves received directly from transmitter

Waves reflected from hill

Waves reflected from building

Receiver At the receiver, waves carrying the same signal arrive at slightly different times, causing interference

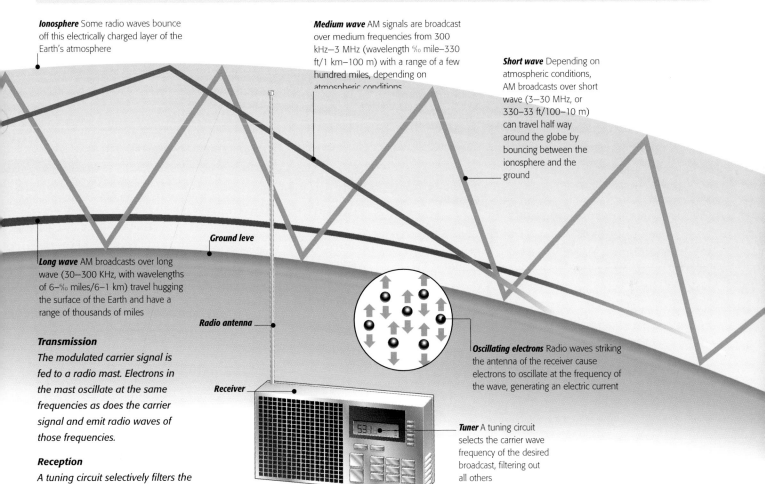

Ionosphere Some radio waves bounce off this electrically charged layer of the Earth's atmosphere

Medium wave AM signals are broadcast over medium frequencies from 300 kHz–3 MHz (wavelength ‰ mile–330 ft/1 km–100 m) with a range of a few hundred miles, depending on atmospheric conditions

Short wave Depending on atmospheric conditions, AM broadcasts over short wave (3–30 MHz, or 330–33 ft/100–10 m) can travel half way around the globe by bouncing between the ionosphere and the ground

Ground leve

Long wave AM broadcasts over long wave (30–300 KHz, with wavelengths of 6–‰ miles/6–1 km) travel hugging the surface of the Earth and have a range of thousands of miles

Radio antenna

Receiver

Oscillating electrons Radio waves striking the antenna of the receiver cause electrons to oscillate at the frequency of the wave, generating an electric current

Transmission

The modulated carrier signal is fed to a radio mast. Electrons in the mast oscillate at the same frequencies as does the carrier signal and emit radio waves of those frequencies.

Reception

A tuning circuit selectively filters the oscillations of electrons in the antenna to allow only frequencies within a certain bandwidth to be received. The small oscillating current produced is a replica of the modulated carrier wave. This is then amplified, rectified, and demodulated, producing an electric equivalent of the original sound wave. A loudspeaker then converts this signal into to sound waves.

Tuner A tuning circuit selects the carrier wave frequency of the desired broadcast, filtering out all others

Amplified AM signal The oscillations of the tiny electric signal within the tuning circuit are amplified

Rectified AM signal Half of the voltage swing of the electric signal is removed

Demodulated signal The carrier-wave frequency is removed from the signal, recreating the original electric signal

Loudspeaker A loudspeaker then converts the electric signal back into sound waves

SEE ALSO: SATELLITE COMMUNICATIONS *p.50* | DIGITAL DATA *p.56* | LIGHT *p.262*

Satellite communications

Since the launch of the first satellite, *Sputnik 1*, in 1957, satellites have revolutionized information broadcast and the collection of data. Communications satellites relay TV pictures, telephone calls, and Internet data from one part of the Earth to another using microwaves that travel at the speed of light. Other types of satellites gather data on the weather, monitor environmental and urban changes, locate mineral deposits, and transmit navigation signals.

In the last 40 years, several thousand satellites have been launched by rockets and reusable spacecraft such as the Space Shuttle. If the satellite is traveling at the appropriate speed for its altitude, the gravitational attraction of the Earth pulls it into a circular or eliptical orbit. Most communications satellites travel in high geostationary orbits, in which they revolve around the Earth at the same rate as the Earth rotates, so they maintain their position above the Earth's surface. These satellites are usually put into a low Earth orbit before being transferred to their final orbit by onboard engines and the final stage of the launch vehicle. Once in the desired orbit, a satellite uses photovoltaic cells fitted on extended solar panels to generate electric power for its operating systems. Gas thrusters and gyroscopes are used to maintain the satellite's orientation and keep its antennae pointed toward Earth.

Communication satellites receive data in the form of radio signals from transmitters on Earth (uplink) and relay the signals to receivers elsewhere on the planet (downlink) via dishlike antennae. Due to the increasing volume of data being transmitted via satellite, the microwave bands currently in use are in danger of being exhausted. To provide extra capacity, NASA is testing a satellite that uses higher-frequency microwaves.

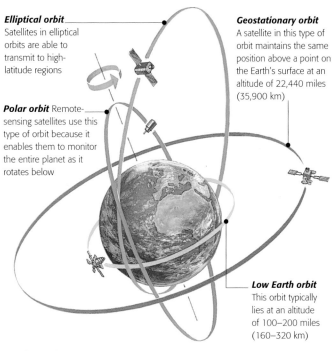

Elliptical orbit
Satellites in elliptical orbits are able to transmit to high-latitude regions

Geostationary orbit
A satellite in this type of orbit maintains the same position above a point on the Earth's surface at an altitude of 22,440 miles (35,900 km)

Polar orbit Remote-sensing satellites use this type of orbit because it enables them to monitor the entire planet as it rotates below

Low Earth orbit
This orbit typically lies at an altitude of 100–200 miles (160–320 km)

Satellite orbits
A satellite's function determines the height, shape, and angle of its orbit. Most communications satellites lie above the equator and complete one orbit in 24 hours, the same time the Earth takes to spin once. In such geostationary orbits, the satellite remains fixed above a single point on the Earth's surface. Remote-sensing satellites orbit several times a day in low polar orbit.

Advanced communications satellite

Launched in 1993, NASA's Advanced Communications Technology Satellite (ACTS) is an experimental spacecraft designed to relay signals at much higher frequencies than conventional satellites do. This allows ACTS to receive and transmit far more data per second than does a satellite employing lower-frequency signals. ACTS is also much more precise in targeting the destination of its downlink transmissions.

Receiving antenna This antenna collects incoming microwave signals

Uplink Audio and visual data are relayed to ACTS in the form of microwave signals

Ground station 1
This ground station uses a large antenna to transmit microwave signals to ACTS

Ground link Much of the data on the ground is passed in digital form via electric or optical-fiber cables

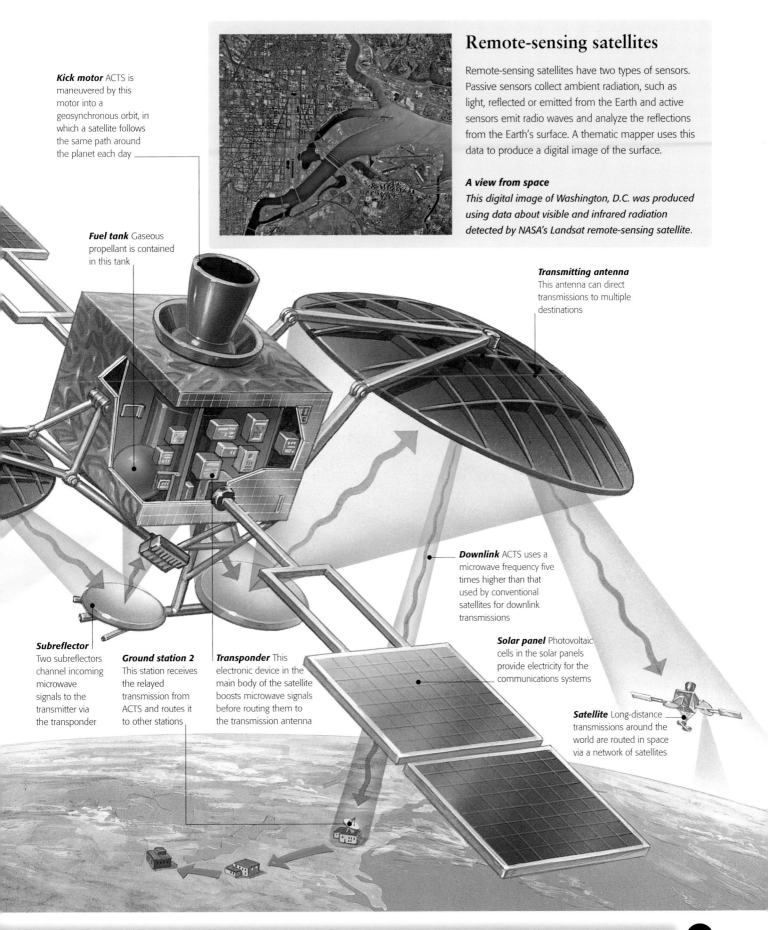

Remote-sensing satellites

Remote-sensing satellites have two types of sensors. Passive sensors collect ambient radiation, such as light, reflected or emitted from the Earth and active sensors emit radio waves and analyze the reflections from the Earth's surface. A thematic mapper uses this data to produce a digital image of the surface.

A view from space
This digital image of Washington, D.C. was produced using data about visible and infrared radiation detected by NASA's Landsat remote-sensing satellite.

Kick motor ACTS is maneuvered by this motor into a geosynchronous orbit, in which a satellite follows the same path around the planet each day

Fuel tank Gaseous propellant is contained in this tank

Transmitting antenna This antenna can direct transmissions to multiple destinations

Downlink ACTS uses a microwave frequency five times higher than that used by conventional satellites for downlink transmissions

Solar panel Photovoltaic cells in the solar panels provide electricity for the communications systems

Satellite Long-distance transmissions around the world are routed in space via a network of satellites

Subreflector Two subreflectors channel incoming microwave signals to the transmitter via the transponder

Ground station 2 This station receives the relayed transmission from ACTS and routes it to other stations

Transponder This electronic device in the main body of the satellite boosts microwave signals before routing them to the transmission antenna

SEE ALSO: THE TELEPHONE NETWORK *p.44* | RADIO *p.48* | GLOBAL POSITIONING SYSTEM *p.136* | SPACE ROCKETS *p.238*

History of computers

The first design for a programmable computer—one that would follow a set of instructions—is usually considered to be the "Analytic Engine" invented by English inventor Charles Babbage in 1832. Babbage's device was designed to perform a sequence of calculations using instructions input on punched cards, and it included both a memory "store" and a processing unit. It was entirely mechanical in design.

Unfortunately, Babbage never assembled his computer, and it was not until the 1900s, with the invention of the electron tube, that components for a viable electronic computer became available. An electron tube is a device that can block, amplify, or act as an on/off switch for an electric current. During the 1920s and 1930s, scientists investigated how to link these devices in arrays that would accept electric signals representing numbers, process the signals according to a program, and output the results. Famous electron-tube computers included the British Colossus, designed to break the German Enigma code during World War II, and the American ENIAC. These machines were huge, and programming them involved changing their circuitry by plugging and unplugging cables, although later machines stored programs in electronically-switched memory areas.

Aside from their large size, computers based on electron tubes had other drawbacks. The heating filaments in the tubes made the computers hot, and the filaments would often "blow." But in 1947, the development of the transistor by scientists at Bell Telephone laboratories transformed the computer landscape.

ENIAC (Electronic Numerical Integrator and Calculator)
Created toward the end of World War II to solve mathematical problems in ballistics, ENIAC was the first all-purpose, stored-program electronic computer. Weighing 30 tons, it worked with decimal rather than binary numbers and needed dozens of operators. ENIAC's 18,000 electron tubes became very hot during operation and it frequently malfunctioned.

1960s mainframe computer
Computers of the 1950s and 1960s ranged from room-sized to cabinet-sized and were based on transistors, which by the 1960s were being grouped on silicon chips. Called mainframes, they were shared by many users in big corporations such as banks but were less powerful than a modern PC. Data and programs for the machines were stored on spools of magnetic tape.

These tiny components were made from crystals of semiconductors such as geranium and silicon, and could do everything an electron tube could do, but were smaller and more reliable. Cheaper, more compact computers were soon in production, although some still occupied a whole room.

Along with the developments in hardware, there were changes in software. Originally, all instructions for computers were written in binary code ("machine code"). In 1951, a programmer named Grace Hopper proposed "reusable software," code segments that could be assembled according to instructions written in a "higher-level language" (something more closely resembling English). Hopper further proposed the concept of a compiler—a program that would translate instructions written in a higher-level language into machine code. FORTRAN, the first fully fledged language, and its compiler were introduced in 1956. During this period, the punched cards and tape used to input data into computers were gradually replaced by magnetic tape and disks.

Integrated circuits

In 1959, engineers at Texas Instruments showed that it was possible to incorporate many transistors, connected by metal tracks, onto one piece of silicon. This innovation became known as an integrated circuit, or "silicon chip," and the trend ever since is summarized in "Moore's Law": the number of transistors that can be put on a chip doubles every 12 to 18 months. Gordon Moore, who formulated this law in 1965, later cofounded the chip manufacturer Intel.

Integrated circuits soon led to the development of yet smaller, cheaper computers, called minicomputers. Although still too expensive for most individuals to afford, these were relatively simple to operate. Other innovations of the 1960s were keyboards for inputting data into computers and monitors for displaying this data and the results of calculations before they were printed out. In 1971, the floppy disk was introduced for data storage.

Microprocessors and microcomputers

Although integrated circuits made computers smaller, the processing units still consisted of a number of circuits on separate chips. In 1971, an engineer working for Intel realized that a set of circuits commissioned for an electronic calculator could all be put onto one chip, and that

Apple Lisa (1983)
The Lisa, a precursor of the Apple Macintosh, was the first personal computer (PC) with a graphical user interface (GUI). With a GUI, commands could be chosen by clicking on pull-down menus or icons (little pictures) on screen, instead of by typing. To work the GUI, the user moved a novel device called a mouse around on the desktop. The Lisa also offered two floppy disk drives, a megabyte of RAM, and a monochrome monitor.

the resulting device could be used as a general-purpose "computer on a chip." The result was the Intel 4004—the world's first microprocessor. Physically, it consisted of a silicon chip in a protective ceramic capsule, with a set of metal pins sticking out that connected it to other components in whatever device it controlled. It contained 2,300 transistors, executed 60,000 operations per second, and could be used for any device—including computers and robots—that required a "brain" for accepting input and following a program of instructions to produce an output. Within five years, many very powerful microprocessors had appeared. The invention of microprocessors set the stage for the arrival of the microcomputer, or personal computer (PC)—an affordable machine for the masses. The first PCs, in kit form, appeared in the mid-1970s, and by the mid-1980s machines such as the Apple Macintosh and those based on a PC first brought out by IBM in 1981 were popular throughout the world. The success of these machines led to an explosion of software, in particular a range of spreadsheet, word-processing, graphic, educational, and games programs. Since the 1980s, a number of strong intertwined themes have driven the computer revolution forward, including a continuing increase in the processing power and decrease in the size and cost of PCs; a switch of emphasis from isolated to linked machines, as evidenced by the growth of local area networks and the Internet; and the spread of computer applications into virtually every aspect of home and business life.

Palm V handheld computer
The continuing trend toward compact computing is exemplified by handhelds. These are typically used as combined organizers, calculators, note-takers, and e-mail devices. Increasingly, they are fitted with devices that provide wireless Internet access and infrared data transfer. Some, such as the Palm handheld, can recognize handwriting.

Digital world

The ability to encode any type of information as a string of binary digits has, since the 1980s, revolutionized the way people communicate with each other, exchange information, buy and sell goods, create works of art and entertainment, and even manufacture objects. Worldwide, vast amounts of digital data are being created and processed every second.

The different products and media that are being created from digital data are vast in range and purpose and include everything from educational and entertainment products to business databases and 3-D-modeled environments. In order to move this huge amount of data around, a global revolution is taking place in the type and scale of hardware used for communications systems, especially cable and wireless networks.

Data types *Information that can be digitized falls into five main categories. For each category, an analog-digital converter (ADC) exists. ADCs include chips found in scanners, computer sound cards, and so on.*

Creating digital data from scratch *In addition, various devices and software exist for creating new digital data.*

Using the data *Some digital data is simply stored on the creator's computer for personal use. But vast amounts are created or made available for use elsewhere—usually for educational, entertainment, or commercial purposes. This "information for dissemination" falls into two broad categories—single media and multimedia products.*

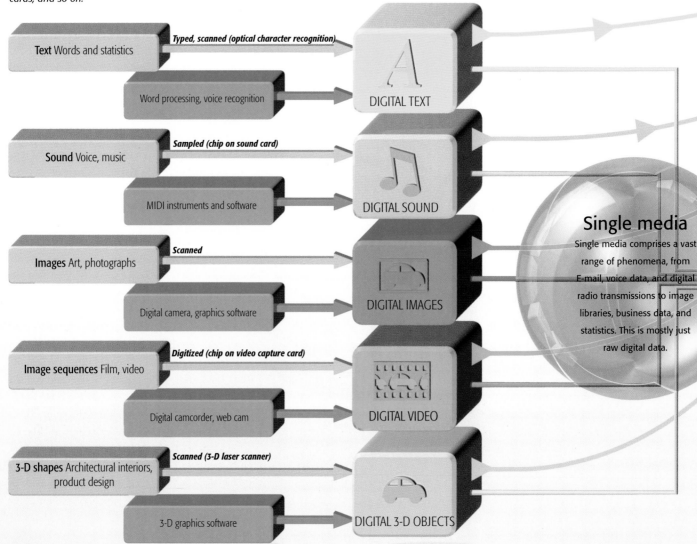

Text Words and statistics

Typed, scanned (optical character recognition)

Word processing, voice recognition

DIGITAL TEXT

Sound Voice, music

Sampled (chip on sound card)

MIDI instruments and software

DIGITAL SOUND

Images Art, photographs

Scanned

Digital camera, graphics software

DIGITAL IMAGES

Image sequences Film, video

Digitized (chip on video capture card)

Digital camcorder, web cam

DIGITAL VIDEO

3-D shapes Architectural interiors, product design

Scanned (3-D laser scanner)

3-D graphics software

DIGITAL 3-D OBJECTS

Single media
Single media comprises a vast range of phenomena, from E-mail, voice data, and digital radio transmissions to image libraries, business data, and statistics. This is mostly just raw digital data.

Multimedia

Increasingly, different data types are combined, edited, and packaged into a single product, often combined with software to provide a choice of "route" through the information. Typical multimedia assemblages include websites, multimedia reference products, and interactive TV programs.

Output

Digital data has to be converted back into analog form—generally sound and/or light output—in order to be understood by humans. Digital-to-analog converters (DACs) include chips on computer sound cards and video graphics cards. The principal output devices currently include computer monitors, games consoles, digital mobile phones, CD-players, digital televisions, and radios. In the future, these may well converge to just two main types of unit—large stationary ones for the home or office and smaller ones for mobile use.

Transmission

Delivery to end users may involve physical media (disks and cartridges), wire cable, optical fiber, or wireless transmission (digital modulation of radio waves). Traditionally, different media types have been associated with particular methods of delivery, but these distinctions are becoming blurred as the bandwidth (data-carrying capacity) of transmission systems increases. In the future, all digital data may be transmitted by ultrafast optical-fiber networks—a true "information superhighway."

End user

The modern consumer is exposed to a continuous stream of digital information and products created by digital processes. But the flow of information is often under the user's control—examples include browsing Internet websites and interactive television. Control is often provided and effected by means of hyperlinks (choices for what to access next) embedded within the data structures.

Server storage

While waiting to be accessed or transmitted, digital data is often stored on the hard disk of a server—a computer set up expressly for the purpose.

Feedback control over flow of data

Manufacture

Digital data is sometimes used to manufacture a tangible product rather than being output directly to consumers. Examples of products are books and newspapers—invariably created digitally today but usually still read in printed form. Objects may also be made by robots sculpting materials such as metal or plastic using digitized shape data, or assembling objects from components using digitized movement data.

SEE ALSO: CABLE TECHNOLOGY *p.46* | DIGITAL DATA *p.56* | INTERNET & E-MAIL *p.70* | WORLD WIDE WEB *p.72*

Digital data

The word digital, when applied to computing or electronics, has a precise meaning: "coded as numbers." The digital revolution, which affects so many aspects of our lives today, relies on the fact that any type of data–text, pictures, sounds, even the shapes of 3-D objects–can be turned into a string of numbers. In this form, remarkable things can be done with the data using the power of computers.

Before the widespread use of computers, there was little purpose to storing data digitally. Most information was kept as printed text and pictures—forms that people could understand directly. Some data (such as sound recordings) were stored as continuous waveforms in materials such as vinyl or magnetic tape and transmitted as modulated electromagnetic waves. Data in these forms is called "analog." Today, digital storage is taking over from analog. One reason is that digital information can be reduced to data strings containing just two values—0 and 1. Such data strings use a counting system called binary in place of the decimal system. Binary numbers are ideally suited to the electronic storage and transmission of information because they require only two alternative states in an electronic circuit—"on" (representing a 1) and "off" (representing a 0.)

A further advantage to digital data is that it is relatively easy to copy and manipulate and can be compressed (recoded to a smaller file size) with little or no loss of information. It can also be stored and transmitted using the same methods, whatever the type of data (text, sound, pictures, or other media). To be used or understood by a human, though, the information has to be reconstructed back into an analog form. This is achieved by special computer processors called digital-to-analog converters (DACs)—they are found on computer sound and graphics cards.

Analog versus digital

```
1010011101111010100001000
0010001111101010000111000
0001110101001110111101010
0001000111000100010101111
```

Analog
An analog signal, such as a sound wave, is continuous. Data was traditionally stored and transmitted in analog form.

Digital
A digital signal is a stream of pulses in a wire, or any other on/off signaling system. The pulses may have only two values–0 or 1.

Decimal versus binary

100s	10s	1s
0	1	3

= 13

8s	4s	2s	1s
1	1	0	1

= 8+4+1 = 13

Decimal number
In a decimal number (reading right to left), the right-hand digit represents ones, the next digit to the left represents 10s, the next digit represents 100s, and so on. So 13 means (1 x 10) + (3 x 1) = 13.

Binary number
In a binary number, the right-hand digit represents ones, the next digit along represents 2s, the next represents 4s, and so on. So in the binary system, 13 becomes 1101, which means (1 x 8) + (1 x 4) + (0 x 2) + (1 x 1).

Bits and bytes

In a binary number, each figure is called a binary digit, or "bit." So a "4-bit" binary number can be anything between 0000 (0 in decimal) and 1111 (15 in decimal). In practice, modern computers handle data in chunks called "bytes" that have a minimum size of 8 bits. Many computers use 16-bit bytes, 32-bit bytes, or even 64-bit bytes. Some examples of 4-bit and 8-bit binary numbers are shown below.

DECIMAL	4-BIT BINARY NUMBERS			
	8s	4s	2s	1s
9 = 8+1=	1	0	0	1
11 = 8+2+1=	1	0	1	1
15 = 8+4+2+1=	1	1	1	1

DECIMAL	8-BIT BINARY NUMBERS							
	128s	64s	32s	16s	8s	4s	2s	1s
22 = 16+4+2=	0	0	0	1	0	1	1	0
197 = 128+64+4+1=	1	1	0	0	0	1	0	1
255 = 128+64+32+16 +8+4+2+1=	1	1	1	1	1	1	1	1

How much space does it take up?

Different types of digital data vary markedly in the space they take up. The approximate quantities of some different media that can be stored on a CD-ROM are shown below. Digital images, sounds, and particularly videos are very space hungry, and this can cause problems with speed of transmission—for example, over the Internet. The main solution to this problem is to compress the data for storage and transmission.

 =   OR  OR OR

| CD-ROM 650 MEGABYTES | SEVERAL LARGE TEXT-ONLY ENCYCLOPEDIAS | 800 HOURS OF MIDI CODE FOR DIGITAL INSTRUMENTS | 1 HOUR OF HIGH-QUALITY STEREOPHONIC SOUND | 600 UNCOMPRESSED 10 X 7 INCH DIGITAL PHOTOS | 1 MINUTE OF UNCOMPRESSED VIDEO |

How information is digitized

There are two different approaches to producing digital data. One is to take an object to be digitized, divide it into tiny pieces, measure each piece, and record this measurement as a number. The more finely the object is fragmented, the more accurate is its digital description. But the accuracy also depends on the precision of the recording method—one that uses 24 bits (three bytes) for each fragment is more accurate than one that uses just 8 bits (one byte). A second method is the "recipe" approach, which records an object such as a tune or drawing as a digitally encoded series of instructions for creating the object. In terms of file size, this is a much more economical method of storing information digitally.

Pixels High quality digital images typically use 24 bits (3 bytes) to record the color of each pixel. This method provides codes for 16 million possible different colors and is termed 24-bit or true color

Digitizing images

A photograph can be digitized by dividing it into thousands of picture elements (pixels) and coding the color of each pixel as a binary number. The digital version of the image is called a bitmap. Typically, each pixel's color is coded as three 8-bit values, representing the relative amounts of red, green, and blue light that, when mixed, would produce that color.

IMAGE ON-SCREEN

	DECIMAL	BINARY
RED	254	1 1 1 1 1 1 1 0
GREEN	200	1 1 0 0 1 0 0 0
BLUE	36	0 0 1 0 0 1 0 0
RED	39	0 0 1 0 0 1 1 1
GREEN	109	0 1 1 0 1 1 0 1
BLUE	175	1 0 1 0 1 1 1 1

```
1111111011001000
0010010000100111
0110110110101111
```
BINARY BITSTREAM

Digital music

A sound can be digitized by fragmenting and measuring its waveform (this is called sampling). Alternatively, digital music can be created by the "recipe" approach using a Musical Instrument Digital Interface (MIDI) device connected to a computer. This method stores music as a series of codes that represent notes, their duration, how forcefully each should be played, and so on.

MIDI KEYBOARD

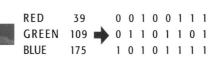

		BINARY
C	Note on	0 0 0 0 0 0 0 1
	Key code	0 0 1 1 1 1 0 0
	Loudness	1 1 0 1 1 1 0 1
	Note off	0 0 0 0 0 0 1 0
C#	Note on	0 0 0 0 0 0 0 1
	Key code	0 0 1 1 1 1 0 1
	Loudness	0 1 0 0 1 1 0 1
	Note off	0 0 0 0 0 0 1 0

```
0000000100111100
1101110100000010
0000000100111101
0100110100000010
```
BINARY BITSTREAM

SEE ALSO: CONTROLLING YOUR COMPUTER *p.62* | MEMORY *p.66* | SOUND RECORDING *p.76* | DIGITAL CINEMA *p.96*

Silicon

Historians divide periods of history into ages that are defined by a particular material used at the time—the Stone Age, the Bronze Age, and the Iron Age. In centuries to come, our descendants will look back on the period that bridged the second and third millennia and refer to it as the "Silicon Age," for it is silicon that defines our times. It is a fundamental material in the key invention of our age, the microprocessor or "microchip," and silicon plays a key role in many other modern technologies, including solar power and the emerging communications technology of photonics.

What is silicon?

After oxygen, silicon is the most abundant material in the Earth's crust, making up nearly 28 percent of its total mass. Silicon does not occur naturally in its pure form but is found in combination with other elements, most frequently with oxygen in the guise of silicon dioxide, the major component of sand and quartz. Silicon compounds have a wide variety of industrial uses, most famously

in glass production, and to make silicones, rubbery materials valued for their water repellent and heat-resistant properties.

Pure silicon is a dark gray shiny solid that feels hard to the touch. Together with some closely related substances, such as germanium, it is a semiconductor. This means that in its pure form, at low temperatures, it is an electric insulator. But when heated or combined with an impurity in a process known as "doping," silicon's conducting properties are altered so that an electric current flows through it under certain conditions. When doped with substances such as phosphorus, silicon develops an excess of free electrons; this is termed n-type silicon. When doped with other substances such as boron, there is an excess of "holes" (spaces that accept electrons), producing p-type silicon. Transistors, the key components in microprocessors and other electronic devices, consist of a sandwich of n-type and p-type silicon.

Silicon chips

When computers first appeared, they were big, slow, and expensive. Matters improved with the replacement of electron tubes with transistors as the means of controlling the flow of current in

Chip-circuit plans

Silicon-chip architects design the electrical circuits of a chip on a computer. The multiple layers of a circuit design are printed on transparent sheets for detailed inspection. These designs are then used to make masks (sheets of opaque material used to shield selected areas of silicon during the photographic printing and etching process). The design masks have to be aligned with extreme accuracy during the fabrication process.

Silicon wafer

This 8-in (20-cm) diameter silicon wafer, seen under an inspection light, has already had the highly complex electric circuits for its component chips etched onto its surface. The wafer is ready to be cut up into 600 or so individual silicon chips.

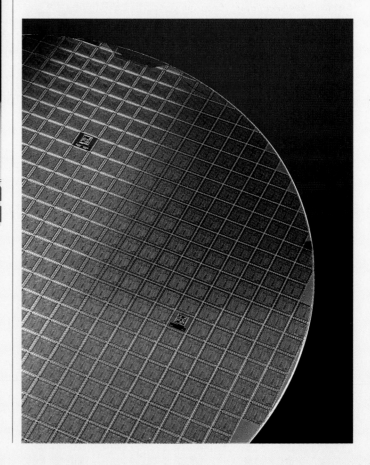

Micromachines
The processes used to produce microprocessors have also been used to produce silicon micromachines—mechanical devices on a similar scale. Components manufactured so far include accelerometers for activating automobile airbags and miniature combination locks.

Directing light
Fabricated from a grid of tiny interlocking silicon bars, each about 1/20,000 in (0.0012 mm) deep, photonic crystals are capable of trapping photons ("packets") of light and of directing the photons along specific and, if necessary, convoluted paths.

circuits. But real progress was not made until 1959, when it was shown that multiple transistors could be formed on the surface of a piece of silicon using various etching techniques. These transistors, composed of tiny areas of doped silicon, could then be connected by metal tracks deposited on the gaps in the silicon surface. The arrangement became known as an integrated circuit, or silicon chip.

Early silicon chips contained relatively simple electronic circuits, but in 1971, an integrated circuit manufacturer named Intel put all the main components for a small computer on a single chip. The result, the 4004 chip, was the world's first microprocessor and was used to power one of the first portable electronic calculators. Since then, several series of increasingly powerful microprocessors have been developed, the trend being to cram more and more transistors onto smaller areas of silicon. The first microprocessor contained 2,300 transistors; by 1999, an average microprocessor contained over 10 million.

How silicon chips are made

The fabrication of silicon chips is a complex process. Silicon must first be obtained in pure form, which is done by melting mined quartz and distilling it until all the impurities have been removed.

A silicon microcrystal is then dropped into the melted quartz and acts as a "seed" around which a larger, cylindrical silicon crystal grows. Using diamond saws, the silicon ingot obtained is sliced into thin disk-shaped "wafers," from which hundreds of chips can be made. Electronic components are created on each wafer in several stages. First, the wafer is coated with an emulsion sensitive to ultraviolet light. The wafer is then exposed to ultraviolet light shone onto it through a mask corresponding to the circuit design, and exposed emulsion is dissolved away. Exposed areas of silicon are "doped" and more components added, using further photographic and chemical methods. Finally, metal contacts and connecting tracks are added to the components.

The etched silicon wafers are cut into individual chips using a diamond saw. Each chip is inserted into a plastic or ceramic package that provides the connections needed for communication with the circuit board in the computer or other device into which it will be mounted. Finally, the chips are thoroughly tested, since the manufacturing process produces significant numbers of faulty chips that must be rejected.

Other uses for silicon

A further use of silicon as a semiconductor is in photovoltaic (solar) cells. Each cell consists of a piece of n-type silicon joined to a piece of p-type silicon. A small movement of electrons across the junction from n-type to p-type creates a small voltage difference across the junction. When light strikes the cell, it dislodges electrons, which flow across the junction, constituting a current. Modern solar cells are almost 70 percent efficient in converting light into electrical energy.

Silicon's hardness, combined with its suitability for fabrication at a microscopic level, has more recently been exploited in the production of micromachines—tiny gears, wheels, and valves carved out of silicon using variations of the techniques used to produce microchips. Micromachines will be used in areas such as medicine and in motion-control devices in spacecraft.

Similar processes are used to make photonic crystals—lattices of tiny interlocking silicon bars that can trap light of particular frequencies. The exact frequency of the light is chosen by altering the distances between the bars. By introducing variations or impurities into the lattice, specific pathways for light to travel though the crystal can be created. As data transfer in the future is likely to be mostly by light pulses rather than electrons, photonic crystals may hold the key to rapid advances in communications.

SEE ALSO: HISTORY OF COMPUTERS *p.52* | COMPUTER MEMORY *p.66* | SOLAR ENERGY *p.174* | ELECTRONICS *p.264*

Personal computers

The term "personal computer"(PC) refers to a self-contained, general-purpose, microprocessor-based computer operated by a single user, as opposed to a large "mainframe" machine shared by users at multiple terminals. Since the 1980s, PCs have become ubiquitous in offices and most homes. Two main types of PCs are IBM-compatible PCs (or, simply, "PCs") and the less widespread Apple Macintoshes.

Personal computers or microcomputers were made possible by two major 20th-century innovations: the integrated circuit and the microprocessor. Together, these two developments have delivered computing power onto a desktop that a few decades ago would have required a machine the size of a room.

The "brain" of a computer is the CPU (central processing unit) or processor, which executes programmed instructions, performs data calculations, and coordinates storage, input, output, and communications devices. The CPU is located on the motherboard, the main circuit board, together with RAM (random-access memory) chips and connectors to the rest of the computer.

The rapid development of CPU technology and operating systems (the basic controlling software) drives computer evolution. One of the major determining factors of computer power is CPU clock speed. The first IBM PC (1981) had an Intel CPU with a clock speed of about 8 MHz. By 2000, seventh-generation CPUs offered speeds in excess of 700 MHz. Modern PCs mostly run versions of Microsoft's Windows operating system. The Apple Macintosh is based on a Motorola CPU that runs Apple's own operating system. Introduced in 1984, it became the first mass-produced machine to have a graphical user interface (GUI), which enabled the user to navigate around a virtual desktop with a mouse. By 1999, the Macintosh had developed to reach supercomputer speeds of one billion operations per second.

Hard disk drive The hard disk is the main storage unit for programs and data, and typically stores several gigabytes (billions of bytes) on a stack of rigid magnetic platters. Unlike RAM, disks retain data even when the computer is turned off

Intel Pentium III processor
At the turn of the 21st century, the most up-to-date IBM-compatible PCs were powered by the Intel Pentium III microprocessor, of which part of the detailed circuitry is shown above. The Pentium III was designed specifically for fast processing of sound, image, and video data.

Internal speaker An internal speaker gives auditory feedback, including warning signals, and reproduces music in the absence of external loudspeakers

Floppy disk drive Floppy drives once used removable 5¼-in magnetic disks, and more recently, 3½-in disks that store 1.4 MB. The advent of E-mail and higher-capacity media such as 250 MB Zip disks have rendered 3½-in floppies obsolete

Apple iMac
A far cry from the first preassembled home computer, the Apple II of 1977, the iMac, launched in 1998, houses the motherboard and monitor in a sculpted translucent case and is available in a range of colors. The latest iMacs and IBM PCs include high-capacity hard drives (often 10 GB or larger), internal modems, DVD-ROM drives, and standardized data communications ports such as universal serial bus (USB) and the high-speed Firewire.

Inside a PC

The major components of a personal computer, excluding the monitor, keyboard, and mouse, are housed in a metal and plastic case. The desktop case of a typical IBM-compatible PC is shown here; computers with more components, such as extra disk drives, are usually housed in free-standing "towers." The case contains a printed circuit board called the motherboard to which all the other major components—such as CPU, hard disk, RAM, and expansion cards—are connected.

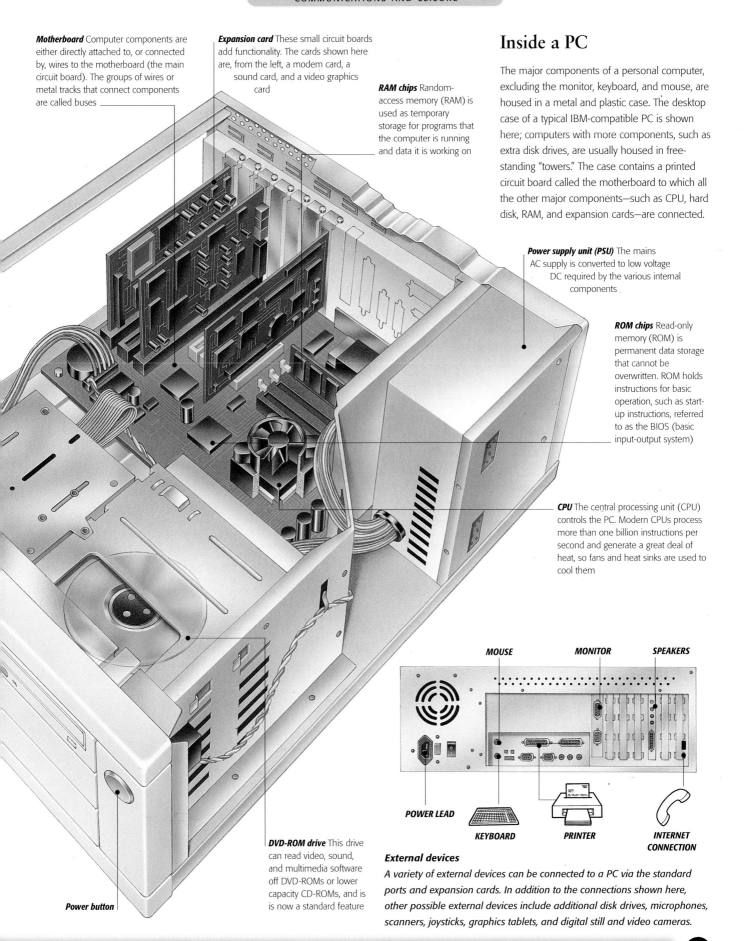

Motherboard Computer components are either directly attached to, or connected by, wires to the motherboard (the main circuit board). The groups of wires or metal tracks that connect components are called buses

Expansion card These small circuit boards add functionality. The cards shown here are, from the left, a modem card, a sound card, and a video graphics card

RAM chips Random-access memory (RAM) is used as temporary storage for programs that the computer is running and data it is working on

Power supply unit (PSU) The mains AC supply is converted to low voltage DC required by the various internal components

ROM chips Read-only memory (ROM) is permanent data storage that cannot be overwritten. ROM holds instructions for basic operation, such as start-up instructions, referred to as the BIOS (basic input-output system)

CPU The central processing unit (CPU) controls the PC. Modern CPUs process more than one billion instructions per second and generate a great deal of heat, so fans and heat sinks are used to cool them

Power button

DVD-ROM drive This drive can read video, sound, and multimedia software off DVD-ROMs or lower capacity CD-ROMs, and is is now a standard feature

MOUSE   MONITOR   SPEAKERS

POWER LEAD   KEYBOARD   PRINTER   INTERNET CONNECTION

External devices

A variety of external devices can be connected to a PC via the standard ports and expansion cards. In addition to the connections shown here, other possible external devices include additional disk drives, microphones, scanners, joysticks, graphics tablets, and digital still and video cameras.

SEE ALSO: CONTROLLING YOUR COMPUTER *p.62* | COMPUTER MEMORY *p.66* | COMPUTER OUTPUT DEVICES *p.68*

Controlling your computer

One of the earliest computer input devices was the keyboard, which was quickly accepted because it resembled a typewriter. But the keyboard has now been partly replaced by the point-and-click actions of the mouse, and in computer games by the joystick. New developments in materials technology have also meant that computer screens can now be made sensitive to fingertip touch.

Communicating with a computer is a matter of transforming an operator's commands into electronic signals that the central processing unit inside the computer interprets and follows. Whether these commands be keystrokes on a keyboard, movements of a mouse, joystick, or trackball, or the pressure of a fingertip on a touch pad or a touch-sensitive screen, they all involve signals relayed as electronic pulses.

Advances in input-device technology are geared toward making these devices more compact and more versatile, and improving on-screen response times. Unlike the mouse, which requires a flat desktop space to be operated, and has a rolling ball mechanism that needs regular cleaning, touch pads can be built into a computer and, being completely sealed, are not affected by dirt and dust so they can operate in any environment.

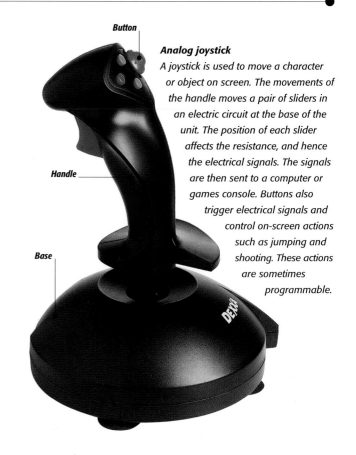

Button

Analog joystick
A joystick is used to move a character or object on screen. The movements of the handle moves a pair of sliders in an electric circuit at the base of the unit. The position of each slider affects the resistance, and hence the electrical signals. The signals are then sent to a computer or games console. Buttons also trigger electrical signals and control on-screen actions such as jumping and shooting. These actions are sometimes programmable.

Handle

Base

Touch-sensitive technologies

Touch-sensitive technology used to manipulate computers is developing in two ways. Firstly, the screen itself can be made into a sensor that detects the presence of a fingertip (or a stylus) and turns it into an electric signal. Touch-screen systems are widely used to operate automated teller machines and other machinery where ease of control is required.

Secondly plastic pads are used as sensors. These touch pads, first widely used as graphic tableaux, have now become a standard device on many laptop computers. Movement of a finger across the rectangular pad causes the cursor to follow the exact same path on screen. In most touch pads the moving finger changes the distribution of electrical charge in the pad's circuits and so produces variations in an electric signal. Touch pads can achieve cursor tracking speeds of up to 40 in (100 cm) per second.

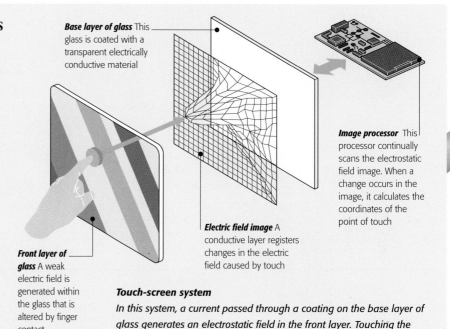

Base layer of glass This glass is coated with a transparent electrically conductive material

Image processor This processor continually scans the electrostatic field image. When a change occurs in the image, it calculates the coordinates of the point of touch

Front layer of glass A weak electric field is generated within the glass that is altered by finger contact

Electric field image A conductive layer registers changes in the electric field caused by touch

Touch-screen system
In this system, a current passed through a coating on the base layer of glass generates an electrostatic field in the front layer. Touching the screen disturbs the field and these changes are plotted by a processor.

The mechanism of the mouse

The mouse allows swift and easy access to the desktop's many icons and operations, such as choosing options from menus, manipulating windows, and moving files. When the mouse is moved, a rubber ball rotates and turns two rollers, each connected to a slotted wheel. A light-emitting diode (LED) shines light through the slots, and transducers (detectors) convert the light signal into electric signals. Pressing a button sends additional signals to the computer.

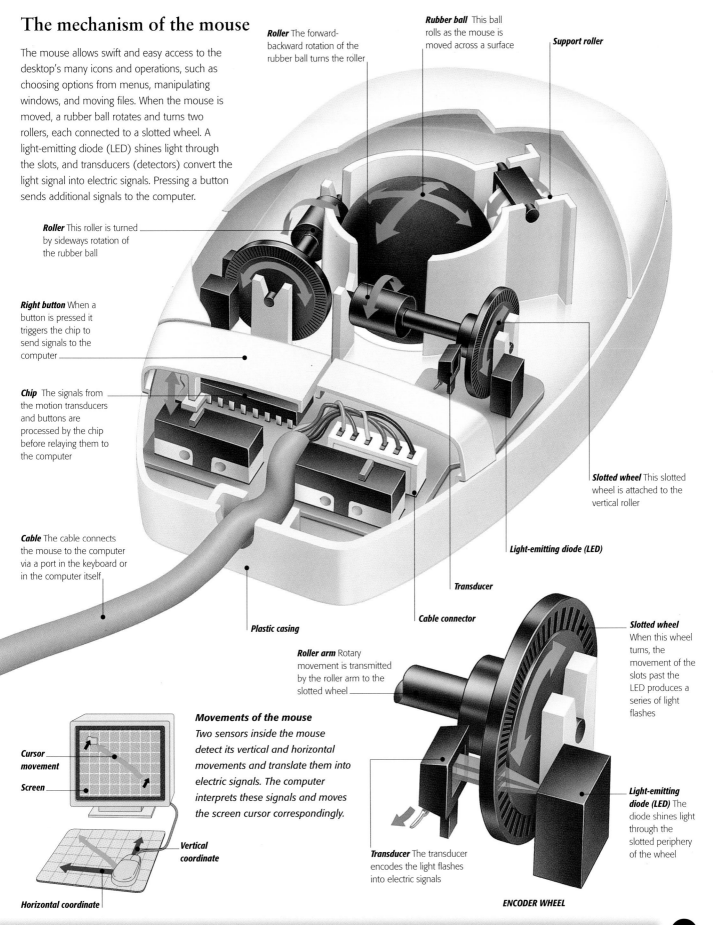

Roller The forward-backward rotation of the rubber ball turns the roller

Rubber ball This ball rolls as the mouse is moved across a surface

Support roller

Roller This roller is turned by sideways rotation of the rubber ball

Right button When a button is pressed it triggers the chip to send signals to the computer

Chip The signals from the motion transducers and buttons are processed by the chip before relaying them to the computer

Cable The cable connects the mouse to the computer via a port in the keyboard or in the computer itself

Plastic casing

Cable connector

Transducer

Light-emitting diode (LED)

Slotted wheel This slotted wheel is attached to the vertical roller

Roller arm Rotary movement is transmitted by the roller arm to the slotted wheel

Slotted wheel When this wheel turns, the movement of the slots past the LED produces a series of light flashes

Cursor movement

Screen

Movements of the mouse
Two sensors inside the mouse detect its vertical and horizontal movements and translate them into electric signals. The computer interprets these signals and moves the screen cursor correspondingly.

Vertical coordinate

Horizontal coordinate

Transducer The transducer encodes the light flashes into electric signals

Light-emitting diode (LED) The diode shines light through the slotted periphery of the wheel

ENCODER WHEEL

SEE ALSO: PERSONAL COMPUTERS *p.60* | SCANNING DEVICES *p.64* | COMPUTER OUTPUT DEVICES *p.68*

Scanning devices

Scanning devices allow visual information to be transferred into a computer. This information ranges from simple black-and-white product bar codes to full-color photographic images. In all cases, light is reflected by or transmitted through the object being scanned to collect data on the brightness of different areas. This data is converted into an electric signal that is then digitally encoded.

One of the most widespread scanning devices today is the flatbed scanner, which was once an expensive professional machine but is now a low-cost, personal-computer accessory. Scanners divide up images into thousands of pixels (picture elements) for display on a computer screen or reproduction by a printer. Low-cost scanners use fluorescent light and a series of mirrors to direct reflected light from a strip of the image onto a charge-coupled device (CCD). Professional drum scanners in the printing industry use a laser beam to scan the surface of the original, and measure the intensity of the reflected light.

Bar-code scanners, which are even more widespread, measure the intensity of light reflected from white and black areas on a bar code. The simplest bar-code readers are pens or guns that are laid directly over the top of the bar code, but more advanced models use sophisticated optics to trace a laser beam through a three-dimensional grid of space, so it will pick up light reflected from a bar code held at almost any angle.

Flatbed scanner

Most scanners work by measuring the light reflected from an original, line by line. The light is focused through a lens onto a charge-coupled device (CCD). The scanner resolution depends on the number of light sensors on the CCD and on the distance it advances with each line. A transparency scanner detects the light passing through, rather than reflecting off, an original.

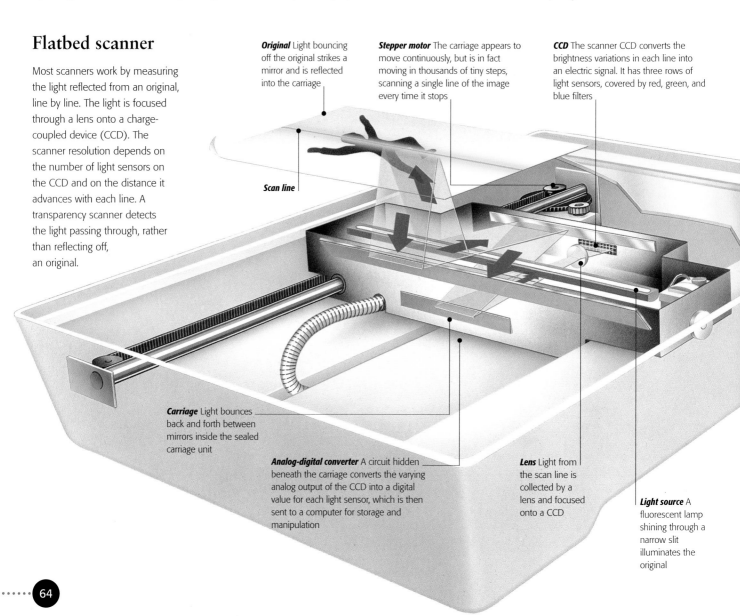

Original Light bouncing off the original strikes a mirror and is reflected into the carriage

Scan line

Stepper motor The carriage appears to move continuously, but is in fact moving in thousands of tiny steps, scanning a single line of the image every time it stops

CCD The scanner CCD converts the brightness variations in each line into an electric signal. It has three rows of light sensors, covered by red, green, and blue filters

Carriage Light bounces back and forth between mirrors inside the sealed carriage unit

Analog-digital converter A circuit hidden beneath the carriage converts the varying analog output of the CCD into a digital value for each light sensor, which is then sent to a computer for storage and manipulation

Lens Light from the scan line is collected by a lens and focused onto a CCD

Light source A fluorescent lamp shining through a narrow slit illuminates the original

Webcams

Webcams are low-resolution digital cameras that capture visual data for use on a computer. They are set up to record at anything from 24 frames per second (for full-motion video) to one frame every few minutes or hours. The webcam is built around a light-detecting CCD chip, which usually has a resolution of 640 x 480 pixels. Webcam pictures are piped directly into the computer through one of its data ports and can then be manipulated on screen. The image files may be uploaded directly onto the World Wide Web by a computer with an Internet connection, or they may be sent as video files attached to E-mails.

Internet videoconferencing

Using the power of the Internet, sound and pictures can be sent between several webcam users at distant locations, allowing them to interact as if they were in a face-to-face meeting. The same technology may soon be used in a stripped-down unit to make domestic videophones a reality.

Bar-code reader

A supermarket bar-code reader has a spinning holographic disk that acts like a rapidly changing lens, focusing a laser beam at different points to draw a 3-D grid in space above it. When the beam is reflected from a bar code, the reader detects a pattern of reflection from dark and light areas.

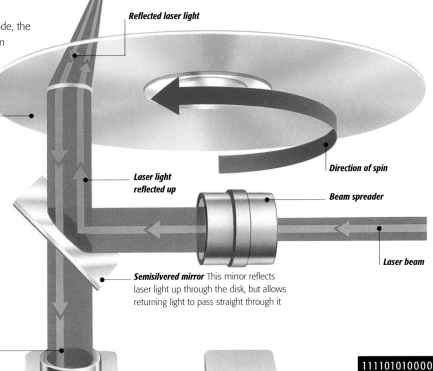

Bar code Dark areas absorb laser light, but light areas reflect it. Some light travels back to reenter the reader. A special code at either end of the bar enables the reader to recognize the beginning and end of the code

Reflected laser light

Holographic disk The disk diffracts and focuses the laser light to a specific point. As the disk rotates, the laser focus traces out a grid

Direction of spin

Laser light reflected up

Beam spreader

Laser beam

Semisilvered mirror This mirror reflects laser light up through the disk, but allows returning light to pass straight through it

Glass plate A glass plate isolates the scanning mechanism from dirt and dust

Laser light received

Photoelectric cell The cell converts the varying levels of laser light into a varying electric current

Analog-digital converter The current is converted into binary electric pulses by this device

Binary code A computer looks up the code from the reader in a database to identify the product

SEE ALSO: DIGITAL WORLD *p.54* | PERSONAL COMPUTERS *p.60* | PHOTOCOPIERS *p.84* | VIDEO TECHNOLOGY *p.104*

Computer memory

All computers require a range of devices for storing binary data, both short- and long-term. During operation, the program a computer is executing, and the data it is working on, are kept in RAM (random-access memory)—transistor-based chips that provide fast data access for the computer's processor. But for long-term storage, most computer systems use a hard disk. Unlike data in RAM, data on a hard disk remains there when the computer is turned off.

There are many types of RAM. Dynamic RAM (DRAM) holds data in arrays of capacitors on a silicon chip. The data has to be refreshed (read from and then rewritten to the capacitor array) about every 15 microseconds, so DRAM is relatively slow, though much quicker than a hard disk. Static RAM (SRAM) is a faster form of RAM that holds binary data in small electric circuits called bistables or flip-flops, which have no need for capacitors. SRAM is more expensive than DRAM.

Besides hard disks and RAM, various other types of memory exist. ROM (read-only memory) chips contain fixed instructions for a computer to run when it starts up and routines for how it should respond to signals from the keyboard or other peripherals. The data in ROM cannot be altered. Flash memory is another type of memory used to carry fixed instructions, but unlike ROM it can be quickly erased and written to in units called blocks; it is used in devices such as digital cellular phones and smart cards. Cache memory is a very fast-access memory region used for data that a processor needs to use frequently.

DRAM

Most of a computer's working memory is in the form of dynamic RAM (DRAM). This comes in the form of plug-in modules called SIMMs or DIMMs (single/double in-line memory modules), each of which contains several chips. A DRAM chip contains an array of transistors, each connected to a capacitor. The 1s and 0s of binary numbers are stored as charges in the capacitors within the chip.

Writing data

To write to DRAM, each address line is opened in turn, switching on the transistors in that line. A binary number is sent down the data lines as a pattern of high- or low-voltage pulses (1s and 0s). In the transistors receiving a pulse, the pulse's voltage charges the capacitor. The number is thus stored as a pattern of charged and noncharged capacitors.

A transistor in DRAM

Each transistor is associated with an address line, a data line, and a capacitor. When the address line to a transistor is opened (green), it switches the transistor on. Any electric current flowing down the data line can then pass through the transistor and build up a charge on its associated capacitor (red).

Reading from DRAM

To read from DRAM, each address line is opened in turn. The electric charges are allowed to flow back out of the capacitors. A pulse of charge coming back out of a data line is taken to represent a 1, and no pulse represents a 0. The DRAM is then "refreshed" to restore the read data.

How a hard disk works

The storage region of a hard disk is a stack of flat platters with magnetizable coatings. Data is stored as patterns of alignment of magnetizable regions within the coating called domains. To read or write data, a mechanism called an actuator moves read-write heads to the correct position relative to the disk while the platters are spun at high speed. Signals are then sent to or received from the read-write heads.

FAT One part of the disk—the file allocation table (FAT)—stores information about the location of all the files on the disk

Read-write head The head reads and writes data, gliding 0.00008 in (0.002 mm) above the platter surface. There is one head on each side of each platter

Actuator arm Each read-write head is carried at the end of a light arm, which turns about a pivot at one end, moving the heads in unison

Data cable Data flows via this cable between the hard disk and a device called the disk controller. The controller governs the spin of the platters and the flow of data to and from the read/write heads and the actuator

Actuator block The actuator receives a continuous flow of instructions for moving the read-write heads. It may move the heads up to 50 times a second

Magnet

Moving coil A coil moves within a permanent magnet at the heart of the actuator. An electric pulse sent to the coil causes it to deflect a precise distance, moving the actuator arms in turn

Platter This stores data on both surfaces

Sector Each track contains several sectors

Track Before first use, the magnetizable coating on each platter is divided into concentric tracks by special signals from the computer, in a process called formatting

Spindle motor This motor spins the platters at several thousand revolutions per minute

Wires These wires carry the electronic data being written or read between the read-write head and the disk controller

Electromagnet When data is being written to the disk, electric pulses sent here produce magnetic fields that align the domains in the underlying track

Read-write head

Once a head is correctly positioned, electric pulses are sent to the electromagnet at its tip in order to write data to a sector. Binary data (1s and 0s), encoded by changes in current direction, are turned into alignment patterns in the domains. Data is read from the disk by the reverse process—passage of the domains under the electromagnet inducing currents in the wire leads.

1 1 0 1 1 1 1 0 1

Data-carrying domains Each domain is aligned in one of two possible directions. A change of alignment from the previous domain signifies a 1, no change signifies a 0

Randomly aligned domains Where the disk has never had data written to it, the domains are aligned randomly

SEE ALSO: THE DIGITAL WORLD *p.54* | PERSONAL COMPUTERS *p.60* | PLASTIC CARD TECHNOLOGY *p.150*

Computer output devices

Computers can produce output in many forms—as static images on paper and moving images on screens, as sound in the form of music or synthesized speech, as mechanical movements of robots or other machines, and as electronic signals for relay to other computers or electronic devices. Common output devices include monitors and printers. Innovative devices such as braille displays for the visually impared cater for those with special needs.

The main output device used with personal computers is the monitor. Monitors for most desktop computers use cathode-ray tubes and work like standard television screens. Laptop computers and flat-screen desktop displays use a different technology called LCD (liquid crystal display). LCDs are slim, light, and consume less power than cathode-ray tubes. Active matrix LCD screens, also called TFT (thin-film transistor) screens, can display high-resolution video images in millions of colors, although the basic technology is the same as that used in most calculators and digital watches.

Early computer printers used a set of pins to hammer patterns onto paper through an inked ribbon. Most modern printers use either laser technology, like that used in some photocopiers, or ink-jet technology. An ink-jet printer has a printhead with up to several hundred nozzles, which squirt droplets of quick-drying inks onto paper in a precisely controlled manner.

Output in braille

Braille displays carry out the function of a monitor screen for computer users with visual impairments. The display unit has a row of cells, each with a grid of small pins that are raised or lowered by piezoelectric signals to represent different braille characters. The main display row represents the line of text being read. Auxiliary cells indicate the position of the cursor on the screen and the length of the text. This unit can be integrated with a speech synthesizer and a braille printer, which embosses braille dots onto paper using small hammers.

BRAILLE DISPLAY ATTACHED TO LAPTOP

Cell Each cell of the display has a grid of pins that is used to display an individual character

Thermal ink-jet color printer

Ink-jet printers use an array of tiny nozzles arranged in a printhead. There are typically 48 nozzles for each of the four basic printing colors. The printhead moves across the paper, squirting droplets onto it under the control of sophisticated driver software. Thermal ink-jet printers use tiny heaters in each nozzle to vaporize some of the ink. The expanding vapor bubble pushes out a droplet of ink, which may be only three picoliters (three-trillionths of a liter) in volume. Piezoelectric ink-jets eject ink using a crystal that distorts when a voltage is applied to it.

1 A signal from the printer hardware turns on a heating element. Ink above the element heats up and quickly forms a vapor bubble.

2 The bubble expands, reaching its maximum size in a few milliseconds. The growing bubble forces ink out of the nozzle.

3 The heater is turned off. The vapor bubble collapses, causing a droplet to separate from the nozzle. Fresh ink moves into the nozzle.

4 As the ink droplet hits the paper, the nozzle is replenished with ink, and the heating and bubble-formation cycle begins again.

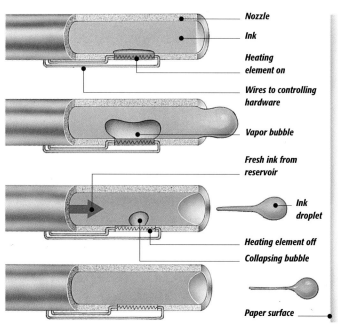

Nozzle

Ink

Heating element on

Wires to controlling hardware

Vapor bubble

Fresh ink from reservoir

Ink droplet

Heating element off

Collapsing bubble

Paper surface

Active matrix LCD screens

Typical LCD screens contain 1,024 x 768 pixels (picture elements), each made of red, green, and blue subpixels. White light from a backlight passes through polarizing filters, a layer of liquid crystal, and color filters to illuminate the subpixels. The liquid crystal behind each subpixel acts as a shutter, determining the amount of light transmitted. An active matrix (a grid of 2.3 million transistors, one for each subpixel) applies voltages to electrodes that affect the transmissive properties of the liquid crystal. The changing pattern of subpixel illumination generates dynamic images.

LAPTOP COMPUTER WITH LCD SCREEN

LCD screen The 12-in (30-cm) wide screen is less than ½ inch (1 cm) thick

Keyboard

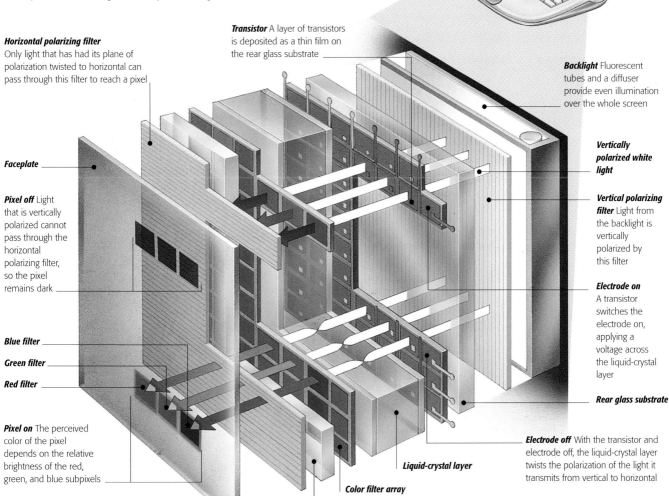

Horizontal polarizing filter
Only light that has had its plane of polarization twisted to horizontal can pass through this filter to reach a pixel

Transistor A layer of transistors is deposited as a thin film on the rear glass substrate

Backlight Fluorescent tubes and a diffuser provide even illumination over the whole screen

Faceplate

Pixel off Light that is vertically polarized cannot pass through the horizontal polarizing filter, so the pixel remains dark

Vertically polarized white light

Vertical polarizing filter Light from the backlight is vertically polarized by this filter

Electrode on A transistor switches the electrode on, applying a voltage across the liquid-crystal layer

Blue filter

Green filter

Red filter

Rear glass substrate

Pixel on The perceived color of the pixel depends on the relative brightness of the red, green, and blue subpixels

Electrode off With the transistor and electrode off, the liquid-crystal layer twists the polarization of the light it transmits from vertical to horizontal

Liquid-crystal layer

Color filter array

CROSS-SECTION OF AN LCD SCREEN

Front glass substrate

How liquid crystals work

Liquid-crystal (LC) molecules flow like liquids, but line up regularly like crystals. The rod-shaped molecules naturally align side by side. This alignment becomes twisted if the LC is trapped between grooved plates called alignment layers. The molecules realign end to end if a voltage is applied across the LC layer. Depending on its alignment, the LC affects the plane of polarized light passing through it.

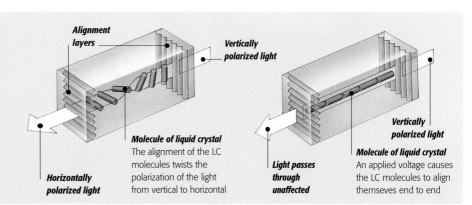

Alignment layers

Vertically polarized light

Molecule of liquid crystal The alignment of the LC molecules twists the polarization of the light from vertical to horizontal

Horizontally polarized light

Light passes through unaffected

Vertically polarized light

Molecule of liquid crystal An applied voltage causes the LC molecules to align themseves end to end

SEE ALSO: PHOTOCOPIERS & FAX MACHINES *p.84* | TELEVISION *p.100* | ELECTRICITY & MAGNETISM *p.260*

Internet and E-mail

The Internet is a global network that links together hundreds of thousands of smaller computer networks and thus millions of computers worldwide. It functions as a way of transferring data and is used especially for E-mail, but also for information gathering, e-commerce, and as a discussion forum. The core of the Internet is a global array of supercomputers connected by high-speed links.

Data to be transmitted over the Internet is first broken down into small "packets," which may travel by different routes. The data is reassembled at its destination. To allow this, each packet is labeled electronically with its destination address and its position in the stream of packets that make up the data. For the system to work, computers on the network must follow a set of rules known as the Internet Protocol (IP). The two most common applications of the Internet today are electronic mail (E-mail) and the World Wide Web—a vast collection of information, commerce, and entertainment centers called "websites," each of which contains links to other related sites. Other Internet features include newsgroups (public forums for discussion) such as Usenet, and on-line chat sites, which involve talking with other Internet users in a "virtual room" by typing messages.

The Internet

At the heart of the Internet is a network of supercomputers. Within each region of the world, the network is connected by devices called nodes to a number of smaller networks, and so, ultimately, to home and office computers. Here, one user is sending data to a remote site from home, while an office worker is downloading files from a remote website.

Internet Service Provider (ISP) An ISP is an organization that provides a gateway to the Internet along with other services, such as fileserver space for installing websites

Home computer Hundreds of millions of home-computer users have modems that allow them to access the Internet via an ordinary phone line linked to an Internet Service Provider (ISP)

Binary information Data travels over the Internet in digital form, as binary code

Router Routers are computers that monitor traffic and find the best route for sending data packets to their destinations

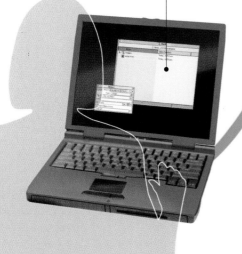

Telephone line The data is usually transmitted as an analog signal, at quite slow rates. But it can also be sent at much higher speeds in digital form using a technology called ADSL (Asymmetric Digital Subscriber Line)

E-mail via the Internet

E-mail is a way of sending messages from one computer to another, anywhere in the world, via telephone lines and the Internet. To send or receive E-mail, you must have an E-mail address, which takes the general form "username@providername.code." For example, the address "dean@scene.com" belongs to someone called "Dean" who works for a commercial organization called "Scene." At one time E-mail was restricted to text messages, but now you can attach images, sounds, and even video clips to messages. The time a message takes to reach its destination—which can be as short as a few seconds—depends on how fast the Internet hardware is working, not on how far the message must travel.

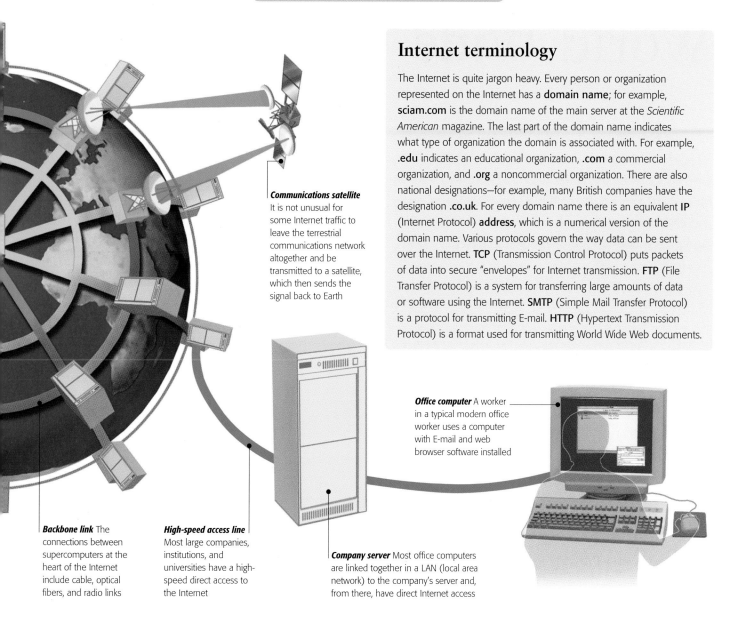

Internet terminology

The Internet is quite jargon heavy. Every person or organization represented on the Internet has a **domain name**; for example, **sciam.com** is the domain name of the main server at the *Scientific American* magazine. The last part of the domain name indicates what type of organization the domain is associated with. For example, **.edu** indicates an educational organization, **.com** a commercial organization, and **.org** a noncommercial organization. There are also national designations—for example, many British companies have the designation **.co.uk**. For every domain name there is an equivalent **IP** (Internet Protocol) **address**, which is a numerical version of the domain name. Various protocols govern the way data can be sent over the Internet. **TCP** (Transmission Control Protocol) puts packets of data into secure "envelopes" for Internet transmission. **FTP** (File Transfer Protocol) is a system for transferring large amounts of data or software using the Internet. **SMTP** (Simple Mail Transfer Protocol) is a protocol for transmitting E-mail. **HTTP** (Hypertext Transmission Protocol) is a format used for transmitting World Wide Web documents.

Communications satellite It is not unusual for some Internet traffic to leave the terrestrial communications network altogether and be transmitted to a satellite, which then sends the signal back to Earth

Office computer A worker in a typical modern office worker uses a computer with E-mail and web browser software installed

Backbone link The connections between supercomputers at the heart of the Internet include cable, optical fibers, and radio links

High-speed access line Most large companies, institutions, and universities have a high-speed direct access to the Internet

Company server Most office computers are linked together in a LAN (local area network) to the company's server and, from there, have direct Internet access

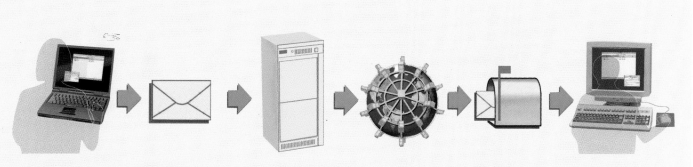

1 The sender types a message, attaches any media files, specifies the E-mail address(es) of the recipients, and then clicks on "send."

2 Client software (for example, Simple Mail Transfer Protocol) encodes the message to enable transmission over a network.

3 A mail server, which is located either at an ISP or on the sender's server, routes the message to the correct electronic address.

4 The Internet carries the encoded message to a server, located at the domain specified in the E-mail address of the recipient.

5 The server converts the message into a form that can be read by the recipient's software and places it in the correct mailbox.

6 The recipient logs onto the server, and all new mail messages are delivered into his or her inbox, along with any attached media files.

SEE ALSO: SATELLITE COMMUNICATIONS *p.50* | COMPUTER OUTPUT DEVICES *p.68* | WORLD WIDE WEB *p.72*

World Wide Web

The World Wide Web consists of millions of digital files stored on computers across the world. A "website" consists of a group of related files, assembled by an individual or a team of web designers. When viewed, a site typically consists of several linked pages, each of which may feature text, images, sounds, and animations. Sites also typically contain many hyperlinks (words, phrases, or buttons that when clicked on take you to other pages or to other websites).

When you access a web page, you are requesting that the files that make up that page are transferred from a remote computer (web server), where they are stored, to your own computer via the Internet. As the files arrive, a program called a web browser assembles them into the page on your screen. To visit a website, it is necessary to be connected to the Internet and have web browser software running on your computer. You type the address of the website into your browser window (for example "www.corporation.com") and press the Enter key.

When the "Web" started, sites consisted mainly of text and a few images. Since then, it has become increasingly media-rich and interactive. Advances that have contributed to this include an increase in the speed of data transmission, better software used for site creation, and developments of browsers and HTML (hypertext markup language), the basic computer language used for creating web pages. A further innovation has been video- and audio-streaming technology—methods for delivering large audio and video files over the Internet in "chunks."

Website structure

Websites vary hugely, but they often have a pyramidal structure. Every website has a "home page," which is like the contents page of a magazine. From the home page, a visitor typically has a choice of several different subsections of the site to explore. Many sites use "frames," which divide each page into content and navigation areas.

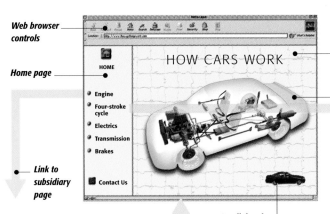

Web browser controls

Home page

Link to subsidiary page

HOW CARS WORK

HOME

- Engine
- Four-stroke cycle
- Electrics
- Transmission
- Brakes

Contact Us

Content area This "frame" is used for the content of each page

Static image Web images have to be stored in a compressed file format (usually JPEG or GIF), otherwise they are too large and take too long to travel over the Internet

Small, looping animation

Link back to home page

Cross-site link

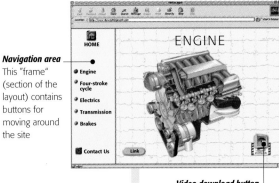

ENGINE

HOME

- Engine
- Four-stroke cycle
- Electrics
- Transmission
- Brakes

Contact Us   Link

Navigation area This "frame" (section of the layout) contains buttons for moving around the site

FOUR-STROKE CYCLE

HOME

- Engine
- Four-stroke cycle
- Electrics
- Transmission
- Brakes

Start
Stop

Contact Us

Video download button Video files are large and may take several minutes to be transmitted

Audio download button Sound files are generally smaller than video files per second of playing time

CONTACT US

Name

E-mail

Message

Link to external website Most sites contain numerous links to other sites. Collectively, these links comprise the "Web" in the World Wide Web

Form Forms are areas within a site requiring user input. They are often programmed using a computer scripting language called JavaScript

Interactive animation Specialized software is needed to build an interactive feature or complex animation into a web page. To view it, a visitor may first have to download a program called a "plug-in"

Creating a website

For each page in a website, the developer must create an HTML document. This contains text for the page, together with tags (codes), which define the layout of the page, the images that will appear on it, how the text will look, the colors used, and everything else about the page. Special web-page creation software can be used for converting a developer's design into a tagged HTML document. Media files, such as images, are often referred to by the HTML document, but are not part of it.

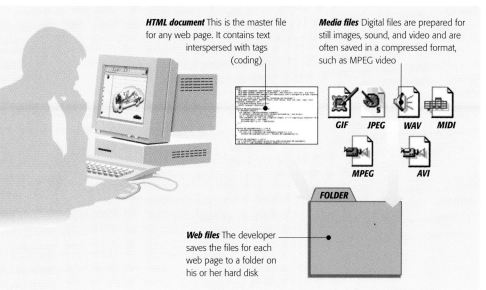

HTML document This is the master file for any web page. It contains text interspersed with tags (coding)

Media files Digital files are prepared for still images, sound, and video and are often saved in a compressed format, such as MPEG video

GIF   JPEG   WAV   MIDI

MPEG   AVI

FOLDER

Web files The developer saves the files for each web page to a folder on his or her hard disk

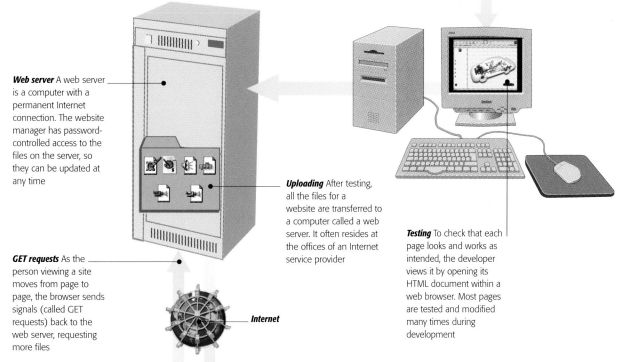

Web server A web server is a computer with a permanent Internet connection. The website manager has password-controlled access to the files on the server, so they can be updated at any time

Uploading After testing, all the files for a website are transferred to a computer called a web server. It often resides at the offices of an Internet service provider

Testing To check that each page looks and works as intended, the developer views it by opening its HTML document within a web browser. Most pages are tested and modified many times during development

GET requests As the person viewing a site moves from page to page, the browser sends signals (called GET requests) back to the web server, requesting more files

Internet

Visiting a web page

To view a web page, a user connected to the Internet must either use a hyperlink or type the page's address into a web browser. Hyperlinks may be found in software such as web pages and E-mails. A copy of the page's HTML file is then transferred from the web server where it resides, to the user's computer. Once the HTML file is loaded into the computer's web browser, signals are sent to the web server requesting that other files comprising the web page be sent.

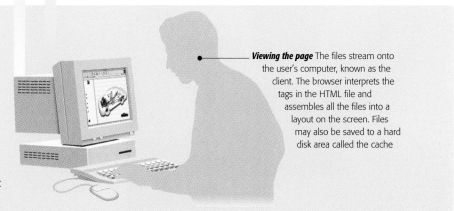

Viewing the page The files stream onto the user's computer, known as the client. The browser interprets the tags in the HTML file and assembles all the files into a layout on the screen. Files may also be saved to a hard disk area called the cache

The virtual world

Virtual reality, until recently the stuff of computer games and science fiction, is fast becoming an important tool with a wide variety of applications. As well as allowing users to enter simulated environments, the technology developed for virtual reality (VR) can be coupled with robotics to give a person the ability to operate in a distant location—a technology know as telepresence.

Virtual basics

Virtual-reality equipment allows a person to enter and interact with a computer-generated three-dimensional (3-D) environment. Movement sensors and powerful computers are needed to create the complete illusion of another world, and to transform the user's movements in the real world into actions in the synthesized reality, often called "cyberspace." The computer must also be linked to a a range of hardware devices that provides constantly updated feedback input to the user.

Flight simulators
Simulators are a form of VR that have been around for several decades. Using a model cockpit supported on hydraulic jacks, and surrounded by large display screens, a flight simulator can recreate landing conditions at any airport in the world, allowing pilots to train in complete safety.

LCD goggles
This headset is a simple form of virtual-reality monitor. A pair of liquid crystal display (LCD) screens inside the goggles produces the same effect as a large monitor viewed from a distance. However, the headset does not have sensors to detect head movement and cannot immerse the wearer in cyberspace.

The most familiar piece of VR technology is the headset, which allows the user to be completely immersed in cyberspace. The headset takes the form of goggles, usually with an LCD screen positioned in front of each eye. The computer sends slightly different images to each screen, and the wearer's brain interprets these images as a view of a real, 3-D environment. The images on the screens are constantly refreshed, so that as the wearer moves his or her head, the view changes. Computing these rapidly changing three-dimensional views is the main reason why VR requires large amounts of computer processing power. In addition to visual information, a VR headset often incorporates stereo headphones, immersing the user in a world of sound as well as 3-D vision. As the headset moves from side to side or tilts up or down, electronic sensors detect the acceleration and deceleration, and feed this information back to the computer, which updates the sound and vision output in the headset. These sensors are similar to those used in automobile safety airbags—movement in the sensor stretches or compresses wires, changing their electric resistance.

Another familiar piece of VR technology is the gauntlet covered in sensors, or data glove, that allows the computer to track the movements of the wearer's hand. Data gloves usually combine acceleration sensors (to track movements of the hand through space), with optical fibers running down the fingers. Light from an interface board on the back of the glove is shone down the fibers, reflected from the end, and detected at the interface board. When the wearer bends his or her fingers, some of the returning light signals are cut off. The computer registers that the optical fibers have been distorted, and adjusts the image of the user's hand in cyberspace accordingly.

For further realism, data gloves are fitted with pressure pads at the fingertips, which can be activated by signals from the computer. Picking up or touching an object in cyberspace results in a realistic pressure being delivered to the fingers. The wearer can also feel how securely they are gripping the object.

Although headsets and data gloves are the most common virtual-reality interfaces, there are many others, offering varying

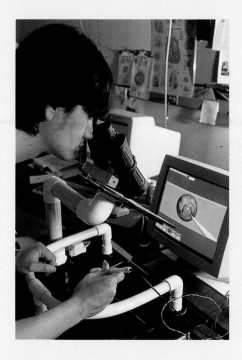

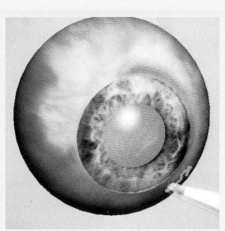

Telepresence

While VR can simulate real situations, telepresence gives people the ability to physically intervene in a real situation from a distance. A person using a telepresence system wears a VR-type headset and gauntlets, which enable them to control remotely the movements of a robot. The robot, know as an "operator," is equipped with sensors that give feedback about its environment to the telepresence system. This feedback provides the wearer with a virtual presence in the robots environment, and allows them to see, hear, and touch things via the robot. Telepresence has a huge range of uses and, for most situations, is far less expensive and more effective than building fully independent robots to carry out tasks. Surgery is already being carried out by telepresence, with a surgeon on one continent conducting delicate operations on another. Bomb-disposal teams use telepresence to defuse explosives without putting themselves in danger, and it can also be used to handle nuclear material. In space, NASA is developing a "robonaut" for the International Space Station—a telepresence operator that will be able to reach a site on the outside of the station in far less time than it takes an astronaut to prepare for a spacewalk.

Virtual surgery
A medical student practices eye surgery using a virtual-reality system. The operator controls a virtual laser with one hand, while peering down an operating microscope. The microscope contains two screens to produce a stereo image of the model eye and laser. Supervising doctors can monitor the student's progress on a larger computer monitor.

degrees of immersion in the VR environment. Gaming joysticks, steering wheels, and even computer mice can now give realistic force feedback to a user playing a game, seats can shake during explosions, and chest padding or even whole-body suits can offer still more realistic experiences.

Putting VR to work

Although the most popular use of VR technology is computer games, VR does have a huge potential for more constructive uses. The earliest VR device was probably the flight simulator, which uses large computer monitors mounted around a rolling, pitching model cockpit to train pilots in flying techniques and how to cope with emergencies. VR can offer realistic training for many dangerous situations without putting real lives or expensive machinery at risk—other applications include training nuclear power plant operators, electricity engineers, and surgeons.

VR systems can also be used to put their wearer into a controlled environment. Architects use VR simulations to demonstrate buildings during their planning stages, while some psychiatrists use VR to cure phobias such as fear of open spaces or fear of heights. Finally, VR vastly expands the ways in which human beings can interact with computer systems. Instead of using a mouse and a monitor, some systems make it possible for a database manager to enter a 3-D representation of the computer file and search for data.

Telepresence at work
This simulation allows people to practice using a robotic arm to dismantle components inside a nuclear reactor. The operator controls the arm (shown white) using a joystick, and manipulates the red and blue radioactive materials just as he or she would in reality.

SEE ALSO: COMPUTER OUTPUT DEVICES *p.68* | AIRLINERS *p.126* | ROBOTICS *p.210* | SPACE STATIONS *p.244*

Sound recording

Sound recording is the process of converting pressure waves in air (sound) into electric signals for storage as magnetic or physical patterns in a medium such as magnetic tape, vinyl record, or compact disc. In analog recording, a copy (analog) of the continuous sound wave is created in the recording medium. Digital recording converts the continuous wave into a binary code of discrete pulses.

During recording, sound waves from a traditional instrument or voice must be converted into an electric signal by a microphone, while electronic instruments, such as synthesizers, can produce an electric signal directly. An analog electric signal is theoretically an exact copy of the original sound wave, while digital code is only a sampled approximation. However, digital audio has many practical advantages. Digital coding systems are designed to carry information in addition to sound, such as error-correction data. In addition, digital copies are identical to digital originals, which means that tracks can be copied repeatedly without deterioration.

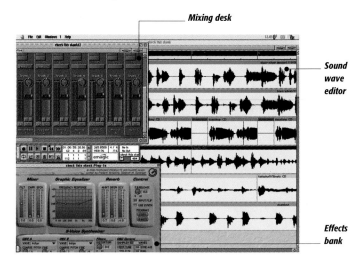

Mixing desk

Sound wave editor

Effects bank

Virtual recording studio
Today, most of the functions of a professional sound recording studio can be carried out by software on a laptop computer, including editing sound waves, arranging and mixing tracks, and adding effects. Recordings can be made from microphones or directly from synthesizers or MIDI instruments.

Microphone

A microphone is a transducer—it converts energy from one form (sound) into another (electric). The varying electric voltage produced by a microphone is a copy (analog) of the varying air-pressure wave associated with sound. There are two main types of microphones: condenser and dynamic. In condenser microphones, a diaphragm (sheet) moves in response to sound waves, producing a varying electric signal. Dynamic microphones use a diaphragm's motion to move a coil inside a magnetic field, which induces a varying electric current in the coil.

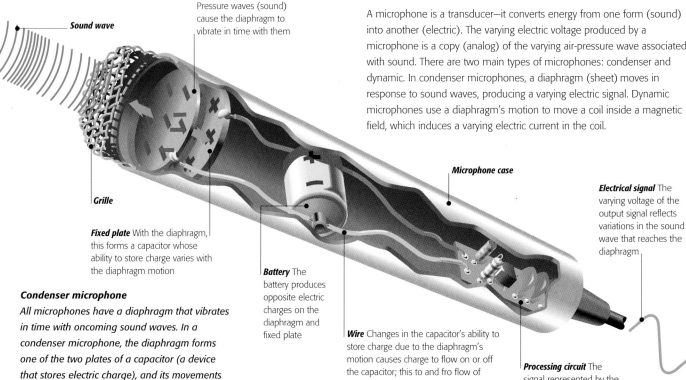

Sound wave

Moving diaphragm Pressure waves (sound) cause the diaphragm to vibrate in time with them

Grille

Fixed plate With the diaphragm, this forms a capacitor whose ability to store charge varies with the diaphragm motion

Microphone case

Electrical signal The varying voltage of the output signal reflects variations in the sound wave that reaches the diaphragm

Condenser microphone
All microphones have a diaphragm that vibrates in time with oncoming sound waves. In a condenser microphone, the diaphragm forms one of the two plates of a capacitor (a device that stores electric charge), and its movements vary the amount of charge stored, causing a varying electric signal in a circuit.

Battery The battery produces opposite electric charges on the diaphragm and fixed plate

Wire Changes in the capacitor's ability to store charge due to the diaphragm's motion causes charge to flow on or off the capacitor; this to and fro flow of charge constitutes an alternating current in the circuit

Processing circuit The signal represented by the varying current undergoes some processing here

Sampling sound

To convert the continuously varying electric-voltage wave produced by a microphone into a digital code, the voltage signal must be sampled. Sampling involves measuring the amplitude (height) of the wave at regular intervals. The code can then be reconverted into an analog signal for playback. The fidelity of the reconverted signal depends on the sampling frequency and the bit depth (which determines the range of possible values of a sample). The standard for CD audio is 16 bit (65,536 levels), 44.1 kHz (measured 44,100 times a second). The 8-bit, 10-kHz sample shown here is adequate for recording speech.

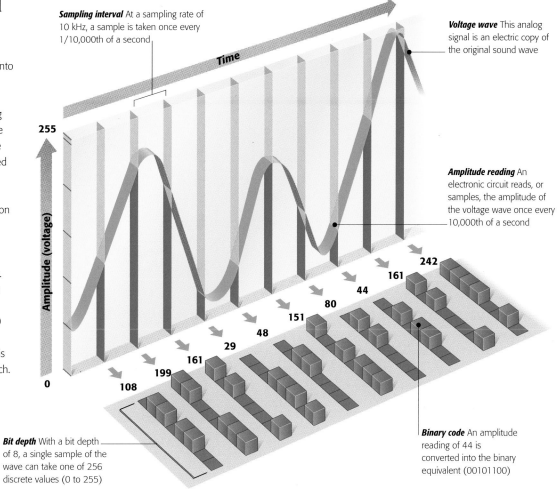

Sampling interval At a sampling rate of 10 kHz, a sample is taken once every 1/10,000th of a second

Time

Voltage wave This analog signal is an electric copy of the original sound wave

255

Amplitude (voltage)

0

Amplitude reading An electronic circuit reads, or samples, the amplitude of the voltage wave once every 10,000th of a second

242
161
44
80
161
151
48
29
199
161
108

Bit depth With a bit depth of 8, a single sample of the wave can take one of 256 discrete values (0 to 255)

Binary code An amplitude reading of 44 is converted into the binary equivalent (00101100)

Digital versus analog

The main advantage of digital systems is that they can easily distinguish between the wanted signal and unwanted noise caused by interference or defects in the recording medium. This is because in digital systems, the signal and noise "look" very different. Errors or missing data can be filled in by a computer, using various error-correction systems that are part of the code. For example, extra bits called parity bits may be added to each sample. To an analog system, signal and noise "look" quite similar, so even sophisticated noise-reduction systems either do not remove all the noise or cause distortion of the signal.

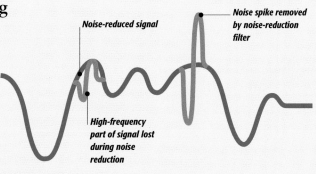

Noise-reduced signal

Noise spike removed by noise-reduction filter

High-frequency part of signal lost during noise reduction

Analog signal

An analog system cannot easily distinguish between a high-frequency noise "spike" and a high-frequency part of the signal. Filters used to remove noise also produce distortions of the signal.

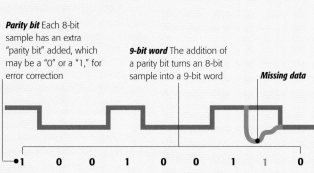

Parity bit Each 8-bit sample has an extra "parity bit" added, which may be a "0" or a "1," for error correction

9-bit word The addition of a parity bit turns an 8-bit sample into a 9-bit word

Missing data

1 0 0 1 0 0 1 1 0

Digital signal

If noise disrupts the reading of a particular part of the code, missing values can be predicted. The coding system shown here specifies that each 9-bit word must have an even number of 1s, so the missing value is deduced to be a "1."

SEE ALSO: THE DIGITAL WORLD *p.54* | DIGITAL DATA *p.56* | SOUND REPRODUCTION *p.80*

Music media

The history of music media is the history of technological development, commercial successes or failures, and format wars. The first mass medium was the gramophone disk (record), which was on the market in the 1890s and steadily improved into the 1950s with electrification, better disk materials, and various size and speed formats. Magnetic recording on wire was developed in the late 1890s, but was not commercially successful until the 1930s. In the late 1940s wire recorders were superseded by open-reel recorders, which used plastic tape coated with metal oxide, and by compact cassettes and 8-track cartridges in the 1960s and 1970s. The major revolution at the close of the 20th century was the development and popularization of digital media. Though some formats such as DAT (digital audio tape) remain confined to professional or semipro markets, the CD (compact disc) and the rewritable MD (minidisc) have brought unprecedented sound quality and durability within reach of domestic consumers with modest budgets. The next major consumer format is likely to be a "solid-state" medium based on flash-memory chips or cards.

From cylinders to discs

The first device that allowed sound recording and playback was Edison's phonograph of 1877, in which sound waves arriving at a diaphragm caused an attached needle to vibrate, inscribing a groove of varying depth into a foil-coated rotating cylinder. The hand-turned cylinder was played back with a needle attached to a diaphragm. In 1887, Emile Berliner developed the gramophone, which used discs instead of cylinders, the great advantage of which was that discs could be mass-produced by stamping from a "master." Discs made of shellac, turning at 78 revolutions per minute (rpm) on hand-wound gramophones gave four minutes of playing time per side, and became popular across this country and Europe in the early 20th century. By the 1920s, electricity was being used for recording and in reproduction, for rotating the turntable and amplifying the vibrations of the needle. In 1948, the long-playing (LP) record turning at 33⅓ rpm, with its finer "microgroove" played with a diamond-tipped pickup, increased sound quality and extended playing time to over 20 minutes per side. A 45-rpm disc was introduced as a direct competitor to the 78 rpm. Both LPs and 45s were made from newly available plastics, or "vinyl." Stereo (two-channel) recordings followed in 1958. Automatic record players and jukeboxes helped popularize vinyl discs. But despite the dramatic fidelity improvements brought by electric recording and reproduction and new disc materials, record sales declined in the 1980s. Initially this was due to the increasing popularity of the cassette, but later because of the digital CD. Now, vinyl LPs remain in production only for specialist markets.

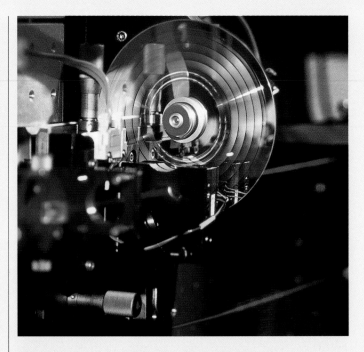

Glass CD master being written by a laser at high speed
Audio code is written by laser from a master tape to a polished glass master plate coated with photoresist. The photoresist is developed and metal deposited on it; "stampers" are then made from the negative metal impression of the disc and used to injection-mold polycarbonate CDs.

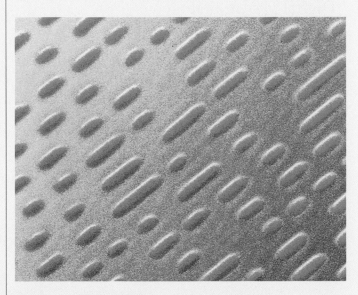

Scanning electron micrograph of CD "pits" and "lands"
High-quality stereo audio data is stored in microscopic "pits" and "lands" on one continuous spiral track. The edges of the pits correspond to "1s" in binary code. The pit size is close to the wavelength of visible light, and they diffract (bend) light to produce the spectral colors seen on CD surfaces.

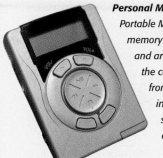

Personal MP3 player
Portable MP3 players using nonvolatile "flash-memory" chips were introduced in the late 1990s and are fast gaining in popularity. Sound files in the compressed MP3 format can be downloaded from the Internet to a computer and loaded into the player. Some MP3 players are smaller than a pen, yet can hold two hours of good-quality stereo sound.

Recording magnetically

Magnetic recording was demonstrated in 1889 by Danish physicist Valdemar Poulsen, whose telegraphon recorded onto steel wire. However, magnetic wire and tape recorders did not achieve adequate fidelity levels until the 1930s, when the first wire recorders for industry were produced. It was not until the late 1940s that a standard tape format using a magnetizable coating on plastic tape was established. From the late 1950s, open-reel tape recorders using ¼-inch tape with two pairs of stereo tracks became popular. By turning over the tape reel, the alternate pair of tracks could be accessed. Open-reel recorders offered various tape speeds—7½, 3¾ and 1⅞ inches per second (ips). A higher speed meant better sound quality (since the same sound was recorded over a longer length of tape and, thus, on more magnetic particles, giving greater detail), but at the expense of recording time.

Magnetic tape received a massive impetus with the introduction of the compact cassette in 1962 and the 8-track cartridge (1965). The latter was specifically developed as an in-car format, but lost ground to the cassette, which has since become a standard fixture in automobile stereos and also in personal stereos. The compact cassette squeezed four tracks (two stereo tracks per side) onto ⅛-inch tape wound on a pair of small spools in a durable plastic case. The narrow tape, poor magnetic medium formulation, and a low speed of 1⅞ ips meant that sound quality was adequate at best. Since then, fidelity has improved dramatically through the introduction of noise-reduction systems and the use of more advanced tape formulations, refining the coating from simple rust to chromium dioxide and then metal particles. The continuing success of the cassette format is also partly due to the convenience of advanced features such as track search and auto-reverse (quick-reverse decks play both sides of the cassette by rotating the playback head, making manual cassette flipping unnecessary) and miniaturization (enabling personal stereo players to be barely larger than a cassette case).

The digital revolution

Digital tape recorders were developed for professionals in the late 1970s in an effort to improve fidelity and duplication quality and to reduce noise. Today, hard disks and tape formats such as DAT set the studio standard. For consumers, the digital revolution arrived in the form of the CD. Unlike analog or digital tape, but like records, CD offers the convenience of random access, meaning that tracks in the middle of the disc can be accessed directly without searching through previous tracks. For home recording, MD (minidisc) and recordable CD are poised to replace analog cassettes. But the future of music media probably lies in solid-state digital devices, with their near-instantaneous track-access times, and compactness and robustness due to lack of moving parts.

Minidisc recording technology

Minidisc, or MD, is a rewritable consumer audio format that provides near-CD digital sound quality from 2½-inch (65-mm) diameter discs encased in durable plastic shells. MD recorders fit into the palm of your hand, and weigh little more than 3.5 oz (100 g). Each MD holds 140 MB of data—about one-fifth as much as a CD—but squeezes 80 minutes of digital stereo sound into this space by compressing the data, removing some frequencies with minimal loss of fidelity. There are two types of disc: prerecorded and rewritable. Prerecorded MDs are like miniature

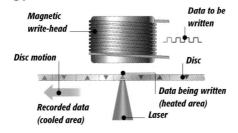

Magneto-optical recording
During recording, a laser on one side of the disc heats the portion of the disc being recorded, while a magnetic write-head on the other side of the disc changes the magnetization pattern.

CDs, with permanently embossed "pits" and "lands," and are read by a laser. Rewritable MDs use a different technology, but work with the same hardware. This technology, called magneto-optical, encodes audio data as a digital code in a magnetizable layer on the disc. This layer is only magnetizable when heated by a laser, so both magnetic and optical components are involved. Rewritable MDs can be repeatedly recorded and erased. Compared with analog cassettes, MD offers much better fidelity, greater compactness and durability, and random (direct) access to tracks.

SEE ALSO: SOUND RECORDING *p.76* | SOUND REPRODUCTION *p.80* | VIDEO TECHNOLOGY *p.104*

Sound reproduction

Reproduction of recorded sound requires a media player to convert the physical or magnetic patterns on the recorded media into an electric signal, an amplifier to boost the signal from the player, and speakers to convert the signal into sound waves. If the recording medium is digital, a digital/analog converter is needed to change the signal into analog form before it can be amplified.

Ideally, sound reproduction equipment recreates the sound exactly as originally recorded. An amplifier must boost the level of the electric signal from the record, tape, or CD player without introducing unpleasant distortion. Most amplifiers accomplish this using transistors, but amplifiers based on older vacuum tube technology are prized by some for the "warmth" of their sound. The final component in the reproduction chain is the loudspeaker. Speakers are transducers–they convert electric energy into sound energy–and work like microphones in reverse. Most speakers use moving-coil technology. "Flat" speaker designs were once usually based on electrostatic technology and operated like condensor microphones in reverse, but the latest flat designs use a new "distributed mode" speaker technology.

Distributed mode loudspeakers
The latest flat-speaker technology, called distributed mode loudspeakers (DML), uses an extremely stiff, thin panel to produce sound. The whole panel vibrates in a complex way, unlike in a conventional speaker, where a cone moves in and out like a piston. DML units do not need an enclosure (box), and they fill a room more evenly with sound than conventional speakers. They can be made from various materials—even glass, as shown—and incorporated into walls and ceilings.

Record player cartridge

A vinyl record stores the two channels of a stereo sound signal as wavelike contours in the walls of a spiral groove. A cartridge at the end of the record player tonearm detects the contour variations with a diamond-tipped stylus, which moves in two perpendicular directions as it follows the inner and outer groove walls. The movements of the stylus cause an attached magnet to move within two sets of coils, inducing separate electric signals in them for the left and right channels. Some cartridges use a moving coil and fixed magnets instead.

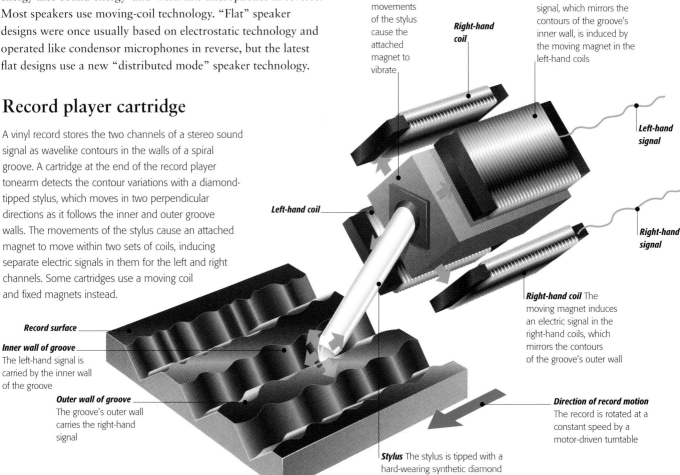

Magnet The movements of the stylus cause the attached magnet to vibrate

Right-hand coil

Left-hand coil An electric signal, which mirrors the contours of the groove's inner wall, is induced by the moving magnet in the left-hand coils

Left-hand signal

Left-hand coil

Right-hand signal

Right-hand coil The moving magnet induces an electric signal in the right-hand coils, which mirrors the contours of the groove's outer wall

Record surface

Inner wall of groove
The left-hand signal is carried by the inner wall of the groove

Outer wall of groove
The groove's outer wall carries the right-hand signal

Stylus The stylus is tipped with a hard-wearing synthetic diamond and sits in the groove

Direction of record motion
The record is rotated at a constant speed by a motor-driven turntable

Speakers

Most hifi speaker units use two moving-coil driver units; a "tweeter" for high frequencies and a "woofer" for low frequencies They are enclosed within specially designed cabinets that incorporate damping material to absorb unwanted vibrations. An alternating electrical current from the amplifier that represents the sound signal is sent to a crossover circuit in the cabinet. This sends high frequencies to the tweeter and low frequencies to the woofer. The varying electrical signals cause the driver unit's coil to move, creating varying pressure (sound) waves.

Moving-coil speaker

A moving-coil speaker converts electrical currents into sound through the magnetization of a wire coil connected to a flexible diaphragm.

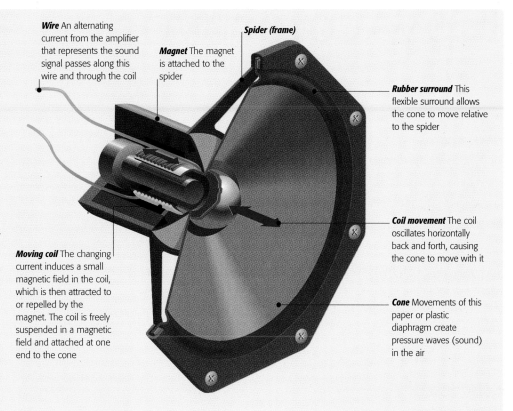

Wire An alternating current from the amplifier that represents the sound signal passes along this wire and through the coil

Spider (frame)

Magnet The magnet is attached to the spider

Rubber surround This flexible surround allows the cone to move relative to the spider

Coil movement The coil oscillates horizontally back and forth, causing the cone to move with it

Cone Movements of this paper or plastic diaphragm create pressure waves (sound) in the air

Moving coil The changing current induces a small magnetic field in the coil, which is then attracted to or repelled by the magnet. The coil is freely suspended in a magnetic field and attached at one end to the cone

Compact disc

The compact disc is a 5-in (12-cm) diameter plastic disk that can hold approximately 74 minutes of 16-bit stereo sound, sampled at 44.1 kHz. The sound data is stored in the form of about 3 billion "pits" in a single spiral track that starts at the center. A typical pit is 1/40,000 in (1 micrometer) long.

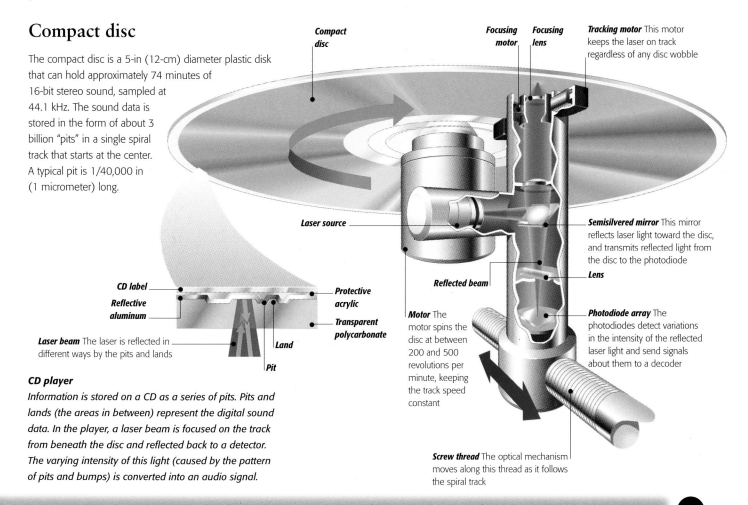

Compact disc

Focusing motor

Focusing lens

Tracking motor This motor keeps the laser on track regardless of any disc wobble

Laser source

Semisilvered mirror This mirror reflects laser light toward the disc, and transmits reflected light from the disc to the photodiode

Reflected beam

Lens

Motor The motor spins the disc at between 200 and 500 revolutions per minute, keeping the track speed constant

Photodiode array The photodiodes detect variations in the intensity of the reflected laser light and send signals about them to a decoder

CD label

Reflective aluminum

Protective acrylic

Transparent polycarbonate

Laser beam The laser is reflected in different ways by the pits and lands

Land

Pit

Screw thread The optical mechanism moves along this thread as it follows the spiral track

CD player

Information is stored on a CD as a series of pits. Pits and lands (the areas in between) represent the digital sound data. In the player, a laser beam is focused on the track from beneath the disc and reflected back to a detector. The varying intensity of this light (caused by the pattern of pits and bumps) is converted into an audio signal.

SEE ALSO: THE DIGITAL WORLD *p.54* | SOUND RECORDING *p.76* | MUSIC MEDIA *p.78* | LASERS *p.220*

How color works

Colors are so familiar to us that it is easy to take them for granted. But the principles of color mixing, which underlie the technology for reproducing a vast range of colors in media such as print, photography, television, and computing, are more complex than they might first appear. They are closely related to the way in which our eyes and brains work together to perceive color.

The phenomenon of color derives from the fact that light, which is a form of electromagnetic radiation, exists in a range of wavelengths. Sunlight and other forms of "white" light contain a mixture of these wavelengths, as can be seen by passing a ray of such light through a glass prism. The prism disperses the light into a continuous range, or spectrum, of colored bands of light, from red (long wavelengths) to violet (short wavelengths).

According to the prevalent "trichromatic" theory of human color perception, the retina of the eye contains three kinds of "cone" cells, each sensitive to a different range of wavelengths (approximately red, green, and blue light). The sensation of any color can be evoked by mixing lights of those three colors—the "primary colors." Color TV works on this principle, giving the impression of a full palette of colors using an array of red, green, and blue phosphor dots. However, a different set of primary colors is used in printing and photography, because colored pigments and dyes mix in a different way to colored light.

Human color perception

When light strikes the retina at the back of the eye, it stimulates one or more classes of receptor cells called cones. Each class of cone responds to different wavelengths of light. When light of a particular wavelength stimulates a class of cones, the cones send electric signals to the brain's visual cortex via the optic nerve and the light's color is processed.

Light ray White light consists of a mixture of wavelengths

Cornea

Lens

Signal transmission Signals are transmitted from eye to brain via the optic nerve

Visual cortex The pattern of signals is perceived here as a color (in this case, white)

Retina The retina lines the back of the eye and contains the sensitive cone cells.

KEY

CONES SENSITIVE TO LIGHT OF LONG WAVELENGTHS (L)
CONES SENSITIVE TO LIGHT OF MEDIUM WAVELENGTHS (M)
CONES SENSITIVE TO LIGHT OF SHORT WAVELENGTHS (S)

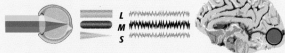

Long wavelengths
A mix of predominantly long light wavelengths (red, orange, and yellow) stimulates the "long wavelength" cones strongly and the "medium" cones only weakly. The color perceived in the brain is orange-red.

Medium wavelengths
A mix of medium light wavelengths (yellow, green, and cyan) stimulates the "medium" cones strongly and the other cones only weakly. The color perceived in the brain from this pattern of signaling is green.

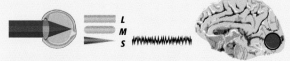

Short wavelengths
A mix of short light wavelengths (blue and violet) stimulates the "short wavelength" cones strongly and the other cones hardly at all. The color perceived in the brain from this pattern of signaling is blue-violet.

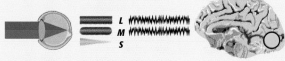

Seeing yellow
An equal mix of long (red) and medium (green) wavelength light produces equal-strength signals from the "long" and "medium" cones. This pattern of signaling is perceived in the brain as the color yellow.

Primary colors of light

Mixing the primary colors of light in equal proportions produces "white" light, whether projected onto a screen or directly into the eyes. By varying the proportions of red, green, and blue lights in the mixture, a very wide range of color sensations can be produced. This phenomenon is exploited in the design of color televisions and computer monitors. These devices have screens made up of millions of phosphor dots that emit red, green, and blue light when hit by cathode rays (beams of fast-moving electrons).

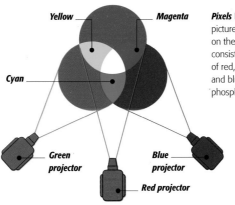

Yellow — Magenta

Cyan

Green projector

Blue projector

Red projector

ADDITIVE COLOR MIXING

Pixels Each picture element on the screen consists of a trio of red, green, and blue phosphor dots

Screen color In a TV or computer monitor, the color of each picture element is determined by the relative intensity of light emitted by the three primary-color phosphors

Primary pigment colors

Just as red, green, and blue lights can be mixed to produce a huge range of color sensations, so pigments and dyes can also be mixed to produce different colors. However, pigments mix in a different way to lights—called "subtractive color mixing"—because pigments gain their color through the wavelengths of light that they reflect back to an observer. The trio of pigments that combine most efficiently to produce the widest range of print colors are yellow, cyan, and magenta—the primary printing or pigment colors.

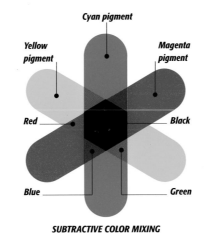

Cyan pigment

Yellow pigment — Magenta pigment

Red — Black

Blue — Green

SUBTRACTIVE COLOR MIXING

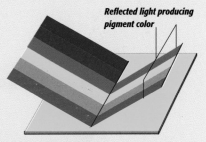

Color printing Print colors are produced by superimposing dots of yellow, cyan, and magenta inks

How subtractive color mixing works

Mixing color pigments differs from mixing colored lights because a pigment is not in itself a source of light. The perceived color of a pigment derives from the light wavelengths it absorbs and those that it reflects from incident light. Mixing pigments on a surface has a subtractive effect—the more pigments used, the greater the range of wavelengths absorbed (subtracted from the incident light), and so the smaller the range of wavelengths reflected.

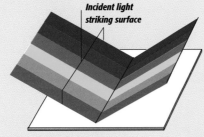

Incident light striking surface

Reflected light producing pigment color

White-painted surface
This color surface absorbs no wavelengths and therefore reflects all wavelengths.

Yellow-painted surface
This color surface absorbs short-wavelength (blue-violet) light and reflects everything else.

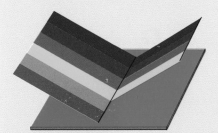

Cyan-painted surface
This color surface absorbs long-wavelength (red-orange) light and reflects everything else.

A mix of cyan and yellow paint
This color surface absorbs wavelengths at both ends of the spectrum and reflects only medium wavelengths, which are perceived as green.

Equal mix of cyan, yellow, and magenta
This color surface absorbs all wavelengths. No wavelengths are reflected, and so the surface is perceived as black.

SEE ALSO: PRINTING *p.88* | PHOTOGRAPHIC FILM *p.90* | TELEVISION *p.100* | LIGHT *p262*

Photocopiers and fax machines

The workplace has been transformed by the introduction of modern photocopying machines that permit multiple copies of textual and graphic material to be made quickly and cheaply. Modern fax machines have taken this revolution one step further by allowing text and graphics to be transmitted through the telephone network to other fax machines that reproduce the information.

Photocopying, or xerography (literally, dry copying), uses the principle of photoconductivity—the fact that certain substances that are relatively resistant to the passage of an electric current become markedly more conductive when exposed to light.

Photocopying a reflection of an original image involves selectively removing an electric charge on a photoconductive drum to recreate the image as a pattern of charges. In order to print the recorded image, the principle of the attraction of opposite charges is brought into play. Electrically charged paper picks up oppositely charged toner (ink powder) that has been attracted to charged areas of the drum. Color photocopiers use four separate ink toners—cyan, magenta, yellow, and black.

Fax (sometimes known as facsimile or telefax) technology began to be developed in the 19th century but did not become widespread until the 1980s. Digital technology used in the Group 3 standard fax launched in 1980 allowed the transmission of image data via ordinary telephone lines.

Fax machine

A Group 3 standard fax machine incorporates a charge-coupled device (CCD) that scans the image in lines and registers light and dark areas as variations in voltage. The voltage variations are digitized and encoded to reduce transmission time. The encoded signal is then transmitted via an analog telephone line. The receiving machine decodes the signal. Early fax machines printed on thermal paper that darkened in response to contact with an array of heated wires, activated by the fax signal. Plain-paper fax machines use an ink-jet or a laser-printing mechanism.

Original image

Light source
The image is illuminated by the light source

Light reflected from white areas of image

CCD The CCD converts brightness variations in each line into electric signals

Analog signal The electrical output from the CCD is an analog of the brightness variations of the original image

READING IMAGE

Laser printer A laser beam turns on and off in response to the incoming signals, discharging areas of the drum that correspond to light areas of the image

Voice-band modem The analog signal is converted into a stream of digital bits and transmitted via the telephone network

FAX TRANSMISSION

Drum-charging electrode

Blank paper   **Revolving drum**

RECEIVING IMAGE

Sending a fax
A fax is sent by digitally sampling the information on a document. The digital information is then sent to the receiving fax machine over the telephone network. A fax machine in receiving mode works in the same way as a printer, using an electrically charged drum to transfer ink onto blank paper.

Pixelated final copy
The digitized image is composed of tiny elements

Black-and-white photocopier

In a photocopier, light passed over a document is reflected from white areas, while the black areas reflect no light. This pattern of light is reflected by mirrors onto a negatively charged drum. Light striking the drum dissipates the electric charge, leaving charged only those areas matching the black parts of the image. Positively charged toner particles attach to the charged areas of the drum and are then transferred onto paper.

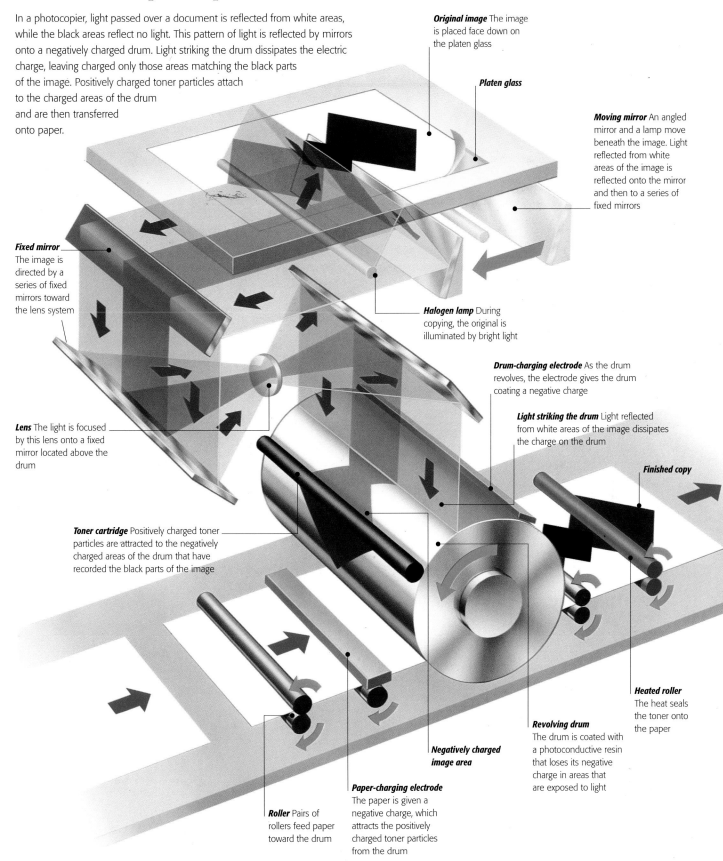

Original image The image is placed face down on the platen glass

Platen glass

Moving mirror An angled mirror and a lamp move beneath the image. Light reflected from white areas of the image is reflected onto the mirror and then to a series of fixed mirrors

Fixed mirror The image is directed by a series of fixed mirrors toward the lens system

Halogen lamp During copying, the original is illuminated by bright light

Drum-charging electrode As the drum revolves, the electrode gives the drum coating a negative charge

Light striking the drum Light reflected from white areas of the image dissipates the charge on the drum

Lens The light is focused by this lens onto a fixed mirror located above the drum

Finished copy

Toner cartridge Positively charged toner particles are attracted to the negatively charged areas of the drum that have recorded the black parts of the image

Heated roller The heat seals the toner onto the paper

Revolving drum The drum is coated with a photoconductive resin that loses its negative charge in areas that are exposed to light

Negatively charged image area

Roller Pairs of rollers feed paper toward the drum

Paper-charging electrode The paper is given a negative charge, which attracts the positively charged toner particles from the drum

SEE ALSO: DIGITAL DATA *p.52* | SCANNING DEVICES *p.64* | COMPUTER OUTPUT DEVICES *p.68* | LIGHT *p.262*

Preparing for print

Today, desktop publishing (DTP) is the most popular way of producing printed material. It allows editors and designers to produce pages quickly, alter text and images easily, and know the finished result will be precisely reproduced by the printer. But transforming images on a computer monitor into ink on a printed page is a complex process, involving several steps of conversion.

The desktop publishing process has three main stages: origination, page layout, and proofing. Origination involves commissioning text and artwork, and converting images into a digital form using a scanner or a digital camera. Alternatively, digital artworks may be created using illustration software. Once stored on the computer, the images may be manipulated using graphics programs to adjust their brightness, color, contrast, etc.

Page layout is the core of desktop publishing. It uses specialized software to position and manipulate images and text on screen. During proofing, this software also acts as an interface between the computer and a desktop printer by converting the layout on screen into a form that the printer can understand and use to print dots of different colors and sizes on a page. The page proofs can then be printed, checked, and corrected.

When the page is ready, the layout program performs color separation. This process breaks down every element on the page into the four colors used for most printing processes. The end result is a set of color layers that will be used to make the actual printing plates.

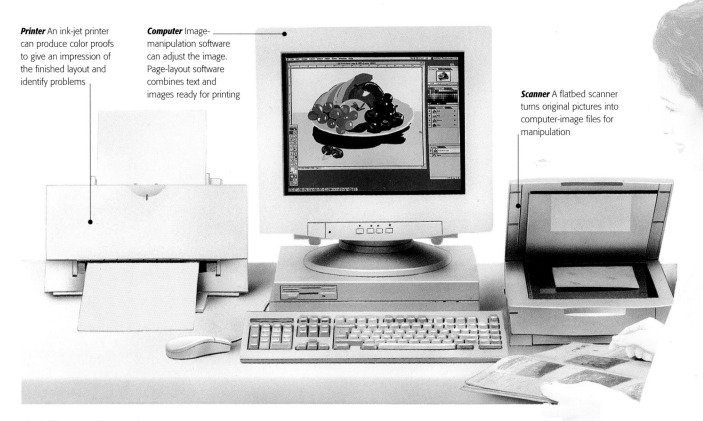

Printer An ink-jet printer can produce color proofs to give an impression of the finished layout and identify problems

Computer Image-manipulation software can adjust the image. Page-layout software combines text and images ready for printing

Scanner A flatbed scanner turns original pictures into computer-image files for manipulation

DTP workstation

In an office DTP setup, a scanner creates low-resolution image files called positionals—a specialist reproduction house then creates high-quality versions of the images, which are suitable for printing. A color printer gives a useful impression of the finished layout, but its colors will not match those of a commercial printing press, so the reproduction house supplies color proofs that are a closer match to the printed product.

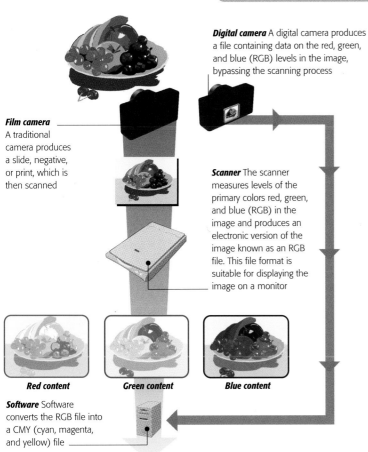

Film camera
A traditional camera produces a slide, negative, or print, which is then scanned

Digital camera A digital camera produces a file containing data on the red, green, and blue (RGB) levels in the image, bypassing the scanning process

Scanner The scanner measures levels of the primary colors red, green, and blue (RGB) in the image and produces an electronic version of the image known as an RGB file. This file format is suitable for displaying the image on a monitor

Red content

Green content

Blue content

Software Software converts the RGB file into a CMY (cyan, magenta, and yellow) file

Cyan content

Magenta content

Yellow content

Black content To sharpen the image, black (K) is added to the color model, producing a CMYK file

Process colors Cyan, magenta, and yellow —the subtractive primary colors—and black will be used as inks on the printed page, so the image must be converted from an RGB file to a CMYK file

Rasterizing Four color films are produced, one for each color in the CMYK file. This process is called raster image processing, or ripping

Platemaking

There are several ways of making printing plates, depending on the printing process to be used. The most popular printing process today—lithography—uses a flat, chemically treated aluminum or plastic plate that attracts ink on some parts and repels it in others. To create the image on the plate, the positive image film is rephotographed to create a negative, and light is then shone through the negative onto the plate. The plate has been previously coated in a light-sensitive chemical, and where it is exposed to light, a greasy coating forms, which attracts ink and repels water. The remaining chemicals are washed off, and the flexible plate is wrapped around a drum ready for printing.

Greasy coating on plate

Ink attracted to greasy coating

Raster dots Each color image is broken down into dots of different sizes. Denser areas of color have larger dots

Magnified color image The combination of dots of different sizes and colors can produce almost any color

Final printed image The CMYK films are used to make plates for each separate color of ink. When overlaid on each other, they give the illusion of full color

Overlaid dots The raster dots for different colors run in different directions, to prevent interference patterns from building up

Four-color process

The image file produced by a digital camera or a scanner contains data about the three primary colors of light—red, green, and blue (RGB). Scanners and cameras work by detecting light through red, green, and blue filters. When mixed in different proportions, these colors can produce almost any color.

Printing colors work by a process known as subtractive color mixing. Inks absorb, or subtract, certain wavelengths of light from light that strikes them. The light reflected from the ink gives it its color. Subtractive color mixing uses the primary colors cyan, magenta, and yellow, which give the greatest range of colors. Black is used to sharpen the image. Before printing, an image file is converted into this four-color, or CMYK, format.

Because the printing process cannot print more than one color of ink at a time, the four colors are separated and rasterized (processed into patterns of different-sized dots on film to show different intensities of color). The resulting films are used to make a set of four separate printing plates.

SEE ALSO: SCANNING DEVICES *p.64* | HOW COLOR WORKS *p.82* | PRINTING *p.88*

Printing

The end result of the complex preprint process is a plate for each color of ink that will appear on the finished page. Usually, this means four plates—one each for cyan, magenta, yellow, and black inks. A wide variety of techniques can be used for transferring the image from plate to paper, but the most commonly used is offset lithography, which relies on chemicals that attract ink and repel water.

The three basic methods of printing in use today are offset lithography, letterpress, and gravure. They use different methods to collect ink on the desired areas of the plate, then transfer the ink by pressing paper onto the plate or a transfer cylinder.

Letterpress, the oldest method of printing, uses a plate with raised letters or patterns on it. When an inked roller runs across the plate, ink sticks to the raised surfaces. In gravure, the letters and images are etched into a metal plate. The ink collects in the recesses, and the surface is wiped clean before printing.

MAGENTA PRINTING PRESS

Ink rollers A series of steel, rubber-coated rollers distribute ink evenly across the width of the plate

Plate cylinder The cylinder spins rapidly to pick up wetting solution and ink

Blanket cylinder The rubber blanket picks up ink from the plate and transfers it to the paper

Wetting solution A mix of water and additives wets the plate

Paper web Presses can run at variable speeds to compensate for changing humidity and other conditions. Maximum speeds are more than 10,000 sheets per hour

Impression cylinder A clean steel cylinder pushes the paper firmly against the blanket

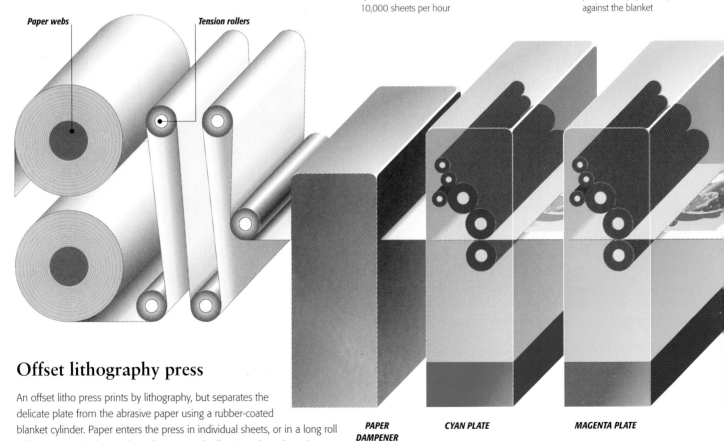

Paper webs

Tension rollers

PAPER DAMPENER

CYAN PLATE

MAGENTA PLATE

Offset lithography press

An offset litho press prints by lithography, but separates the delicate plate from the abrasive paper using a rubber-coated blanket cylinder. Paper enters the press in individual sheets, or in a long roll called a web, which runs through a series of rollers in order to keep it taut. After wetting, the paper passes over four impression cylinders, where cyan, magenta, yellow, and black inks are added. Finally, it is dried, cut, and folded.

Direct-digital printing

Printing direct from a digital file is becoming increasingly common, especially for short print runs. A direct-digital printer works just like a laser printer. The machine's software splits the image into cyan, magenta, yellow, and black parts, a process called color separation. The color-separated image is then converted into a pattern of dots (or rasterized) and the dot pattern is printed directly onto the page.

Digital technology is also affecting conventional printing—plates can now be made direct from a page -ayout file, without passing through the stages of printing to film and photographic transfer.

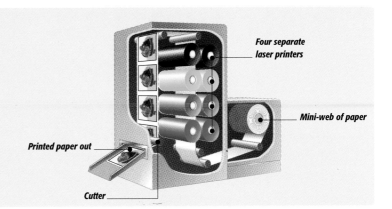

Four separate laser printers

Mini-web of paper

Printed paper out

Cutter

Lithography in action

A lithographic plate is made using a photographic process that gives letters and other image areas on the plate a greasy coating. Because the offsetting and printing process involves reversing the image twice, the plate is identical to the printed page. It is usually made of plastic or very thin aluminum.

Offset lithography involves collecting ink on the desired areas of the plate, transferring it onto a rubber roller, and then pressing this against a sheet of paper so the ink is transferred.

Wetting The plate, on which image areas are covered with a greasy coating, is covered in a wetting solution. This solution collects on uncoated areas of the plate

Inking An inked roller now runs across the plate. The oil-based ink is attracted to the coated areas of the plate, but repelled from the wet parts, since it will not mix with water

Offsetting A rubber-coated blanket cylinder is now rolled across the plate, picking up ink and water from it, and forming a reversed copy of the original image.

Printing Paper is pressed against the blanket cylinder by an impression cylinder and absorbs the ink. The image is reversed again, so the printed image is identical to that on the plate

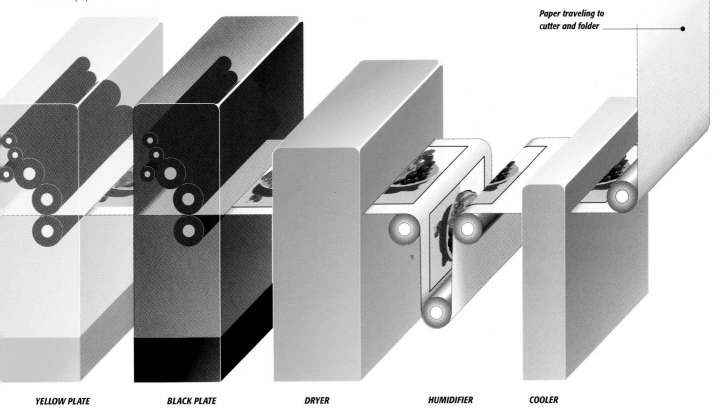

Paper traveling to cutter and folder

YELLOW PLATE   **BLACK PLATE**   **DRYER**   **HUMIDIFIER**   **COOLER**

SEE ALSO: HOW COLOR WORKS *p.82* | PREPARING FOR PRINT *p.86* | PHOTOCOPIERS & FAX MACHINES *p.84*

Photographic film

Photographic film is produced today in a huge assortment of forms. These include black-and-white and color, slide, negative (print), and "instant-print" films. Films also come in different sizes and formats, including APS (Advanced Photo System) and 35-mm cassettes and huge spools of film for professional cinematographers. Most types of films have different speeds (degrees of light sensitivity).

At the heart of any film is a layer or layers of gelatin emulsion that contain billions of grains of light-sensitive silver compounds called silver halides. The grains are sensitized to undergo a chemical change when exposed to light of any wavelength (black-and-white film) or a particular range of wavelengths (color). During development of negative films, chemical reactions produce silver (in most black-and-white films) or dyes (color films) in areas where grains were exposed. A similar process is then used to produce a print on photographic paper from the developed negative.

In color-slide film, dyes appear only in unexposed areas of the emulsion during development. This process is called color reversal. The result is a tri-layered transparency containing yellow, magenta, and cyan dyes that combine to produce the original scene. A patch of green in the original scene, for example, exposes only the middle layer of the transparency, producing no dye, while in the other, unexposed layers, yellow and cyan dyes are released and these combine to produce green.

EXPOSING THE FILM

Color-negative roll film

36EXP

Photographed scene

Abrasion coating

UV filter

Blue/violet sensitive layer

Silver-halide crystals

Blue-blocking filter

Green/yellow/ cyan sensitive layer

Red/orange sensitive layer

Film base

Nonreflective coating

Exposing and developing color-negative film

Color-negative film contains several layers, of which three contain silver-halide crystals that are sensitive to light of different wavelengths. When exposed to light of a wavelength to which they are sensitive, the crystals are chemically altered, and the pattern of alteration leaves a latent image in the film. During development, reactions occur between the altered crystals and chemicals called couplers in the film layers, leading to dye production. The film is then bleached and the dyes are fixed in place to produce a color negative.

DEVELOPMENT

Yellow dye Yellow dye is produced where short (blue/violet) wavelengths have been absorbed

Magenta dye Magenta dye is produced where medium (green/yellow/cyan) wavelengths have been absorbed

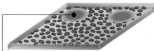

Cyan dye Cyan dye is produced where long (red/orange) wavelengths have been absorbed

BLEACH/FIX

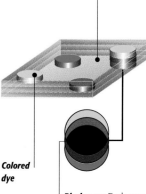

Colored negative Bleaching leaves clear areas on the color negative, while fixing with a solution permanently preserves the position of the dyes

Colored dye

Black areas Dark areas occur on the negative where all three layers have been exposed

Film speed and grain

Photographic film comes in a variety of speeds (International Standards Organisation or ISO ratings), from 50 ISO or lower (slow films) up to 3200 ISO or higher (fast films). The average size of the silver-halide grains is bigger in the fast films and they capture more of the incoming light for a given camera shutter speed and aperture setting. Fast films are best for taking photographs in low lighting conditions or for capturing fast-moving objects (when a high shutter speed must be used), but they result in "grainier" photographs. For portrait photography in good lighting conditions, slower films with finer grains are ideal.

Photographic film emulsion

The silver-halide grains in a typical slow film vary in size but on average are about one micrometer (thousandth of a millimeter) in diameter. They are held fixed in place by a substance called gelatin.

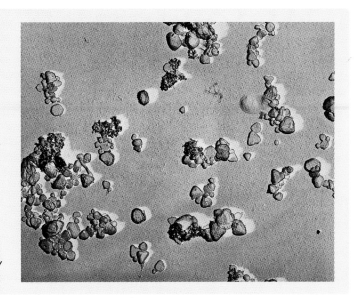

EXPOSING THE PAPER

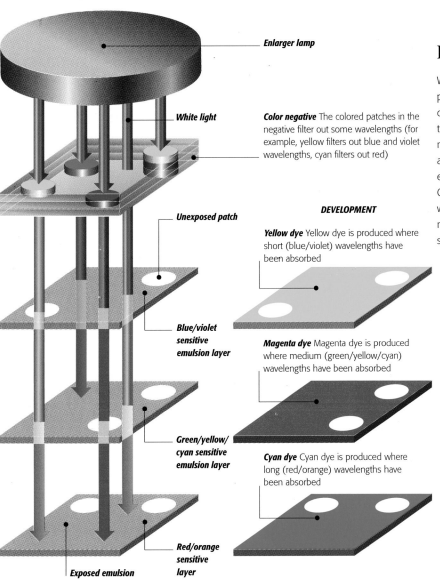

Enlarger lamp

White light

Color negative The colored patches in the negative filter out some wavelengths (for example, yellow filters out blue and violet wavelengths, cyan filters out red)

Unexposed patch

Blue/violet sensitive emulsion layer

Green/yellow/ cyan sensitive emulsion layer

Red/orange sensitive layer

Exposed emulsion

Printing from a color negative

White light is shone through the negative onto color print paper to form a large image, using a machine called an enlarger. Like color film, the paper contains three emulsion layers that are sensitive to short, medium, and long wavelengths of light. Transparent areas in the negative let through all wavelengths, exposing the emulsion in all three layers of the paper. Colored areas in the negative filter out some wavelengths, leaving nonexposed patches in one or more of the paper layers. The development and fixing stages then follow.

DEVELOPMENT

Yellow dye Yellow dye is produced where short (blue/violet) wavelengths have been absorbed

Magenta dye Magenta dye is produced where medium (green/yellow/cyan) wavelengths have been absorbed

Cyan dye Cyan dye is produced where long (red/orange) wavelengths have been absorbed

FIXING

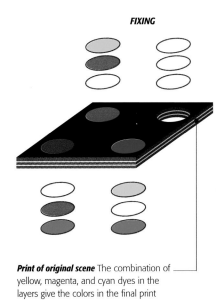

Print of original scene The combination of yellow, magenta, and cyan dyes in the layers give the colors in the final print

SEE ALSO: HOW COLOR WORKS *p.82* | PREPARING FOR PRINT *p.86* | CAMERAS *p.92* | LIGHT *p.262*

Cameras

Although modern cameras are computerized to allow photographs to be taken in difficult conditions by simply pointing and shooting, a camera is quite simple in principle. It comprises a light-tight body to hold the film and transport mechanism (or light-sensitive chip, in a digital camera), a lens to focus the image, an adjustable aperture (window) to admit light, and a shutter that opens for a preset time.

The two main types of cameras are the single lens reflex (SLR) and the compact. Compacts are light, small, highly automated, and easy to use. Modern SLRs are highly automated too, but offer the photographer more control over the picture because the aperture,

shutter speed, focus, and other settings can be manually overriden. Additionally, a wide range of lenses can be fitted to an SLR body.

Modern cameras are microprocessor-based devices. Sensors detect the amount of light entering the camera. This data is used to set an appropriate aperture size and shutter speed, and charge a flash unit (if present) when extra illumination is needed. Autofocus systems measure the distance to the subject using infrared or ultrasonic beams and move a motorized lens to focus the image. Many conventional film cameras have a motor to wind the film.

Digital cameras use an array of light-sensitive elements on a chip to record images, which can then be output to a computer for manipulation and archiving. Most digital cameras are compacts, though digital SLRs are increasingly being used by professionals.

Basic SLR camera

An SLR camera has a hinged mirror that reflects light entering the lens toward the viewfinder, but flips up to allow this light to reach the film during an exposure. This ensures that the image on the viewfinder is exactly the same as the image that will be captured on film. The camera body can be fitted with one of dozens of different lenses, including wide-angle for landscapes, telephoto for capturing distant objects, macro lenses for close-up work, and multipurpose zooms that provide a range of focal lengths.

Winder Turning this lever advances the film to bring the next frame into position behind the shutter

Shutter speed control Turning this dial alters the length of time for which the shutter opens to allow light onto the film

Pentaprism This glass prism directs light into the viewfinder

Light sensor This sensor detects light levels and helps obtain a correctly-exposed image

Viewfinder

Shutter release Pressing this button flips the mirror up and opens the shutter for a preset interval

Aperture control

Mirror This mirror reflects light from the lens into the pentaprism, but flips up during an exposure to allow light to reach the film

Focusing ring Turning this ring moves the lens elements, changing the overall focal length and bringing nearby or more distant objects into focus

Aperture The size of this hole is adjusted by the aperture control ring to vary the amount of light that reaches the film while the shutter is open

Lens This interchangeable lens unit contains several individual lens elements made from high-quality optical glass that focus an image onto the film

Light from scene to be photographed

Iris diaphragm Overlapping blades moves to change the size of the aperture

Shutter The shutter consists of a pair of metal blinds that open for a preset interval to allow a controlled amount of light to reach the film

Digital cameras

Instead of capturing images on film, digital cameras use light-sensitive chips, usually of a type called the CCD (charge-coupled device). A CCD is an array of millions of light-sensitive cells called pixels, which produce electric signals that vary with the amount of light that hits them. An analog-to-digital converter transforms these signals into digital form for storage. These digital images are then reconverted into analog form for display on the camera's electronic viewfinder or on a television screen.

Charge-coupled device (CCD) This array of light-sensitive cells converts light into an electric signal, which is relayed to the color screen and to the analog-to-digital converter

Lens The lens focuses the image onto the CCD as light enters the camera

Light from scene to be photographed

Microprocessor This microchip relays digitized image information to the memory chip and to the television and computer output sockets

Memory chip This chip temporarily stores images while the camera is switched on

"Flash memory" card A removable electronic card holds images for long-term storage

Television output socket A cable connection enables pictures to be displayed on a television set

Digital-to-analog converter This device reconverts the digitized image into analog form so that it can be viewed on the camera's color screen or on a television set

Viewfinder screen Instead of using a normal optical viewfinder, a digital camera may display an image on a small color LCD (liquid-crystal display)

Analog-to-digital converter This device reads the image information from the CCD and converts it into digital form so that it can be stored in the camera's memory or on a computer

Computer output port Images can be downloaded onto a computer, where they can be manipulated or sent on to a printer

Film rewinder This is used to rewind the exposed film back into the cassette before removal for development

Film When the film is exposed to light, changes occur in layers of light-sensitive emulsion and a latent photographic image forms

Film cassette After each shot is taken, film is unwound from the cassette and a new frame moves into position behind the shutter

Field of view

The area visible through a camera lens depends on the angle of light entering the lens, and is known as the field of view. Wide-angle lenses receive light at angles above 46° (the angle of view of the human eye). Telephoto lenses have a narrower angle of view, but produce a magnified image.

Lens arrangement
A camera lens consists of a combination of convex lenses (which bend light inward) and concave lenses (which bend light outward). The relative distance between the component lenses determines the angle of view.

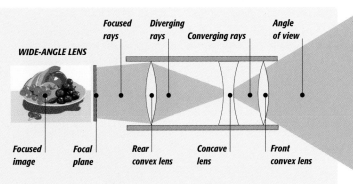

WIDE-ANGLE LENS

Focused rays — Diverging rays — Converging rays — Angle of view

Focused image — Focal plane — Rear convex lens — Concave lens — Front convex lens

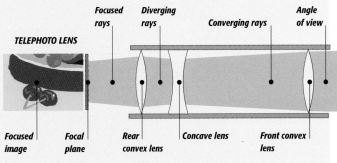

TELEPHOTO LENS

Focused rays — Diverging rays — Converging rays — Angle of view

Focused image — Focal plane — Rear convex lens — Concave lens — Front convex lens

SEE ALSO: DIGITAL WORLD *p.54* | COMPUTER OUTPUT DEVICES *p.68* | PHOTOGRAPHIC FILM *p.90* | LIGHT *p.262*

Movie cameras and projectors

Motion pictures exploit a feature of human sight called persistence of vision, whereby images are retained momentarily even after the object being viewed has changed. Movie cameras and projectors use the same principles and technology to record and display a succession of still images, or frames, at a rate of 24 frames per second—fast enough to generate the impression of smooth, live movement.

In order to capture or project a succession of still images, movie cameras and projectors advance a length of film frame by frame past a device called a gate. Each frame is held still in the gate for a fraction of a second, in which time it is exposed to light by a rotating shutter. In both cameras and projectors, the shutter allows light to reach the film frame only when it is in position in the gate, shutting out light while the film advances. A claw mechanism engages sprocket holes in the film to synchronize frame advance with the rotation of the shutter. The film soundtrack is recorded as an optical or magnetic pattern along the edges of the film during editing and is read by sound heads in the projector.

Modern trends in movie cameras and projection include digital cinema and large-format films, such as IMAX. An IMAX projector is powered by a 15,000-watt short-arc xenon lamp, which is about five times brighter than a conventional projector lamp, in order to create sharper and brighter images.

Movie camera

A movie camera records a succession of images onto the frames of a reel of film. After each frame is exposed, a motor-driven claw mechanism brings the next frame into place behind the shutter. Each second, 24 frames are exposed.

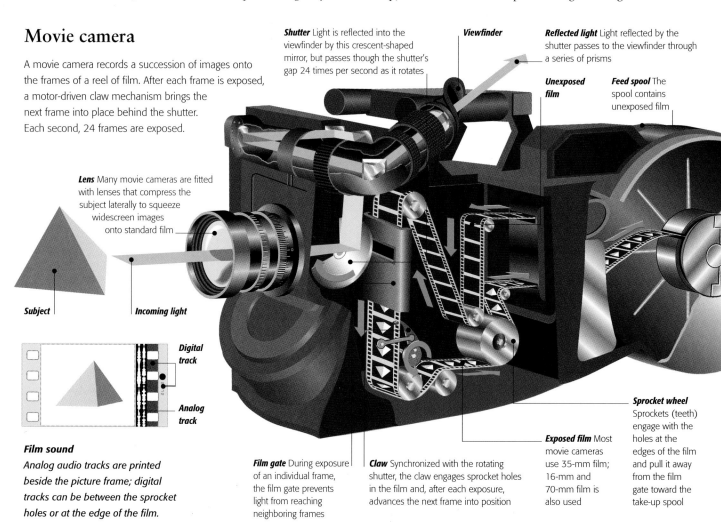

Shutter Light is reflected into the viewfinder by this crescent-shaped mirror, but passes though the shutter's gap 24 times per second as it rotates

Viewfinder

Reflected light Light reflected by the shutter passes to the viewfinder through a series of prisms

Unexposed film

Feed spool The spool contains unexposed film

Lens Many movie cameras are fitted with lenses that compress the subject laterally to squeeze widescreen images onto standard film

Subject

Incoming light

Digital track

Analog track

Sprocket wheel Sprockets (teeth) engage with the holes at the edges of the film and pull it away from the film gate toward the take-up spool

Exposed film Most movie cameras use 35-mm film; 16-mm and 70-mm film is also used

Film sound
Analog audio tracks are printed beside the picture frame; digital tracks can be between the sprocket holes or at the edge of the film.

Film gate During exposure of an individual frame, the film gate prevents light from reaching neighboring frames

Claw Synchronized with the rotating shutter, the claw engages sprocket holes in the film and, after each exposure, advances the next frame into position

IMAX 3-D

IMAX movies are shot using film that is ten times normal size, projected onto giant screens that extend beyond the range of the viewer's peripheral vision, to give them the impression of being "inside" the film action. IMAX 3-D further enhances the illusion of reality. The system uses two films shot simultaneously by a binocular camera, which has two lenses set the same distance apart as a pair of eyes. The two films are then projected simultaneously through separate polarizing filters, one of which polarizes light vertically and the other horizontally. A 3-D effect is produced when the projected images are viewed through glasses that have one vertically- and one horizontally-polarizing lens.

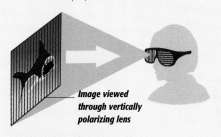

Image viewed through vertically polarizing lens

Film A: vertically polarized
The vertically polarizing lens allows vertically polarized light through to one eye but blocks out the horizontally polarized projection.

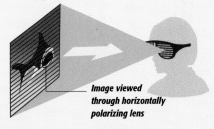

Image viewed through horizontally polarizing lens

Film B: horizontally polarized
The horizontally polarizing lens transmits horizontally polarized light to the other eye but blocks out the vertically polarized projection.

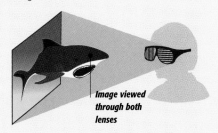

Image viewed through both lenses

Films A and B combined
Each eye sees an image recorded by only one of the camera lenses. The perspective difference between the two images creates a 3-D effect.

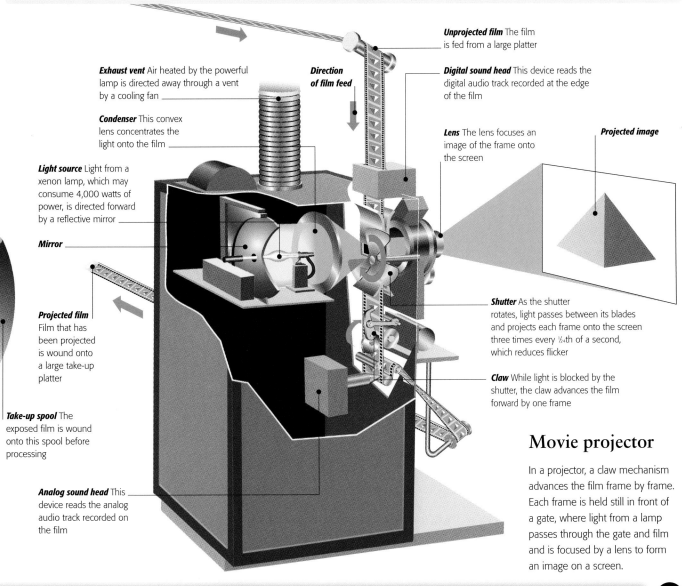

Unprojected film The film is fed from a large platter

Digital sound head This device reads the digital audio track recorded at the edge of the film

Exhaust vent Air heated by the powerful lamp is directed away through a vent by a cooling fan

Direction of film feed

Condenser This convex lens concentrates the light onto the film

Lens The lens focuses an image of the frame onto the screen

Projected image

Light source Light from a xenon lamp, which may consume 4,000 watts of power, is directed forward by a reflective mirror

Mirror

Shutter As the shutter rotates, light passes between its blades and projects each frame onto the screen three times every 1/24th of a second, which reduces flicker

Projected film Film that has been projected is wound onto a large take-up platter

Claw While light is blocked by the shutter, the claw advances the film forward by one frame

Take-up spool The exposed film is wound onto this spool before processing

Analog sound head This device reads the analog audio track recorded on the film

Movie projector

In a projector, a claw mechanism advances the film frame by frame. Each frame is held still in front of a gate, where light from a lamp passes through the gate and film and is focused by a lens to form an image on a screen.

SEE ALSO: DIGITAL CINEMA *p.96* | SPECIAL EFFECTS *p.98* | VIDEO TECHNOLOGY *p.104* | LIGHT *p.262*

Digital cinema

Digital cinema technology does away with celluloid film and traditional methods of projection. Along with the loss of the film go flicker, scratches, focus flutter, color deterioration, and a whole range of degenerative processes that reduce the quality of a film the more it is projected. Future filmgoers will enjoy a clearer, brighter, more colorful picture, bringing them closer to the vision of the filmmaker.

Traditional cinema-projection methods shine bright light through a transparent film medium that holds the image information. Digital cinema technology stores image information in digital form rather than on film and must therefore process light inside the projector. This has been accomplished with the development of Digital Light Processing (DLP) Cinema technology. DLP uses arrays of microscopic mirrors, called Digital Micromirror Devices (DMDs), that move in response to the digital image

information. Bright light is shone onto these arrays and the light reflected forms an image. A succession of image frames shown at a rate of 24 per second gives the the impression of a moving picture. Digital-format movies are either recorded in electronic form using a digital camera or converted from film into digital code. The conversion process involves taking each frame of a film and splitting it into red, green, and blue components. Each frame is divided into small pixels (picture elements) and the intensity of each color in each pixel is measured. The intensity values are recorded as digital data and saved onto hard disk. Digital motion pictures can be sent to movie theaters along broadband cable connections or beamed globally by satellite.

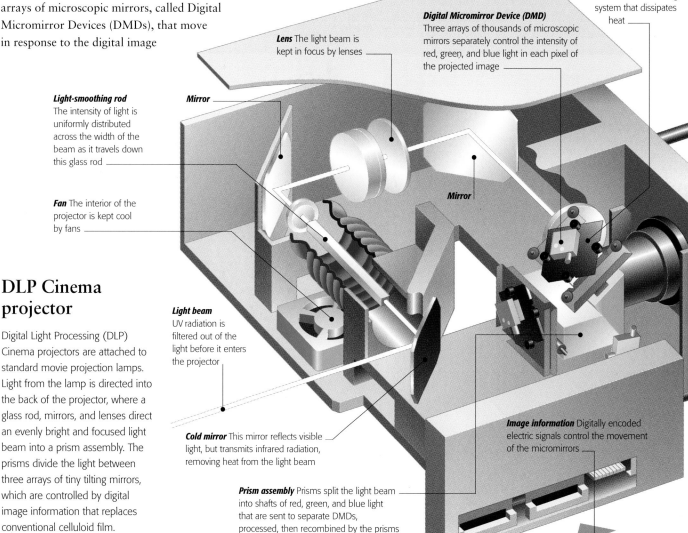

Heat sink Each DMD is attached to a fluid-cooling system that dissipates heat

Digital Micromirror Device (DMD) Three arrays of thousands of microscopic mirrors separately control the intensity of red, green, and blue light in each pixel of the projected image

Lens The light beam is kept in focus by lenses

Light-smoothing rod The intensity of light is uniformly distributed across the width of the beam as it travels down this glass rod

Mirror

Mirror

Fan The interior of the projector is kept cool by fans

DLP Cinema projector

Digital Light Processing (DLP) Cinema projectors are attached to standard movie projection lamps. Light from the lamp is directed into the back of the projector, where a glass rod, mirrors, and lenses direct an evenly bright and focused light beam into a prism assembly. The prisms divide the light between three arrays of tiny tilting mirrors, which are controlled by digital image information that replaces conventional celluloid film.

Light beam UV radiation is filtered out of the light before it enters the projector

Cold mirror This mirror reflects visible light, but transmits infrared radiation, removing heat from the light beam

Prism assembly Prisms split the light beam into shafts of red, green, and blue light that are sent to separate DMDs, processed, then recombined by the prisms

Image information Digitally encoded electric signals control the movement of the micromirrors

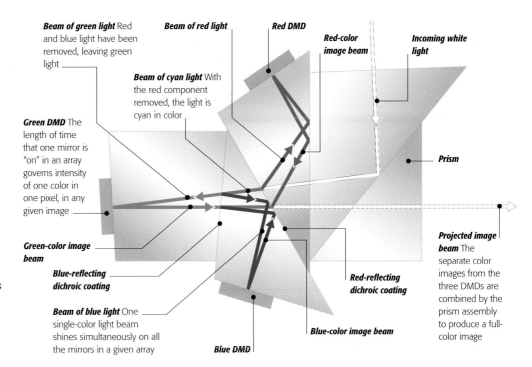

Beam of green light Red and blue light have been removed, leaving green light

Beam of red light

Red DMD

Red-color image beam

Incoming white light

Beam of cyan light With the red component removed, the light is cyan in color

Green DMD The length of time that one mirror is "on" in an array governs intensity of one color in one pixel, in any given image

Prism

Prism assembly

White light entering the prism assembly is split into its three component colors by dichroic coatings on the prisms, which transmit certain frequencies of light but reflect others. Each single-color light beam is shone over the whole of a Digital Micromirror Device (DMD). The DMDs convert each beam into single-color images every ¹⁄₂₄th second. These images are recombined by the prisms to produce a full-color, image.

Green-color image beam

Blue-reflecting dichroic coating

Beam of blue light One single-color light beam shines simultaneously on all the mirrors in a given array

Red-reflecting dichroic coating

Blue-color image beam

Blue DMD

Projected image beam The separate color images from the three DMDs are combined by the prism assembly to produce a full-color image

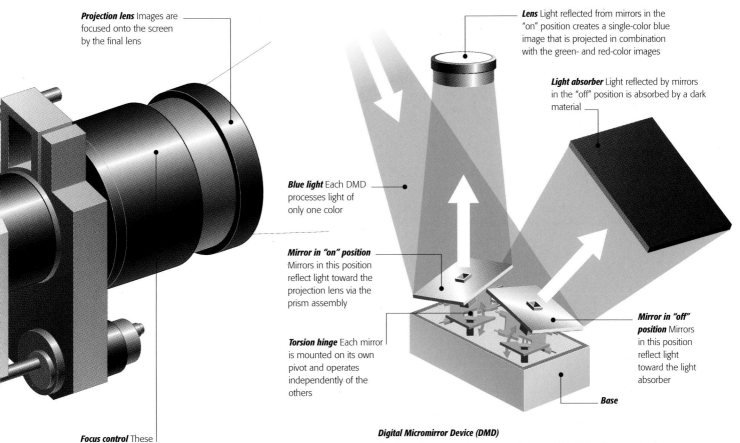

Projection lens Images are focused onto the screen by the final lens

Lens Light reflected from mirrors in the "on" position creates a single-color blue image that is projected in combination with the green- and red-color images

Light absorber Light reflected by mirrors in the "off" position is absorbed by a dark material

Blue light Each DMD processes light of only one color

Mirror in "on" position Mirrors in this position reflect light toward the projection lens via the prism assembly

Mirror in "off" position Mirrors in this position reflect light toward the light absorber

Torsion hinge Each mirror is mounted on its own pivot and operates independently of the others

Base

Focus control These controls enable the image to be focused onto a screen

Digital Micromirror Device (DMD)

A micromirror array is composed of over 500,000 microscopic mirrors on an area the size of a postage stamp. Each micromirror represents one pixel, or picture element, of the projected image and operates entirely independently of the others. A single micromirror can switch between "on" and "off" positions thousands of times during a 1/24th-second frame to produce the correct color intensity for each pixel in the image.

SEE ALSO: DIGITAL DATA *p.56* | HOW COLOR WORKS *p.82* | MOVIE CAMERAS & PROJECTORS *p.94* | LIGHT *p.262*

Special effects

Film and television special effects make the impossible seem real, and allow filmmakers to translate their visions into moving pictures. The history of special effects is as old as cinema itself, but in the past few years it has been revolutionized by the arrival of computer-generated imagery (CGI).

Traditional effects

The first special effects motion pictures were made by French director Georges Méliès starting in 1896. Méliès demonstrated that motion picture special effects could fool the audience in many ways. On film, there is no way of telling the scales of different objects, so miniature models can be turned into huge objects on screen. Film can also be shot frame by frame, moving models slightly between frames, creating the illusion of movement when the film is played back at full speed (stop-motion animation). Exposing two images on the same frame of film can be used to combine action sequences shot at different times (allowing an actor to appear in different guises in a single scene). Early filmmakers created these composite pictures by blocking out part of a scene (either by covering part of the lens with a matte (mask) or

Computer-generated animation

The Utahraptor *for the BBC-TV series* Walking With Dinosaurs *was created in several stages. Left to right: a sculpted model, scanned by lasers, is reduced by computer to thousands of tiny polygons; a basic skeleton is created for animators to plot the creature's movements; an artist designs textures to add realism to the finished model; then the detailed model is "enveloped" around the animation skeleton and rendered to produce the final result.*

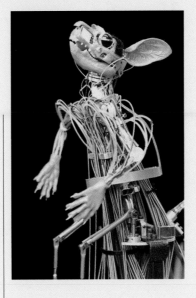

Animatronic model
The latest animatronics models are cabable of performing a wide range of movements. These movements are activated electronically, and a complex network of wires is needed to carry electric signals and move the joints. In the finished model these wires are hidden by an outer skin.

by placing a sheet of glass with a blacked-out area in front of the lens) so that a portion of film in each frame was not exposed. The film was then rewound and a new sequence was shot on the unexposed portion of film. This technique, known as matte, limited the action because actors could not cross the boundary between the area exposed in the first sequence and the area exposed in the second sequence. The invention of the optical printer in the 1920s made the task of combining separate elements much easier. Optical printers consist of a projector on which several films can be projected at any one time, and a movie camera, which shoots the composite picture onto a new piece of film in the camera. As no part of the scene has to be blocked out, this allows the actors far more freedom of movement and interaction.

Combining action and background

The simplest way of blending moving action seamlessly with a separately shot background is rear projection—actors perform in front of a screen, onto which a background sequence is projected. Like all effects, rear projection relies on meticulous planning—the precise perspectives of the action sequence and the background must be worked out in advance, or they will clash when combined. With the arrival of color photography, another method became available—color separation or "blue screen." In this technique, actors perform in front of a blue or green screen. This backdrop is then removed by optical or digital processes, so that the action can be isolated and composited with a different background. In order to make sure that movements and camera angles match for all the elements that appear in a finished scene, robotic "motion-control" cameras are used. The movements of such a camera are computer-controlled and can be repeated exactly to film several elements in a series of "takes" in the same way so that the correct perspective is achieved.

CGI and animatronics

The most recent revolution in special effects has been the arrival of CGI. Computers have been used "behind the scenes" since the 1970s, but their processing power was limited, and CGI could only create flat-looking artificial backgrounds. In the early 1990s, computers finally became powerful enough to create lifelike 3-D images. Films such as *Gladiator* used CGI to merge live action with a digital recreation of ancient Rome on a scale that would have been unimaginable for earlier directors.

Creating an image that blends properly with live-action shots means that a CGI shot has to be composed and "lit" inside the computer in just the same way as a real one. The final images have to be produced, frame by frame, using "ray-tracing" software to plot the paths of light rays from various light sources around the virtual set, creating realistic reflections and highlights.

CGI is often combined with animatronics—detailed models with electronic mechanisms that give them complex movements. Animatronics are frequently used for close-up shots and for scenes where a remote-controlled model has to interact with its environment—for example, in *Walking With Dinosaurs*, the shots of dinosaurs feeding used computerized models.

Blue screening

Blue screening involves filming models or people against a background color (usually blue), and then replacing the blue elements on the film with a different background, so that anything not blue in the original image is blended seamlessly with the new background. For many years this was done using optical printers, but today the process is carried out by computer software—entire films are increasingly transferred into digital format on a computer for editing and addition of special effects, before being printed back onto film for projection.

Foreground image

Background image

Combined image

SEE ALSO: THE VIRTUAL WORLD *p.74* | MOVIE CAMERAS & PROJECTORS *p.94* | DIGITAL CINEMA *p.96* | LIGHT *p.262*

Television

Television creates the illusion of moving color pictures by exploiting two properties of human vision—that a quick succession of still images of a moving scene looks like a moving image, and that a mixture of three primary-color images appears to be full color. These principles, combined with the technological developments of cathode-ray tubes and radio transmission, deliver TV pictures to our homes.

Light entering a TV camera is split into red, green, and blue color components, which form primary-color images in three separate tubes. The tubes scan the images in a series of 525 horizontal lines, recording the brightness along each line as an electrical signal. A synchronization signal ensures that the three tubes scan together, so that when the three electrical signals are combined, they give color and brightness data about the same part of the image at the same time. Thirty full images, or frames, are collated from scans each second. The video (picture) signals from the tubes are combined with the synchronization signal and an audio signal before transmission.

Most TV receivers are based on the cathode-ray tube (CRT), a sealed, evacuated glass tube that contains three electron guns at one end and a phosphor-coated screen at the other. Each electron gun emits a beam of fast-moving electrons (cathode rays), which scans across the screen with varying intensity, according to the brightness of the video signal. Phosphor stripes or dots on the screen emit red, green, or blue light when hit by electrons. As the electron beams scan across the phosphor screen, they build up each transmitted frame line by line. The eye combines the rapidly changing lines of lit phosphor dots into moving full-color images.

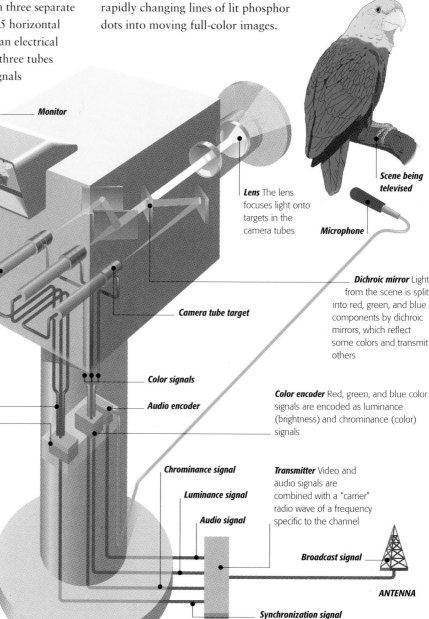

Monitor

Lens The lens focuses light onto targets in the camera tubes

Microphone

Scene being televised

Camera tube The primary color image formed in each tube is converted into an electric signal by a scanning electron beam. Some cameras use charge-coupled devices (CCDs) instead of tubes

Camera tube target

Dichroic mirror Light from the scene is split into red, green, and blue components by dichroic mirrors, which reflect some colors and transmit others

Synchronization signal

Color signals

Synchronization unit This generates a signal that synchronizes scanning in the three camera tubes

Audio encoder

Color encoder Red, green, and blue color signals are encoded as luminance (brightness) and chrominance (color) signals

Television camera

Light entering the camera is split into three primary-color components by special filters called dichroic mirrors. Red, green, and blue images are focused onto light-sensitive targets in three camera tubes. Synchronized electron beams in the tubes scan the images formed and record variations in brightness as varying voltages. This information is encoded, together with synchronization and audio signals, and combined with a radio wave for broadcast.

Chrominance signal

Luminance signal

Audio signal

Transmitter Video and audio signals are combined with a "carrier" radio wave of a frequency specific to the channel

Broadcast signal

ANTENNA

Synchronization signal

IMAGE ON TV SCREEN

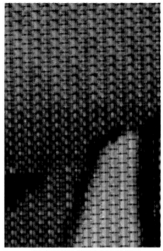

MAGNIFIED VIEW OF SCREEN

Phosphor screen

The points of red, green, and blue light that make up a TV picture become visible when highly magnified. This light is emitted by phosphor dots that are arranged in stripes on the inside of the screen and glow when struck by electrons. As the electron beams scan the screen, they pass through a grille that allows electrons from each beam to strike only phosphor stripes that emit light of the same color as the color signal of the beam.

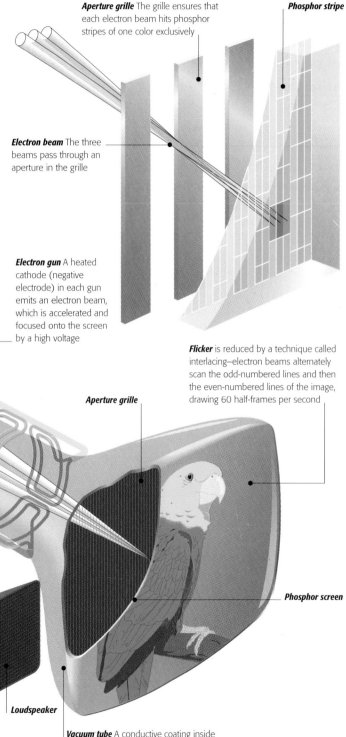

Aperture grille The grille ensures that each electron beam hits phosphor stripes of one color exclusively

Phosphor stripe

Electron beam The three beams pass through an aperture in the grille

Electron gun A heated cathode (negative electrode) in each gun emits an electron beam, which is accelerated and focused onto the screen by a high voltage

Flicker is reduced by a technique called interlacing—electron beams alternately scan the odd-numbered lines and then the even-numbered lines of the image, drawing 60 half-frames per second

Aperture grille

Antenna

Synchronization signal

Tuner A channel is chosen by using the tuner to select the appropriate carrier wave frequency and receive the signal

Synchronization unit The synchronization signal is split into horizontal and vertical deflection signals to apply to the coils

Chrominance signal

Luminance signal

Color signal

Audio detector This decodes and amplifies the audio signal and sends it to the speaker

Color decoder The video (chrominance and luminance) signals are converted into red, green, and blue colour signals

Audio signal

Vertical deflection coil

Horizontal deflection coil

Phosphor screen

Loudspeaker

Vacuum tube A conductive coating inside the glass tube carries electrons that arrive at the screen back to the electric supply

Television receiver

Chrominance, luminance, synchronization, and audio signals are separated from the received radio wave and decoded. The two video signals are split into three component color signals and applied to three electron guns. The intensity of the electron beam emitted by each gun is modulated by the corresponding color signal. The three beams are scanned across the screen by the changing magnetic fields of deflection coils, which are controlled by the synchronization signal.

SEE ALSO: RADIO *p.48* | HOW COLOR WORKS *p.82* | THE FUTURE OF TELEVISION *p.102*

The future of television

Television pictures in the future will be wider, sharper, clearer, and digital. New technologies display high-resolution, cinema-format images on large, flat screens and envelope the viewer in 3-D "surround sound." In addition to picture and sound quality improvements, digital TV provides greater channel choice, delivery of programs on demand, and services such as E-mail, Internet access, and shopping. The TV set may even become the primary gateway into the world of electronic media.

Widescreen TV is perhaps the most apparent of the new developments in TV technology. Conventional TV has an aspect ratio (ratio of screen width to height) of 4:3. Widescreen TV has an aspect ratio of 16:9 (equivalent to 5.3:3), a shape that better fills the human field of vision and is more appropriate for movies and sports coverage. The 16:9 aspect ratio is also part of a new high-definition TV system called HDTV. Standard definition TV (SDTV) has a resolution of 525 to 625 lines horizontally, depending on the regional system. HDTV has about 1,100 lines of resolution, giving a much sharper, more detailed picture. HDTV pictures are best appreciated on very large screens, and new screen technologies, such as plasma displays, are emerging to meet this demand.

A parallel development is digital TV, which uses computers to compress signals for transmission in order to make efficient use of the limited bandwidth (portion of the electromagnetic spectrum) available for each channel. Compression enables four separate SDTV programs, or one HDTV program, to be transmitted digitally in the bandwidth of a single analog channel.

Digital compression

Digital compression makes much of the information in an analog TV signal redundant. Interframe compression exploits the presence of repeated patterns within each still frame of the program. Interframe compression (shown below) works by transmitting only the changes between successive frames. In contrast, analog TV systems encode the entire content of each frame in the signal, even if the image is unchanged between frames.

Frame 1
The moving image is of a ball falling against a static background. All the information in the first frame is transmitted.

Interframe change
The only interframe change is in the position of the ball, so only the data for the new ball position and the newly revealed area of sky are transmitted.

Frame 2
To construct the second frame, a decoder in the TV receiver combines the data for the first frame with the data for the change between frames.

Widescreen TV

The aspect ratio of conventional TV is 4:3. Movie theater screens are widescreen and have a much higher aspect ratio, closer to 16:9. When movies are broadcast for conventional TV sets, the picture must be either "panned and scanned" or "letterboxed" as shown below. Widescreen TVs have a high aspect ratio of 16:9, making such compromises unnecessary.

16:9 IMAGE ON 16:9 SCREEN

Widescreen TV
A widescreen TV displays a 16:9 movie image on the entire screen without any cropping or letterboxing.

PANNED AND SCANNED IMAGE (4:3 SCREEN)

Pan and scan
Here, the important part of the movie image is captured through a 4:3 aspect ratio "window" before broadcast. The remainder is cropped out.

LETTERBOXED IMAGE (4:3 SCREEN)

Letterboxed image
Letterboxing displays the full widescreen image at reduced size leaving the screen black above and below the image.

Flat-screen technology

HDTV is best appreciated on large screens. Conventional TV sets, which produce an image on the screen using a device called a cathode-ray tube (CRT), can incorporate large screens but require extremely large and heavy CRTs to do so. Alternatives to CRTs are needed in order to make large screens practicable. New flat-screen technologies under development include LCD (liquid crystal display), but most commercial flat screens use plasma display panels (PDPs). PDPs may be over a yard wide, but only a few inches deep, so they can be hung on a wall like a painting.

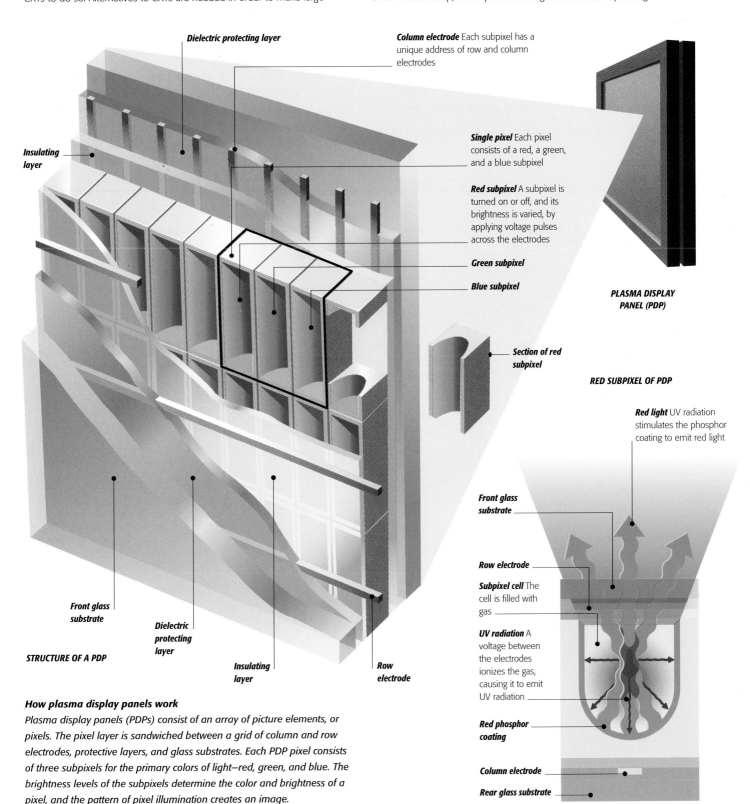

Dielectric protecting layer

Column electrode Each subpixel has a unique address of row and column electrodes

Insulating layer

Single pixel Each pixel consists of a red, a green, and a blue subpixel

Red subpixel A subpixel is turned on or off, and its brightness is varied, by applying voltage pulses across the electrodes

Green subpixel

Blue subpixel

PLASMA DISPLAY PANEL (PDP)

Section of red subpixel

RED SUBPIXEL OF PDP

Red light UV radiation stimulates the phosphor coating to emit red light

Front glass substrate

Row electrode

Subpixel cell The cell is filled with gas

UV radiation A voltage between the electrodes ionizes the gas, causing it to emit UV radiation

Red phosphor coating

Column electrode

Rear glass substrate

Front glass substrate

Dielectric protecting layer

Insulating layer

Row electrode

STRUCTURE OF A PDP

How plasma display panels work

Plasma display panels (PDPs) consist of an array of picture elements, or pixels. The pixel layer is sandwiched between a grid of column and row electrodes, protective layers, and glass substrates. Each PDP pixel consists of three subpixels for the primary colors of light—red, green, and blue. The brightness levels of the subpixels determine the color and brightness of a pixel, and the pattern of pixel illumination creates an image.

SEE ALSO: DIGITAL WORLD *p.54* | COMPUTER OUTPUT DEVICES *p.68* | HOW COLOR WORKS *p.82* | TELEVISION *p.100*

Video technology

Video technology involves the electronic capture, recording, and playback of a sequence of still images and sound. Most video devices process and store this information in analog form, as continuously varying electric currents and magnetic patterns on tape. Digital video equipment encodes images and sound as discrete binary pulses for manipulation by computer and storage on tape, disk, or microchips.

Video recording involves converting moving images captured by a camera into a signal that comprises 30 frames (complete still images) for each second recorded. The electric signal can be recorded on magnetic tape. Each frame is divided into several hundred horizontal lines. A single frame is stored as two interlaced fields (half frames), each of which contains data for alternate lines of the image. A television set draws 60 fields per second, line by line, giving the illusion of a moving picture.

Traditional TV cameras use filters to split the image into three primary-color images. These images are scanned line by line using photomultiplier tubes, which generate an electric current in proportion to the brightness of the incident light. Today, images are more likely to be captured with CCDs instead of tubes.

Video signals are recorded in a range of formats on various media, including VHS, 8-mm, and DV (digital video) tapes and DVD (digital versatile disks). Digital video recordings must be converted into analog form for display on current TV sets.

Video recording and playback

When a cassette is inserted into a videocassette recorder (VCR), a loop of tape is pulled out of it by the loading mechanism. The tape winds around a set of rollers and heads, which are coils of wire that record a magnetic pattern on the tape when an electric signal passes through them, or produce an electric signal as the magnetized tape moves past during playback. Separate heads are used for audio and video recording and playback.

Recording and playback heads These video heads are mounted on the drum

Sound track

Sound head This head records or reads audio signals in a horizontal track along the tape edge

Guide roller

Tracks on a videotape
On analog videotape, the magnetic particles in each track form wavy patterns in various directions. On digital videotape, they align in only two ways, representing 1s and 0s.

Drum The drum spins at 1,800 revolutions per minute in the opposite direction to the tape's movement

Guide roller

Erase head The tape passes by a head with a constant strong magnetic field, which erases any information on it

Loading pole

Capstan This roller pulls the tape through the machine

Video track

Pinch roller These rollers push the tape against the capstan

Take-up reel

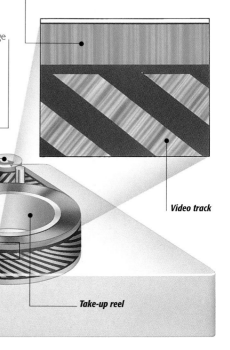

Tape A typical VHS cassette contains about 800 ft (240 m) of ½-in (12.5-mm) wide mylar (plastic) tape coated with magnetizable iron oxide

Supply reel The tape unwinds into the machine at a rate of 1⅜ in (33.3 mm) per second; slower speeds are used for extended recording times

Inside a VHS videocassette recorder (VCR)
A system called helical scanning is used to record picture information in diagonal tracks across the tape. Two or more recording heads on a rapidly spinning drum cause magnetic particles on the tape to line up. As one head reaches the edge of the tape, another begins to write a new track parallel to the last. The sound signal is recorded along the tape edge.

Electronic images

The charge-coupled device, or CCD, lies at the heart of the modern video camera. It is a silicon semiconductor chip that collects electric charges in millions of tiny squares called pixels, in proportion to the amount of light falling on it. The charges can then be flushed out of the CCD in sequence, transforming the picture information into a varying electric current. A typical CCD repeats this process 30 times every second.

1 Capturing light

When photons of light strike silicon atoms in the CCD's outer layer, they knock out negatively charged electrons, which collect in individual pixels, attracted by positive charges behind an insulator layer. The number of electrons in a pixel is proportional to the amount of light striking it.

2 Flushing down

The light source is now cut off and no more electrons are generated. The positive charge on the electrode is moved downward. The rippling movement of the positive charge drags electrons collected on the conducting layer down, and the readout row is filled with electrons from the row above.

3 Reading out

As the positive charges on the upper pixels switch back to normal, all the electrons collected by the CCD have shifted down by one pixel. Electrons on the readout row of the CCD are now flushed along by moving the positive charge to the right. Electrons emerging at the end of the row comprise a varying electric current.

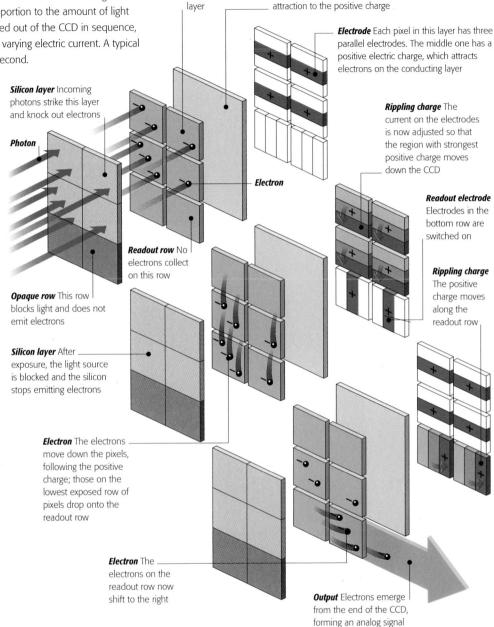

Conducting layer The electrons collect on pixels in the conducting layer

Insulating layer The electrons cannot travel through the insulating layer, so are trapped on the conducting layer by their attraction to the positive charge

Electrode Each pixel in this layer has three parallel electrodes. The middle one has a positive electric charge, which attracts electrons on the conducting layer

Rippling charge The current on the electrodes is now adjusted so that the region with strongest positive charge moves down the CCD

Readout electrode Electrodes in the bottom row are switched on

Rippling charge The positive charge moves along the readout row

Silicon layer Incoming photons strike this layer and knock out electrons

Photon

Electron

Readout row No electrons collect on this row

Opaque row This row blocks light and does not emit electrons

Silicon layer After exposure, the light source is blocked and the silicon stops emitting electrons

Electron The electrons move down the pixels, following the positive charge; those on the lowest exposed row of pixels drop onto the readout row

Electron The electrons on the readout row now shift to the right

Output Electrons emerge from the end of the CCD, forming an analog signal

DVD

Digital Versatile Disks store video data in digital form—the varying analog signals are converted into a code comprising of binary 1s and 0s. "Pits" and "lands" on the DVD record the sequence of these binary numbers, and can be read with a laser in the same way as a compact disc (CD).

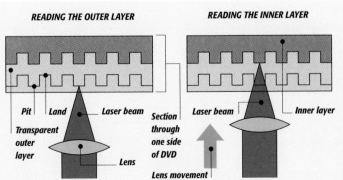

READING THE OUTER LAYER

READING THE INNER LAYER

Pit **Land** **Laser beam**

Transparent outer layer

Lens

Section through one side of DVD

Laser beam **Inner layer**

Lens movement

High capacity

DVDs can store up to 17 GB—as much as 25 CD-ROMs—because the pits and lands used to store data are much smaller than those on a CD. Data may be stored in two layers on each of the two sides of a DVD. The laser beam is focused by a lens onto the upper or lower layer, depending on which one is being read.

SEE ALSO: SOUND REPRODUCTION *p.80* | CAMERAS *p.92* | TELEVISION *p.100* | ELECTRICITY & MAGNETISM *p.260*

Transportation

Since the invention of the steam engine at the end of the 18th century, continuing developments in transportation have made the world seem like a smaller place. Vehicles have become faster, cheaper, and more efficient, shrinking the time taken to travel by land and sea and conquering new realms in the air and beneath the waves. Today, mass transportation such as bullet trains and aircraft allow people to make journeys in hours that would once have taken days or weeks, while private vehicles—most significantly the automobile—give millions new freedom to travel. This chapter reveals the technology behind many familiar, and a few not so familiar, forms of transportation.

Cars

Around 40 million cars are manufactured every year, the majority of which are powered by gasoline-fueled internal combustion engines. To function, these engines must operate in conjunction with the fuel supply, fuel injection, cooling, air filtration, lubrication, and exhaust systems. The rotation produced by the engine is transmitted to the wheels by a system of linked gears and shafts, while control of the vehicle is provided by interdependent steering, braking, and suspension systems. The electric system controls an array of other operations, including starting, ignition, lighting, and heating.

A standard layout for many of these operating systems was established in 1891 by Panhard-Leveassor in France and is still employed in many rear-wheel drive cars today. In recent years, however, there has been a rapid increase in the use of microprocessors and other electronic controls to improve the efficiency of the fuel supply, braking, and other operating systems. Manufacturers over the last century have also made huge advances in improving the aerodynamics and safety features of car bodies, as well as transforming the performance and fuel efficiency of automobile engines. The introduction of catalytic converters and lead-free fuels has reduced pollution levels.

Engine A front-mounted, four-cylinder engine, which is lubricated by oil pumped around its moving parts, drives the rear wheels

Battery The lead-acid battery provides power for the starter, and other electrical systems

Air filter Air drawn into the engine for combustion is first cleaned by filters to prevent clogging by dust particles

Air bag Protective bags fitted in front of the driver and passenger inflate automatically in the event of a crash

Spark plug These plugs generate sparks that ignite fuel in the cylinders

Engine fan This fan blows air around the radiator and engine to cool them

Radiator Water-cooled engines use a water-filled radiator to dissipate heat generated by combustion

MacPherson Strut This suspension unit consists of a shock absorber mounted inside a coil spring

Alternator The engine turns the alternator, which generates electricity to recharge the battery and power the electrical systems once the engine is running

Brake disks Hydraulically activated pads squeeze this metal disk to slow the wheel down

Steering rack The steering wheel turns a pinion gear wheel, which moves this toothed rack from side-to-side to steer the wheels

Gearbox This automatic gearbox converts the high-speed engine rotation into a less rapid but more powerful turning force to drive the wheels

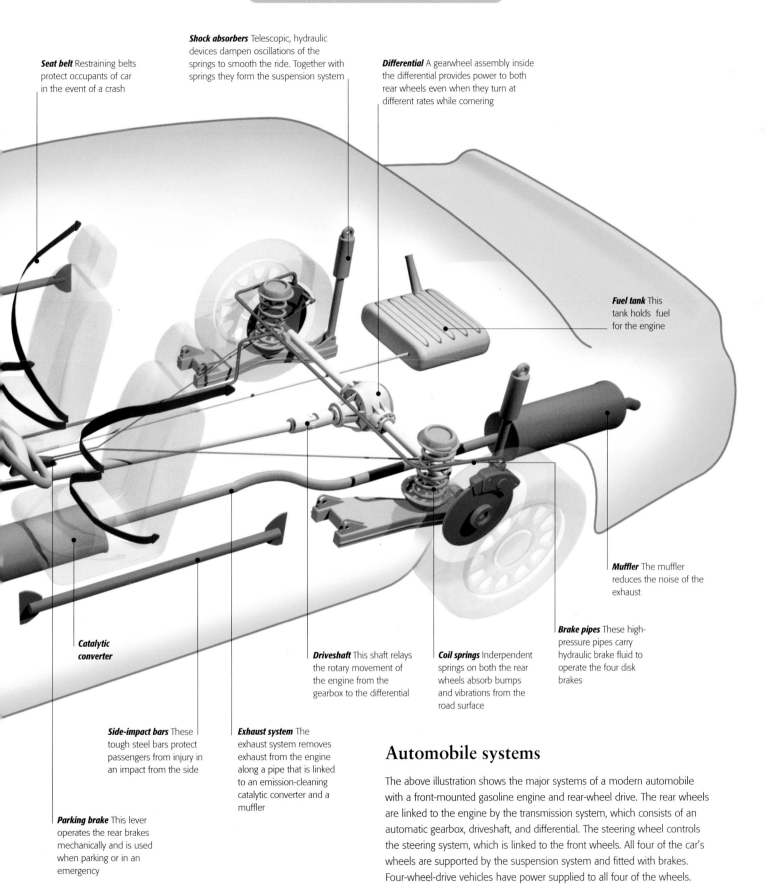

Shock absorbers Telescopic, hydraulic devices dampen oscillations of the springs to smooth the ride. Together with springs they form the suspension system

Seat belt Restraining belts protect occupants of car in the event of a crash

Differential A gearwheel assembly inside the differential provides power to both rear wheels even when they turn at different rates while cornering

Fuel tank This tank holds fuel for the engine

Muffler The muffler reduces the noise of the exhaust

Brake pipes These high-pressure pipes carry hydraulic brake fluid to operate the four disk brakes

Catalytic converter

Driveshaft This shaft relays the rotary movement of the engine from the gearbox to the differential

Coil springs Inderpendent springs on both the rear wheels absorb bumps and vibrations from the road surface

Side-impact bars These tough steel bars protect passengers from injury in an impact from the side

Exhaust system The exhaust system removes exhaust from the engine along a pipe that is linked to an emission-cleaning catalytic converter and a muffler

Parking brake This lever operates the rear brakes mechanically and is used when parking or in an emergency

Automobile systems

The above illustration shows the major systems of a modern automobile with a front-mounted gasoline engine and rear-wheel drive. The rear wheels are linked to the engine by the transmission system, which consists of an automatic gearbox, driveshaft, and differential. The steering wheel controls the steering system, which is linked to the front wheels. All four of the car's wheels are supported by the suspension system and fitted with brakes. Four-wheel-drive vehicles have power supplied to all four of the wheels.

Piston engines

A huge range of machines—including automobiles, lawn mowers, chain saws, and pumps—rely on piston engines for their power. Piston engines harness energy from a burning fuel-air mixture to move a piston up and down within a cylinder. This "reciprocating" movement is converted into rotary motion by a crankshaft. The modern internal combustion engine was invented by Karl Benz in 1886, but steam piston engines were around much earlier.

A typical internal combustion engine burns a highly combustible mixture of gasoline and air, ignited by a spark plug at the top of the cylinder. The composition of the mixture is controlled by a carburetor, or increasingly by electronic fuel injection systems, and enters each cylinder through an intake port at the top. As the mixture combusts, it drives the piston down producing power. As each cylinder only produces power during this downstroke, an engine needs several cylinders with out-of-phase stroke cycles to produce power continuously. Diesel engines work according to the same basic principle, but the fuel-air mixture is highly compressed before it combusts, leading to greater fuel efficiency.

Four-stroke cycle

Most auto engines use a four-stroke cycle. One cycle comprises induction, compression, power, and exhaust strokes, making two complete revolutions of the crankshaft. Inlet valves admit the fuel-air mixture and exhaust valves allow exhaust gases to be expelled. Valve opening and closing is precisely controlled and timed by a camshaft. Four or more cylinders firing in sequence produces smooth, continuous power.

1 Induction stroke A fuel-air mixture is sucked through the inlet valve into the cylinder when the turning crankshaft pulls the piston downward.

2 Compression stroke The inlet valve closes and the mixture is compressed as the piston rises. Just before it reaches the top, the spark plug fires.

3 Power stroke The mixture ignites; the hot expanding gases formed in the explosion drive the piston downward again, turning the crankshaft.

4 Exhaust stroke On the fourth stage of the cycle, the exhaust valve opens and the exhaust gases are expelled by the ascending piston.

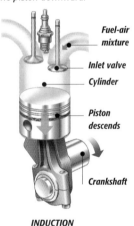

Fuel-air mixture
Inlet valve
Cylinder
Piston descends
Crankshaft

INDUCTION

Spark plug
Mixture compressed
Piston rises

COMPRESSION

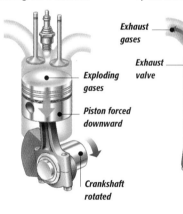

Exhaust gases
Exploding gases
Exhaust valve
Piston forced downward
Piston rises
Crankshaft rotated

POWER

EXHAUST

GDI engine

Gasoline direct injection (GDI) engines differ from standard injection engines in that fuel is injected directly into the cylinder rather than into an intake port. This results in greater fuel efficiency (more fuel is actually burned, and less wasted) and increased power and response, so GDI engines can burn "lean" fuel-air mixtures that produce fewer harmful emissions.

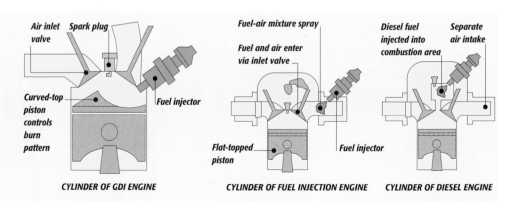

Air inlet valve **Spark plug**
Curved-top piston controls burn pattern
Fuel injector

CYLINDER OF GDI ENGINE

Fuel-air mixture spray
Fuel and air enter via inlet valve
Flat-topped piston
Fuel injector

CYLINDER OF FUEL INJECTION ENGINE

Diesel fuel injected into combustion area
Separate air intake

CYLINDER OF DIESEL ENGINE

V-12 engine

Cylinders in an engine are usually arranged "in-line," as groups of four or six in a straight line. Other engines are "flat," with cylinders horizontally opposed, or in "V" configurations. Generally, the more cylinders an engine has, the more powerful it is, and the smoother its power output. Eight-, ten-, and twelve-cylinder engines usually have a "V" layout.

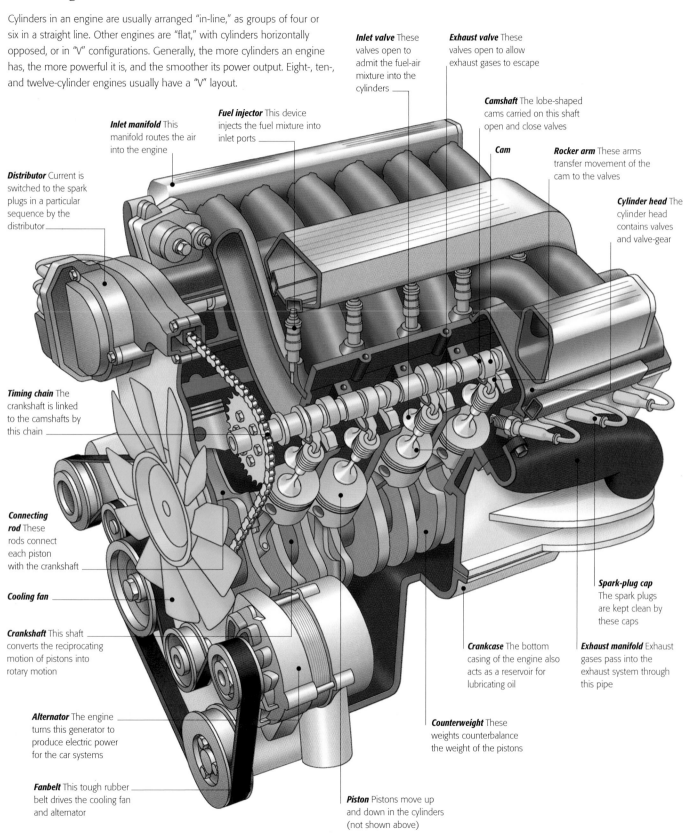

Inlet valve These valves open to admit the fuel-air mixture into the cylinders

Exhaust valve These valves open to allow exhaust gases to escape

Camshaft The lobe-shaped cams carried on this shaft open and close valves

Cam

Rocker arm These arms transfer movement of the cam to the valves

Cylinder head The cylinder head contains valves and valve-gear

Fuel injector This device injects the fuel mixture into inlet ports

Inlet manifold This manifold routes the air into the engine

Distributor Current is switched to the spark plugs in a particular sequence by the distributor

Timing chain The crankshaft is linked to the camshafts by this chain

Connecting rod These rods connect each piston with the crankshaft

Cooling fan

Crankshaft This shaft converts the reciprocating motion of pistons into rotary motion

Alternator The engine turns this generator to produce electric power for the car systems

Fanbelt This tough rubber belt drives the cooling fan and alternator

Piston Pistons move up and down in the cylinders (not shown above)

Counterweight These weights counterbalance the weight of the pistons

Crankcase The bottom casing of the engine also acts as a reservoir for lubricating oil

Spark-plug cap The spark plugs are kept clean by these caps

Exhaust manifold Exhaust gases pass into the exhaust system through this pipe

SEE ALSO: CARS *p.108* | SPEED & SAFETY *p.112* | ALTERNATIVE CARS *p.114* | JET ENGINES *p.128*

Speed and safety

Automobile manufacturers have always strived to better the performance of their motor cars. By increasing the combustion rate of engines, faster and more powerful cars have been developed. These cars utilize advanced braking and suspension systems to improve their road-handling, and safety systems that can prevent injury to passengers in the event of an accident.

Most automobile engines burn a mixture of gasoline and air. If the amount of air entering the engine can be increased, the amount of fuel that can be mixed with it also increases, resulting in more power. To achieve this, pumps are used to compress air before it

enters the engine. Early "supercharger" pumps were belt or chain driven by the engine crankshaft. A more modern type of pump, the turbocharger, uses exhaust gases to drive a compressor turbine.

Safety considerations are paramount when designing modern automobiles. In order to reduce the chances of the driver losing control of the car, suspension systems maximize the grip between a car's wheels and the road, while antilock brakes decrease the risk of skidding. However, accidents still occur. A family car typically weighs two tons and has a large amount of kinetic energy even when traveling at relatively slow speeds. This energy must be safely dissipated in the event of a crash. Crumple zones, passenger safety cells, side-impact bars, air bags, inertia-reel safety belts, and safety glass all help to protect passengers.

Turbocharger

Turbochargers boost the performance of internal combustion engines by compressing air before it enters the carburetor or fuel-injection system. By forcing more air into the parts of the engine that mix fuel with air, more fuel can be fed into the engine cylinders. Turbochargers use exhaust gases, which are rapidly expelled from the engine, to drive a turbine assembly that compresses the air. The turbocharger is not operational all the time, but kicks in under hard acceleration, increasing the fuel combustion capacity of the engine.

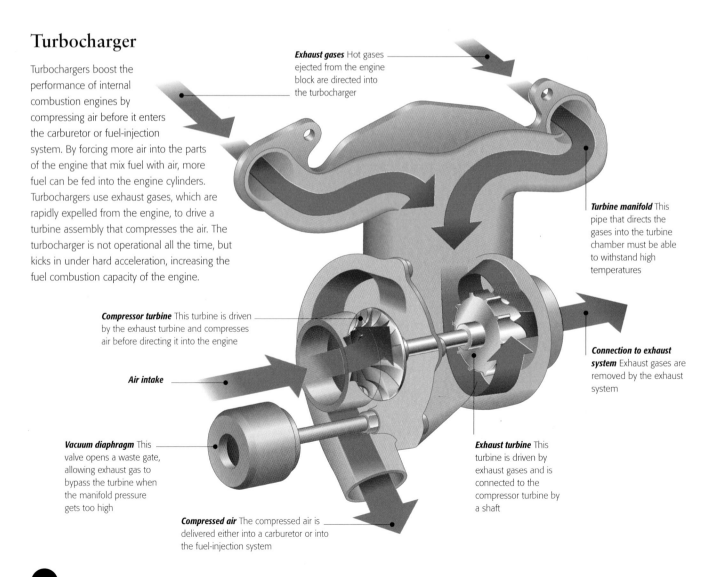

Exhaust gases Hot gases ejected from the engine block are directed into the turbocharger

Turbine manifold This pipe that directs the gases into the turbine chamber must be able to withstand high temperatures

Compressor turbine This turbine is driven by the exhaust turbine and compresses air before directing it into the engine

Air intake

Connection to exhaust system Exhaust gases are removed by the exhaust system

Vacuum diaphragm This valve opens a waste gate, allowing exhaust gas to bypass the turbine when the manifold pressure gets too high

Exhaust turbine This turbine is driven by exhaust gases and is connected to the compressor turbine by a shaft

Compressed air The compressed air is delivered either into a carburetor or into the fuel-injection system

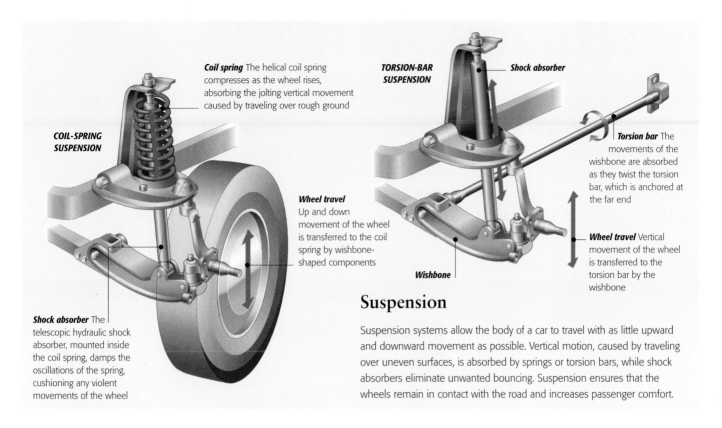

Coil spring The helical coil spring compresses as the wheel rises, absorbing the jolting vertical movement caused by traveling over rough ground

TORSION-BAR SUSPENSION

Shock absorber

COIL-SPRING SUSPENSION

Torsion bar The movements of the wishbone are absorbed as they twist the torsion bar, which is anchored at the far end

Wheel travel Up and down movement of the wheel is transferred to the coil spring by wishbone-shaped components

Wheel travel Vertical movement of the wheel is transferred to the torsion bar by the wishbone

Wishbone

Shock absorber The telescopic hydraulic shock absorber, mounted inside the coil spring, damps the oscillations of the spring, cushioning any violent movements of the wheel

Suspension

Suspension systems allow the body of a car to travel with as little upward and downward movement as possible. Vertical motion, caused by traveling over uneven surfaces, is absorbed by springs or torsion bars, while shock absorbers eliminate unwanted bouncing. Suspension ensures that the wheels remain in contact with the road and increases passenger comfort.

Safety systems

Two strategies are employed to protect people involved in a car crash from harm. The car itself absorbs much of the energy of an impact and thus reduces the forces acting on the passengers. A protective cocoon of devices also restrains and shelters the passengers themselves. The chance of accidents occuring is reduced by improved braking systems, which provide drivers with more control of their cars when they brake suddenly.

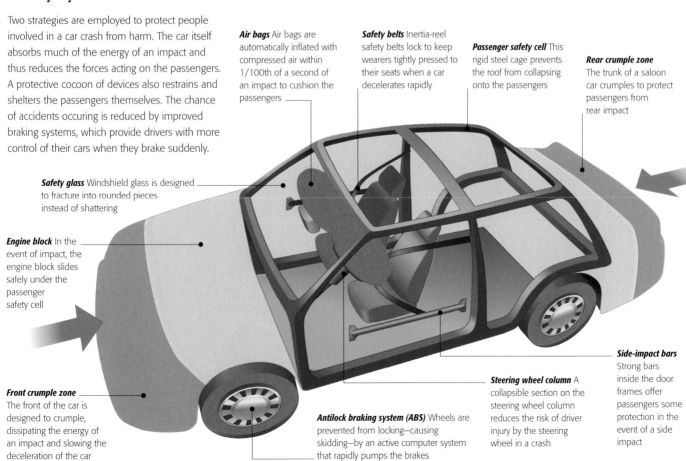

Air bags Air bags are automatically inflated with compressed air within 1/100th of a second of an impact to cushion the passengers

Safety belts Inertia-reel safety belts lock to keep wearers tightly pressed to their seats when a car decelerates rapidly

Passenger safety cell This rigid steel cage prevents the roof from collapsing onto the passengers

Rear crumple zone The trunk of a saloon car crumples to protect passengers from rear impact

Safety glass Windshield glass is designed to fracture into rounded pieces instead of shattering

Engine block In the event of impact, the engine block slides safely under the passenger safety cell

Side-impact bars Strong bars inside the door frames offer passengers some protection in the event of a side impact

Front crumple zone The front of the car is designed to crumple, dissipating the energy of an impact and slowing the deceleration of the car

Antilock braking system (ABS) Wheels are prevented from locking—causing skidding—by an active computer system that rapidly pumps the brakes

Steering wheel column A collapsible section on the steering wheel column reduces the risk of driver injury by the steering wheel in a crash

SEE ALSO: CARS *p.108* | PISTON ENGINES *p.110* | ALTERNATIVE CARS *p.114* | RACING CARS *p.116*

Alternative cars

The need to preserve natural resources and to reduce pollution of the environment has led car manufacturers to seek alternatives to standard gasoline and diesel combustion engines. Gas turbine, battery, and solar-powered engines have been advanced as possible solutions. Currently, the motor industry is investigating two additional alternatives: gasoline-electric hybrids and fuel cells.

From steam to gasoline engines

During the early years of motoring, cars powered by steam engines outnumbered those using internal combustion engines. In 1902, there were only 909 automobiles registered in New York State. Of those, 485 were steam powered. When horseless carriages first took to the highway, internal combustion engines were still in their infancy—the first practical gasoline engine was built by Karl Benz in 1885. In contrast, steam engine technology was already sophisticated after a century of powering railroad locomotives. Simple to build and operate, steam engines were easily adaptable for automotive use. Steam cars of the early 20th century were fast, efficient, and easy to drive. A Stanley steam car driven by American F. H. Marriott set a land speed record of 127 mph (205 kph) at Ormond Beach, Florida, in 1906 and held this record for three years.

Improvements in the design of internal combustion engines in the early 1900s persuaded the industrialist Henry Ford to adopt

Fuel-cell solutions
Daimler-Benz has developed an electric vehicle which is powered by a fuel cell that generates its own hydrogen fuel from methanol.

gasoline-fueled piston engines for his highly successful Ford Model T, which he began mass-producing before the outbreak of World War I. More than any other single car design, the Model T established the popularity of the gasoline engine and marked the decline of steam cars, which were last made in the 1930s.

Gas turbine engines

The next significant challenge to gasoline- and diesel-fueled piston engines came from gas turbine engines, which were tested for use in automobiles by the United States and Britain in the 1950s and 1960s. A derivation of the jet engine, these engines used a power turbine to turn the vehicle's wheels rather than provide thrust directly from exhaust gases. Gas turbines burned kerosene at high temperatures and had environmental advantages because they burned the fuel cleanly and efficiently. However, the early gas turbine cars were not economical, only managing a maximum of 12 miles to the gallon (4 km/liter) of kerosene. Although experiments with fuel-saving heat exchangers improved consumption, the industry abandoned the gas turbine engine, since it was at its most efficient when running at or near full power—modern piston car engines only need to run at about 20–30 percent of full power for much of the time. Gas turbine engines have survived as electricity-generating units in airplanes and trains, and used to power some trucks, ships, and locomotives. Although gas

Solar power across a continent
This car, which has a curved array of photovoltaic (solar) cells on its roof, was one of the solar-powered cars to compete in the Pentax World Solar Challenge Race held in Australia between Darwin and Adelaide in 1987.

Multiple battery power
The performance of electric-powered cars has been hampered by the weight and limited power capacity of the batteries they use, which need several hours to recharge.

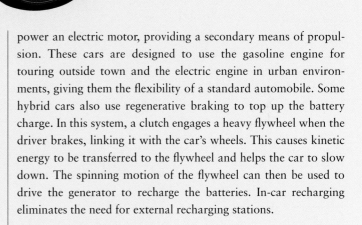

Solar city car
This two-seat electric car has solar cell on its roof that convert sunlight into electric energy to supplement its batteries. The solar-rechargeable battery car could become a viable solution to combating automobile-generated pollution.

turbine technology has been shelved, many experts believe the gas-turbine/electric hybrid car could provide a fuel-efficient means of transportation.

Electric cars

Electric motors have powered vehicles of various types for over a century. The two major problems with electric motors are their inefficiency and the need for heavy batteries. Short operating range, low speeds, and long charging periods make them impracticable for many applications. However, for some commercial vehicles such as local delivery vans and forklifts, where speed and range are unimportant, electric motors are ideal. Future advances in battery design should provide economies in both cost and weight that could make family electric cars a viable proposition. While electric cars produce no harmful emissions themselves, pollution is created indirectly by the power stations that generate the electricity to charge their batteries.

Solar-powered cars

In a bid to create "pollution-free" cars that are cheap to run, prototypes of solar-powered vehicles have been in development for decades. Such vehicles either use photovoltaic (solar) cells to convert sunlight into electricity that directly powers a motor, or they run on chemical batteries that are recharged by the solar cells. The major problem with solar cells is that each cell can only produce a small amount of electricity, so hundreds of cells are needed to generate enough power. Because solar cells are expensive to produce, the initial cost of solar-powered cars is high. However, improvements in solar cell and battery technology should reduce the number of cells needed and allow cheaper, commercially viable solar-powered vehicles to be developed.

Hybrid cars

A new type of car is the dual fuel gasoline-electric hybrid. A high-efficiency gasoline engine drives the car and also drives a generator that recharges batteries carried on board. These batteries can power an electric motor, providing a secondary means of propulsion. These cars are designed to use the gasoline engine for touring outside town and the electric engine in urban environments, giving them the flexibility of a standard automobile. Some hybrid cars also use regenerative braking to top up the battery charge. In this system, a clutch engages a heavy flywheel when the driver brakes, linking it with the car's wheels. This causes kinetic energy to be transferred to the flywheel and helps the car to slow down. The spinning motion of the flywheel can then be used to drive the generator to recharge the batteries. In-car recharging eliminates the need for external recharging stations.

Fuel-cell-powered cars

Perhaps the most technologically appealing of the alternative cars is the fuel-cell-powered vehicle. In one design of fuel cell, hydrogen reacts with atmospheric oxygen to produce electric energy, with water and heat as the only waste products. Hydrogen-based fuel cells produce no emission of carbon dioxide—one of the greenhouse gases. However, hydrogen is highly flammable and storing it is hazardous, so some designs of fuel cells use methanol as a source of hydrogen. Although generating hydrogen by this method produces some pollutants, methanol engines still produce far fewer emissions than combustion engines. Fuel cells are efficient and durable because no combustion is required to generate electricity and they have no moving parts. Experimental fuel-cell cars are already in operation, and mass-produced fuel-cell cars should be available within a few years.

SEE ALSO: CARS *p.108* | PISTON ENGINES *p.110* | SOLAR ENERGY *p.174* | BATTERIES *p.178*

Racing cars

Automobiles have been raced since their invention at the end of the 19th century. Formula 1 is now the foremost high-performance motor racing event, watched by billions of television viewers around the world every year. The modern Formula 1 car is a triumph of applied science, packed with technology to improve its speed, reliability, and safety. However, all the car's systems must conform to strict rules. As a result most Formula 1 cars are evenly matched, and it is driver skill that separates the winner from the rest.

The most distinctive feature of a Formula 1 car is its low, wide shape. Although engines are being continually improved, air resistance will hold back the most powerful vehicle, so aerodynamics are as important as power in today's car. Front and rear airfoils create downforce, pressing the car down onto the road and improving the grip of the tires. However, the car must also be as light as possible, in order to obtain the maximum acceleration from the engine. The shell of a Formula 1 car is therefore made from an extremely light carbon-fiber composite.

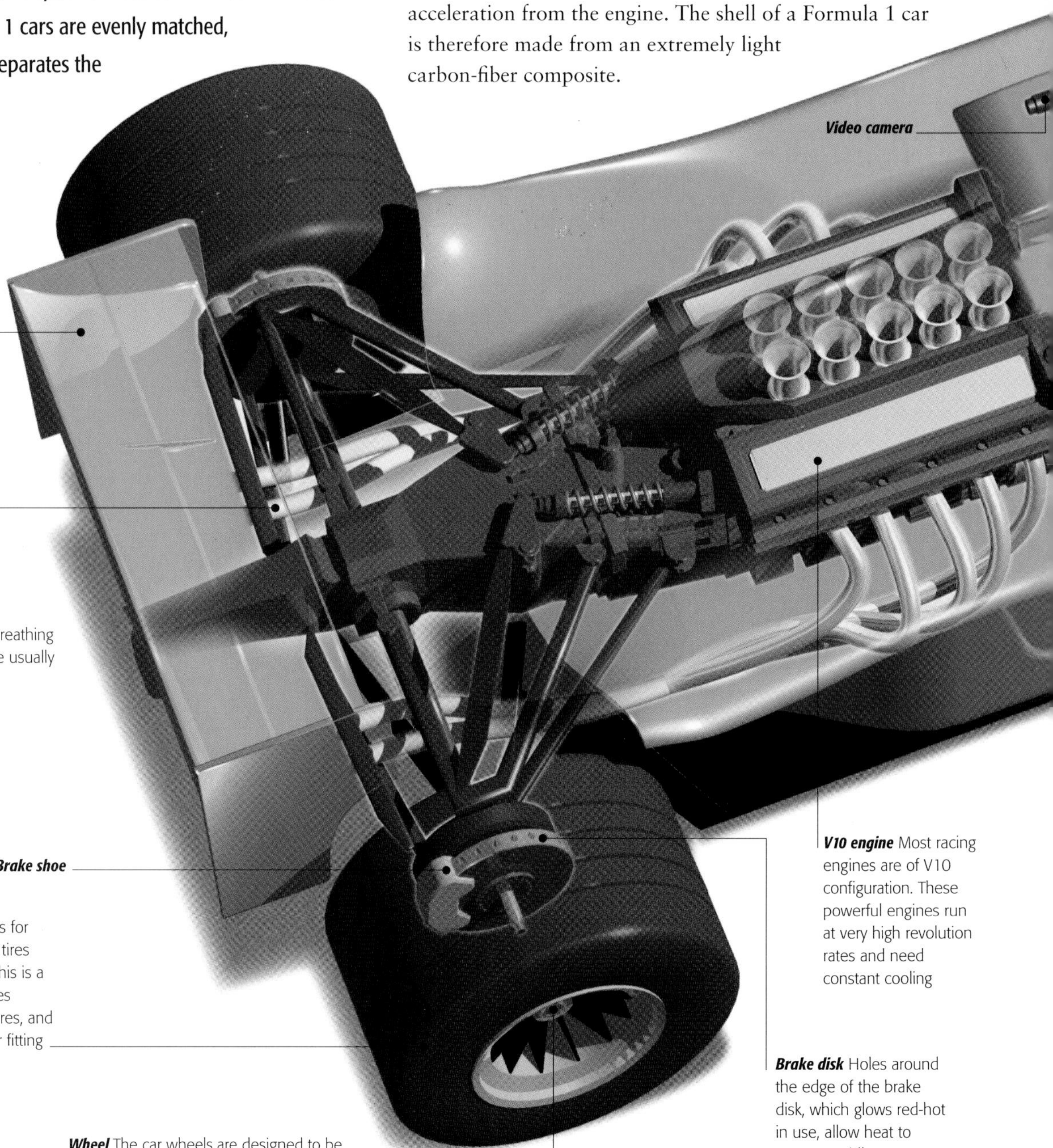

Video camera

Rear wing The airfoil produces a downward force that increases rear-wheel grip. It is set at a steeper angle for increased downforce on circuits with numerous tight curves

Tailpipe The exhaust system is designed to maximize engine breathing and minimize drag. Tailpipes are usually hidden beneath the rear wing

Brake shoe

Tire Race teams fit different tires for different weather—wet weather tires have more treads for grip, but this is a disadvantage on a dry track. Tires perform best at high temperatures, and take some time to heat up after fitting

V10 engine Most racing engines are of V10 configuration. These powerful engines run at very high revolution rates and need constant cooling

Brake disk Holes around the edge of the brake disk, which glows red-hot in use, allow heat to escape rapidly

Wheel The car wheels are designed to be released and refitted quickly during pitstops lasting just seconds

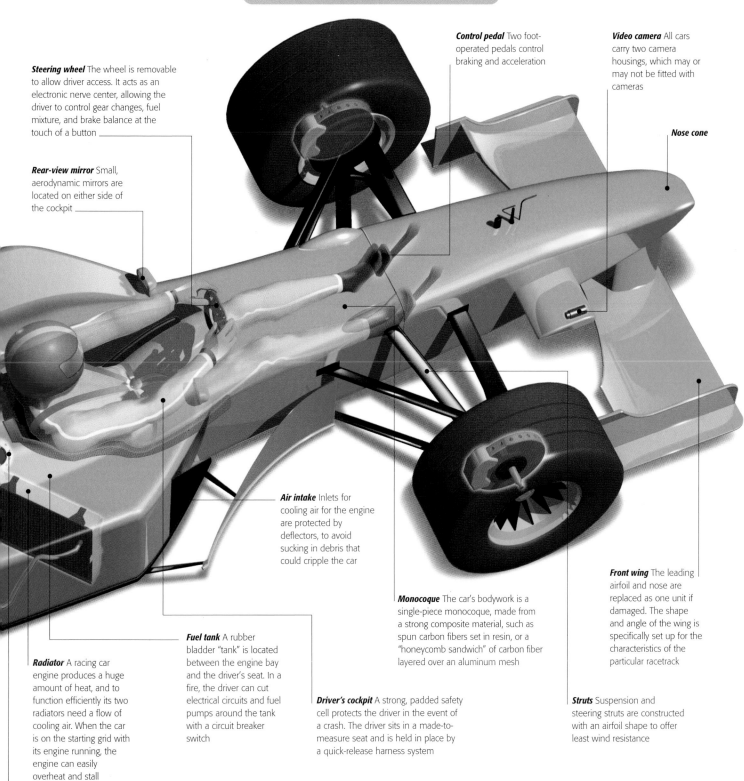

Control pedal Two foot-operated pedals control braking and acceleration

Video camera All cars carry two camera housings, which may or may not be fitted with cameras

Nose cone

Steering wheel The wheel is removable to allow driver access. It acts as an electronic nerve center, allowing the driver to control gear changes, fuel mixture, and brake balance at the touch of a button

Rear-view mirror Small, aerodynamic mirrors are located on either side of the cockpit

Air intake Inlets for cooling air for the engine are protected by deflectors, to avoid sucking in debris that could cripple the car

Front wing The leading airfoil and nose are replaced as one unit if damaged. The shape and angle of the wing is specifically set up for the characteristics of the particular racetrack

Radiator A racing car engine produces a huge amount of heat, and to function efficiently its two radiators need a flow of cooling air. When the car is on the starting grid with its engine running, the engine can easily overheat and stall

Fuel tank A rubber bladder "tank" is located between the engine bay and the driver's seat. In a fire, the driver can cut electrical circuits and fuel pumps around the tank with a circuit breaker switch

Monocoque The car's bodywork is a single-piece monocoque, made from a strong composite material, such as spun carbon fibers set in resin, or a "honeycomb sandwich" of carbon fiber layered over an aluminum mesh

Driver's cockpit A strong, padded safety cell protects the driver in the event of a crash. The driver sits in a made-to-measure seat and is held in place by a quick-release harness system

Struts Suspension and steering struts are constructed with an airfoil shape to offer least wind resistance

Fuel filler A fuel cap allows access to the car's gas tanks. Fuel is pumped in at very high rates—vents in the filler pipes allow air to escape as the tank fills up

Formula 1 racing car

Formula 1 racing cars have evolved from the relatively simple racers of early last century to the computer-aided, high-performance cars of today. A modern racing team involves a huge number of people behind the scenes, ranging from engineers and aerodynamics experts to the pit mechanics who keep the car running during the race. From the driver's point of view, the Formula 1 car is now safer than ever before. The car illustrated is typical of the technologically advanced cars racing in Grand Prix around the world.

SEE ALSO: CARS *p.108* | PISTON ENGINES *p.110* | SPEED & SAFETY *p.112* | COMPOSITE MATERIALS *p.144*

High-speed trains

High-speed trains push technology to the limit to get the maximum speed and performance out of traditional railroad and locomotive designs. In the 1960s, most people believed that steel wheels and track would disappear and be replaced by maglev and hovertrains. However, the huge financial investment needed to build completely new types of tracks and trains was not forthcoming, and engineers had to look again at the potential of traditional railroads.

Modern high-speed trains, such as the Japanese Shinkansen and the French Train à Grande Vitesse (TGV), share many of the same technological approaches for increasing speed without sacrificing safety and run on dedicated tracks. These trains have aerodynamic design and are powered by electric motors that use pantographs to pick up electricity from overhead cables. The trains are lightweight because the power car and trailer bodies are made from aluminum and the power cars do not carry heavy diesel engines onboard. The TGV reduced the weight of train sets even further by pioneering a truck (wheel housing) that is shared between two trailers. The high-speed railroads these trains run on are constructed so that all the curves have a radius of at least 3 miles (5 km), and the rails are banked to keep the train safely on the track. Where it is not possible to build dedicated tracks, so-called "tilting trains" are used. Trains such as the U.S. Acela and the Italian Superpendolino incorporate suspension systems that tilt the train as it goes around corners, enabling greater speeds to be attained on conventional tracks.

Eurostar

The Eurostar made its first journey through the Channel Tunnel in 1994. It incorporates many features from the French TGV, but has been adapted for the multistandard power supplies of France, Britain, and Belgium. Eurostar has 20 cars and weighs 900 tons (816 tonnes). It can carry up to 800 passengers, reaching a top speed of 185 mph (300 km/h) on journeys between Paris and London or Brussels. The train has 12 electric motors mounted on power trucks that provide over 12,000 kilowatts of power.

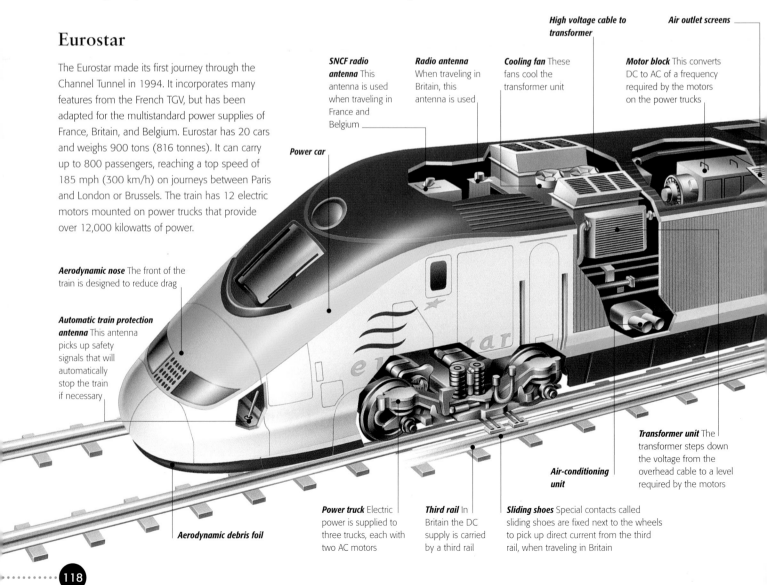

High voltage cable to transformer

Air outlet screens

SNCF radio antenna This antenna is used when traveling in France and Belgium

Radio antenna When traveling in Britain, this antenna is used

Cooling fan These fans cool the transformer unit

Motor block This converts DC to AC of a frequency required by the motors on the power trucks

Power car

Aerodynamic nose The front of the train is designed to reduce drag

Automatic train protection antenna This antenna picks up safety signals that will automatically stop the train if necessary

Transformer unit The transformer steps down the voltage from the overhead cable to a level required by the motors

Air-conditioning unit

Aerodynamic debris foil

Power truck Electric power is supplied to three trucks, each with two AC motors

Third rail In Britain the DC supply is carried by a third rail

Sliding shoes Special contacts called sliding shoes are fixed next to the wheels to pick up direct current from the third rail, when traveling in Britain

Trucks

The power trucks of high-speed trains carry the motors, gears, and braking and suspension systems in a strong steel frame. The Eurostar has three such trucks at either end—two on each power car and one on the end trailer nearest each power car. The other trailers are mounted on trucks that carry only suspension systems. These trucks are shared between neighboring trailers on the Eurostar, reducing the weight of the train and the noise inside, as well as improving its aerodynamics.

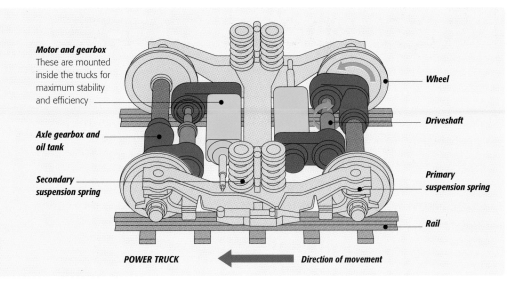

Motor and gearbox These are mounted inside the trucks for maximum stability and efficiency

Axle gearbox and oil tank

Secondary suspension spring

Wheel

Driveshaft

Primary suspension spring

Rail

POWER TRUCK   **Direction of movement**

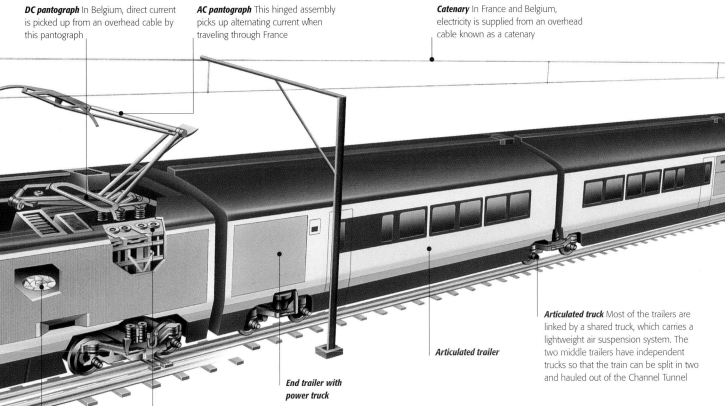

DC pantograph In Belgium, direct current is picked up from an overhead cable by this pantograph

AC pantograph This hinged assembly picks up alternating current when traveling through France

Catenary In France and Belgium, electricity is supplied from an overhead cable known as a catenary

Articulated truck Most of the trailers are linked by a shared truck, which carries a lightweight air suspension system. The two middle trailers have independent trucks so that the train can be split in two and hauled out of the Channel Tunnel

Articulated trailer

End trailer with power truck

Inverter The inverter converts the French AC supply to DC and passes it on to the motor blocks

Cooling fans These fans cool the motor blocks

Diesel electric trains

The world's fastest diesel train, the British Intercity 125, has reached a top speed of 145 mph (235 km/h), using a diesel engine as a generator to drive the train's electric motors. Alternating current generated by the diesel engine is converted by a rectifier into the direct current power supply needed by the motors. The motors themselves are mounted inside the trucks of the locomotive.

SEE ALSO: TUNNELS p.20 | MAGLEV & MONORAILS p.120 | ELECTRICITY & MAGNETISM p.260

Maglev and monorails

The development of high-speed trains has kept railroads competitive with other modes of transportation and popular with travelers. Less conventional technologies, however, offer faster, safer, and more environmentally friendly solutions for intercity journeys and urban transit. The challenge for these technologies is to attract investment to ensure they become viable modes of transportation.

Maglev (short for "magnetic levitation") is a unique mode of transportation that uses magnetic fields to levitate a vehicle above a specially constructed guideway. Maglev vehicles, operating at speeds of about 280 mph (450 km/h), could compete with short-haul flights between cities by offering comparable journey times. Maximum speeds of 343 mph (550 km/h) have already been achieved at test tracks in Germany and Japan. Such high speeds are possible due to the lack of contact between the guideway and the vehicle when it is in motion, so contact friction, a major impediment for conventional trains, does not have to be overcome. The main source of resistance to a maglev vehicle is air resistance, which can be reduced by streamlining. Unlike conventional trains, maglev vehicles do not carry propulsion units on board; instead, the propulsion system is fitted in the guideway. The innovative guiding and propulsion systems eliminate the need for wheels, brakes, motors, and devices for collecting, converting, and transmitting electrical power. Maglev vehicles are therefore lighter, quieter, less prone to wear than conventional trains, and use about five times less fuel than aircraft over the same distance. Despite offering the possibility of rapid, and energy-efficient transportation, maglev has not yet made the transition from experimental to commercial technology, because of the high cost of constructing new networks of dedicated guideways.

However, conventionally-propelled monorail trains, which also run on specially constructed guideways, are gaining popularity as a mode of urban transit. Quiet, efficient, and relatively nonpolluting, they are seen as a solution to the increasing traffic congestion in many modern cities.

Maglev vehicles

Maglev vehicles are constructed in two parts. The carriage body in which passengers travel is mounted on an undercarriage that houses the levitation and guidance magnets. The undercarriage wraps around the guideway and systems controlling the magnets ensure that the vehicle remains very close to the guideway, but does not touch it.

Levitation and propulsion
Cable coils wound underneath the guideway generate a traveling magnetic field that moves down the length of the guideway. Magnetic forces of attraction between this field and electromagnets on the vehicle levitate the vehicle and drag it along behind the advancing field. Thus, the guideway acts as both levitation and the propulsion unit. The speed at which the guideway field travels can be varied for acceleration or braking.

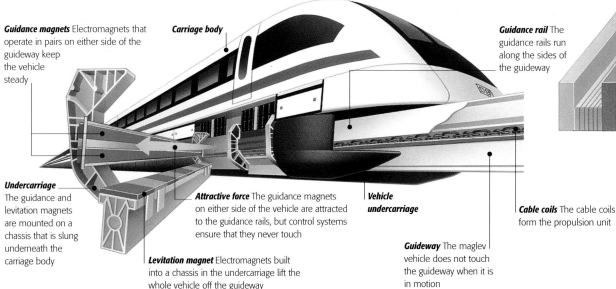

Guidance magnets Electromagnets that operate in pairs on either side of the guideway keep the vehicle steady

Carriage body

Guidance rail The guidance rails run along the sides of the guideway

Undercarriage The guidance and levitation magnets are mounted on a chassis that is slung underneath the carriage body

Attractive force The guidance magnets on either side of the vehicle are attracted to the guidance rails, but control systems ensure that they never touch

Vehicle undercarriage

Cable coils The cable coils form the propulsion unit

Levitation magnet Electromagnets built into a chassis in the undercarriage lift the whole vehicle off the guideway

Guideway The maglev vehicle does not touch the guideway when it is in motion

Monorail trains

Monorails run on single, precast concrete beams that are usually elevated, but can also be at or below ground level. The trains may travel on top of the beam or hang underneath it. Electrically driven horizontally and vertically aligned rubber wheels propel the train and keep it stable on the beam. Rubber tires running directly on the smooth beams suffer little wear, and maintenance is low compared to metal-wheeled trains that run on metal rails.

Urban chic
The Darling Harbour monorail in Sydney, Australia, provides a rapid link between the city center and the modern harborfront development.

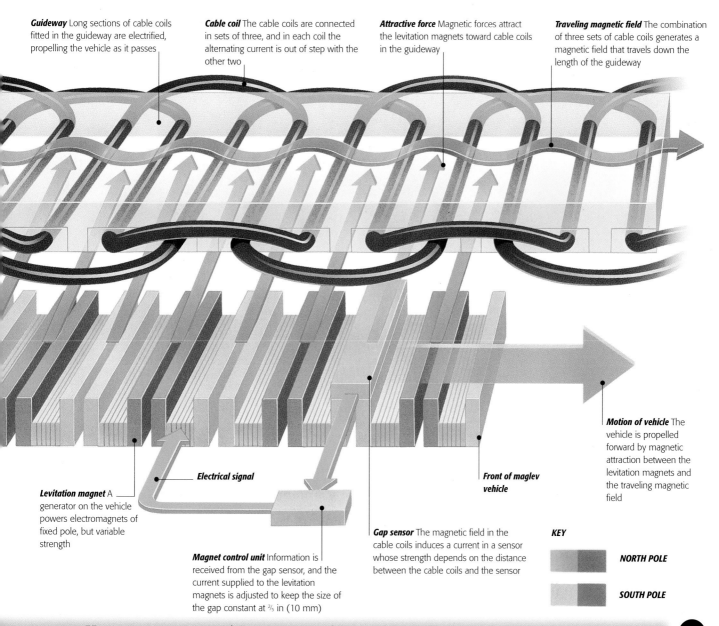

Guideway Long sections of cable coils fitted in the guideway are electrified, propelling the vehicle as it passes

Cable coil The cable coils are connected in sets of three, and in each coil the alternating current is out of step with the other two

Attractive force Magnetic forces attract the levitation magnets toward cable coils in the guideway

Traveling magnetic field The combination of three sets of cable coils generates a magnetic field that travels down the length of the guideway

Levitation magnet A generator on the vehicle powers electromagnets of fixed pole, but variable strength

Electrical signal

Magnet control unit Information is received from the gap sensor, and the current supplied to the levitation magnets is adjusted to keep the size of the gap constant at ²⁄₅ in (10 mm)

Gap sensor The magnetic field in the cable coils induces a current in a sensor whose strength depends on the distance between the cable coils and the sensor

Front of maglev vehicle

Motion of vehicle The vehicle is propelled forward by magnetic attraction between the levitation magnets and the traveling magnetic field

KEY

	NORTH POLE
	SOUTH POLE

SEE ALSO: HIGH-SPEED TRAINS *p.118* | AIRLINERS *p.126* | ELECTRICITY & MAGNETISM *p.260*

Balloons and airships

The first recorded manned flight was achieved in a hot-air balloon designed by Montgolfier brothers in 1783. Balloons are now used for recreational purposes and by meteorologists for gathering data. In the early 20th century, powered balloons, known as airships, rivaled the airplane as a means of air transportation, but they are now mostly used as platforms for television cameras and for advertising purposes.

The "envelope" of a hot-air balloon or an airship displaces a large volume of air, generating a force called upthrust. If the total weight of the balloon or airship is less than the upthrust, it will rise. Hot-air balloons rise when the air in the envelope is heated because this causes it to become less dense than the cooler air in the surrounding atmosphere. Most gas balloons and airships use helium, a nonflammable gas with very low density, to produce lift. Airships have engines, rudders, and elevators that give control over the speed and direction of flight, but balloons have no means of propulsion and drift with the wind.

Conventional hot-air balloons
Hot-air balloons, such as the ones seen here floating over New Mexico, rise upward when heat generated by a propane burner causes the air in the envelope to expand. This forces air out of the bottom of the envelope, reducing the overall weight of the balloon to less than the upthrust on it.

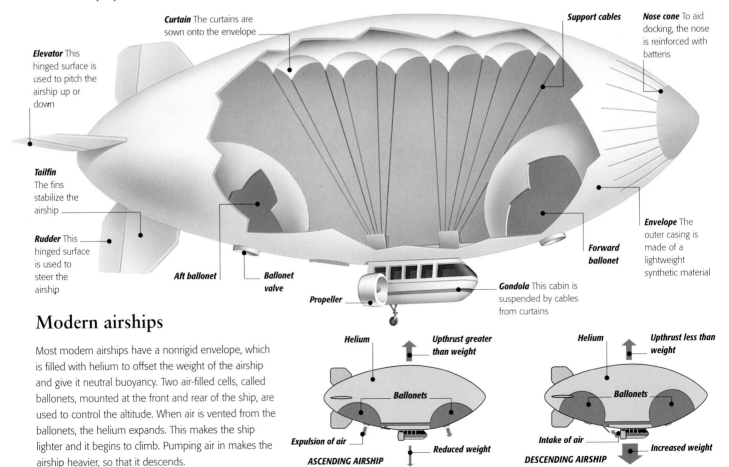

Curtain The curtains are sown onto the envelope

Support cables

Nose cone To aid docking, the nose is reinforced with battens

Elevator This hinged surface is used to pitch the airship up or down

Tailfin The fins stabilize the airship

Rudder This hinged surface is used to steer the airship

Aft ballonet

Ballonet valve

Propeller

Gondola This cabin is suspended by cables from curtains

Forward ballonet

Envelope The outer casing is made of a lightweight synthetic material

Modern airships

Most modern airships have a nonrigid envelope, which is filled with helium to offset the weight of the airship and give it neutral buoyancy. Two air-filled cells, called ballonets, mounted at the front and rear of the ship, are used to control the altitude. When air is vented from the ballonets, the helium expands. This makes the ship lighter and it begins to climb. Pumping air in makes the airship heavier, so that it descends.

Helium

Upthrust greater than weight

Ballonets

Expulsion of air

Reduced weight

ASCENDING AIRSHIP

Helium

Upthrust less than weight

Ballonets

Intake of air

Increased weight

DESCENDING AIRSHIP

Breitling Orbiter 3

The 200-ft (60-m) high *Breitling Orbiter 3* is a hybrid balloon that flew nonstop around the world in March 1999. The main lift was provided by helium, which could be expanded to create extra lift by heating air at the base of the envelope. The crew of two traveled in an airtight gondola, which was pressurized by a computer-controlled supply of oxygen and nitrogen. Carbon dioxide produced by the crew was removed using lithium filters.

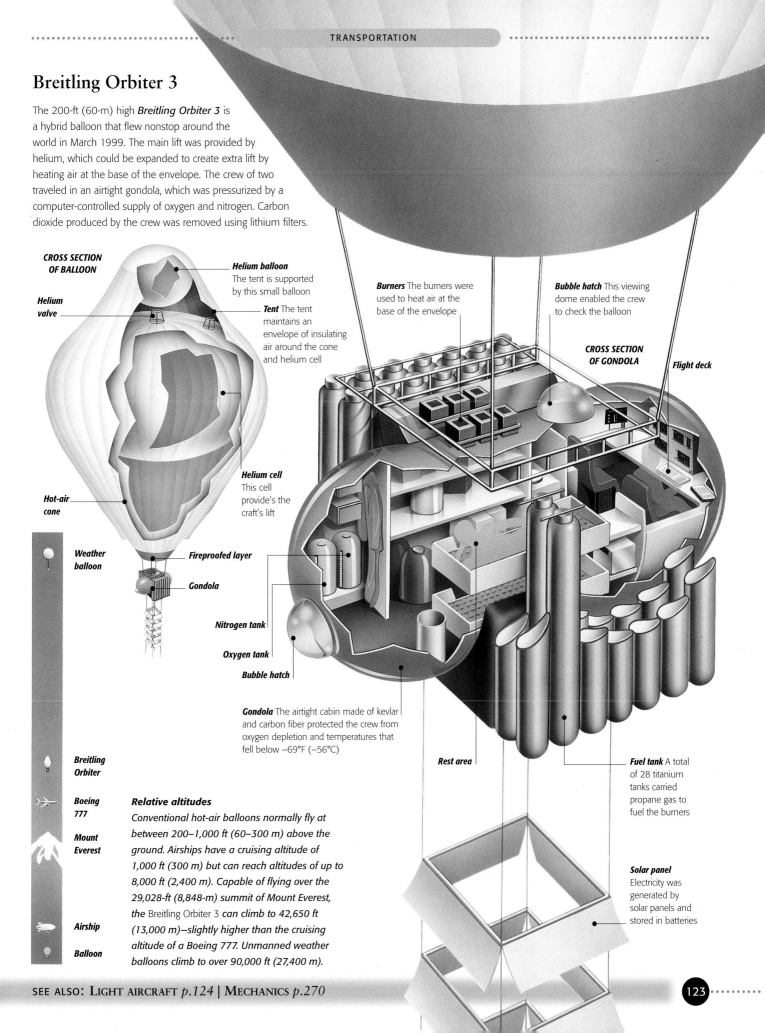

CROSS SECTION OF BALLOON

Helium valve

Helium balloon The tent is supported by this small balloon

Tent The tent maintains an envelope of insulating air around the cone and helium cell

Hot-air cone

Helium cell This cell provide's the craft's lift

Weather balloon

Fireproofed layer

Gondola

Nitrogen tank

Oxygen tank

Bubble hatch

Gondola The airtight cabin made of kevlar and carbon fiber protected the crew from oxygen depletion and temperatures that fell below −69°F (−56°C)

Burners The burners were used to heat air at the base of the envelope

Bubble hatch This viewing dome enabled the crew to check the balloon

CROSS SECTION OF GONDOLA

Flight deck

Rest area

Fuel tank A total of 28 titanium tanks carried propane gas to fuel the burners

Solar panel Electricity was generated by solar panels and stored in batteries

Breitling Orbiter

Boeing 777

Mount Everest

Airship

Balloon

Relative altitudes

Conventional hot-air balloons normally fly at between 200–1,000 ft (60–300 m) above the ground. Airships have a cruising altitude of 1,000 ft (300 m) but can reach altitudes of up to 8,000 ft (2,400 m). Capable of flying over the 29,028-ft (8,848-m) summit of Mount Everest, the Breitling Orbiter 3 *can climb to 42,650 ft (13,000 m)—slightly higher than the cruising altitude of a Boeing 777. Unmanned weather balloons climb to over 90,000 ft (27,400 m).*

SEE ALSO: LIGHT AIRCRAFT p.124 | MECHANICS p.270

Light aircraft

An airplane relies on airflow over its wings to generate a lifting force, which overcomes the plane's weight. This airflow is provided by the forward motion of the plane caused by thrust from an engine. A fin and tail plane stabilize the aircraft in flight. Most light aircraft use mechanical linkages to transmit the pilot's steering commands to control surfaces on the wings, fin, and tail plane.

Light aircraft are small, relatively simple planes that are used for recreation and business travel. They are usually powered by piston engines that drive propellers, although some have jet engines. Unlike large, high-speed aircraft, which have undercarriage (wheels) that retracts during flight for streamlining, light aircraft usually have fixed undercarriage to save weight and cost. With the development of new composite materials, aircraft have become stronger, lighter, and capable of flying greater ranges.

Changing course

An airplane has a set of three basic control surfaces that change its attitude (orientation)—the elevators, the ailerons, and the rudder. Pairs of elevators on the horizontal tail plane force the nose up or down, a maneuver known as pitching. Ailerons are situated on the rear edge of each wing and roll the plane from side to side. The rudder on the vertical fin points the nose of the plane to port or starboard, which is known as yawing. The ailerons and rudder are usually used together to turn the aircraft smoothly, causing the plane to yaw and roll (or bank) simultaneously.

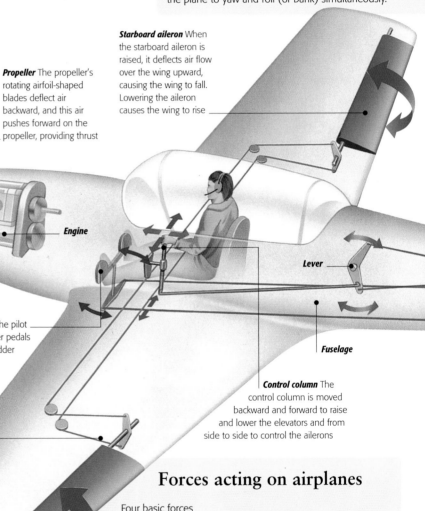

Propeller The propeller's rotating airfoil-shaped blades deflect air backward, and this air pushes forward on the propeller, providing thrust

Starboard aileron When the starboard aileron is raised, it deflects air flow over the wing upward, causing the wing to fall. Lowering the aileron causes the wing to rise

Engine

Lever

Fuselage

Rudder pedal The pilot uses the rudder pedals to pivot the rudder to one side or the other

Control column The control column is moved backward and forward to raise and lower the elevators and from side to side to control the ailerons

Aileron cable A sideways shift in the control column is passed by cables, pulleys, and levers to both ailerons, deflecting them in opposite directions

Port aileron When the port aileron is lowered, it deflects air flow under the wing downward, causing the wing to rise. Raising the aileron has the opposite effect

Flight controls

In a simple light airplane, moving a sticklike control column and a pair of rudder pedals causes deflections of the hinged control surfaces. These deflections cause the plane to climb, dive, or turn to port (left) or starboard (right).

Forces acting on airplanes

Four basic forces act on a plane— thrust from the engine, drag (air resistance), which opposes thrust, lift from the airflow over the wings, and the plane's weight. In steady, level flight, the opposing pairs of forces (thrust and drag, and lift and weight) are kept in balance by adjusting the engine power and control surfaces.

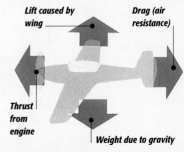

Lift caused by wing

Drag (air resistance)

Thrust from engine

Weight due to gravity

FORCES OF FLIGHT

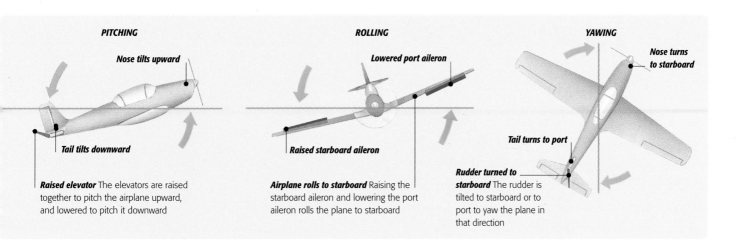

PITCHING

Nose tilts upward

Tail tilts downward

Raised elevator The elevators are raised together to pitch the airplane upward, and lowered to pitch it downward

ROLLING

Lowered port aileron

Raised starboard aileron

Airplane rolls to starboard Raising the starboard aileron and lowering the port aileron rolls the plane to starboard

YAWING

Nose turns to starboard

Tail turns to port

Rudder turned to starboard The rudder is tilted to starboard or to port to yaw the plane in that direction

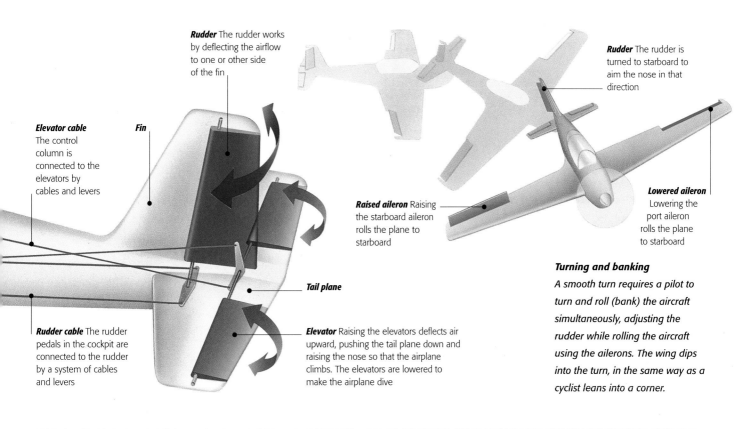

Rudder The rudder works by deflecting the airflow to one or other side of the fin

Elevator cable The control column is connected to the elevators by cables and levers

Fin

Rudder cable The rudder pedals in the cockpit are connected to the rudder by a system of cables and levers

Tail plane

Elevator Raising the elevators deflects air upward, pushing the tail plane down and raising the nose so that the airplane climbs. The elevators are lowered to make the airplane dive

Raised aileron Raising the starboard aileron rolls the plane to starboard

Rudder The rudder is turned to starboard to aim the nose in that direction

Lowered aileron Lowering the port aileron rolls the plane to starboard

Turning and banking

A smooth turn requires a pilot to turn and roll (bank) the aircraft simultaneously, adjusting the rudder while rolling the aircraft using the ailerons. The wing dips into the turn, in the same way as a cyclist leans into a corner.

Generating lift

The airfoil shape of a wing is crucial for providing lift. When a plane flies level, the flow of air over its wings generates lift. If the plane's nose tilts upward, the angle that the wings meet the airflow increases and more lift is produced: the aircraft climbs. But if the wing meets the oncoming air too steeply, the plane stalls: lift is no longer generated and the aircraft falls.

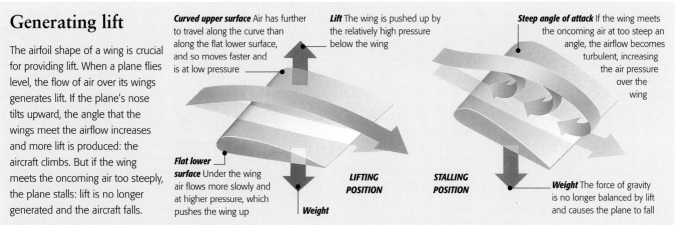

Curved upper surface Air has further to travel along the curve than along the flat lower surface, and so moves faster and is at low pressure

Lift The wing is pushed up by the relatively high pressure below the wing

Flat lower surface Under the wing air flows more slowly and at higher pressure, which pushes the wing up

Weight

LIFTING POSITION

STALLING POSITION

Steep angle of attack If the wing meets the oncoming air at too steep an angle, the airflow becomes turbulent, increasing the air pressure over the wing

Weight The force of gravity is no longer balanced by lift and causes the plane to fall

SEE ALSO: AIRLINERS *p.126* | PISTON ENGINES *p.110* | JET ENGINES *p.128* | AIRCRAFT GUIDANCE *p.134*

Airliners

Modern airliners depend on the same aerodynamic principles as light aircraft. However, in order to meet contemporary standards of safety and comfort and to carry much heavier payloads over far greater distances, this medium of mass transportation has required huge developments in instrumentation systems, wing and engine design, and in the materials used in their construction.

The largest passenger-carrying aircraft under development is the Airbus A3XX, which will carry up to 840 passengers. It will have a takeoff weight of 640 tons (580 tonnes) and a range of up to 10,000 miles (16,000 km). In order to propel an aircraft of this size, powerful engines have been developed. These engines must be fuel efficient in order to meet antipollution standards. The need for fuel economy is one reason why fuel-hungry supersonic aircraft, such as Concorde, have not become more widely used. It has also meant that modern aluminum alloys and carbon fiber composites are increasingly used in the construction of aircraft because they markedly reduce weight and increase strength. Perhaps the most significant innovation in airliner design has been in control systems. All modern airliners rely on "fly-by-wire" technology that substitutes the mechanical linkages of traditional control systems with electronic data lines, and enables complex aircraft to be flown by a crew of two.

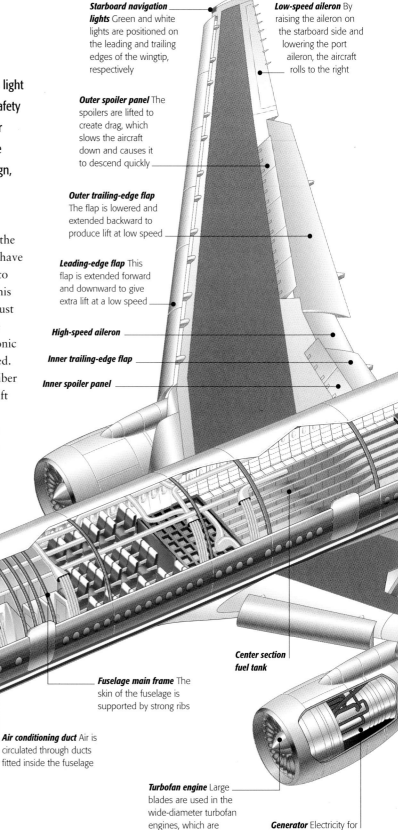

Starboard navigation lights Green and white lights are positioned on the leading and trailing edges of the wingtip, respectively

Outer spoiler panel The spoilers are lifted to create drag, which slows the aircraft down and causes it to descend quickly

Outer trailing-edge flap The flap is lowered and extended backward to produce lift at low speed

Leading-edge flap This flap is extended forward and downward to give extra lift at a low speed

High-speed aileron

Inner trailing-edge flap

Inner spoiler panel

Low-speed aileron By raising the aileron on the starboard side and lowering the port aileron, the aircraft rolls to the right

Fuselage The skin of the fuselage is made of aluminum alloy

Cockpit

Weather radar scanner

Nose landing gear

Baggage compartment Cargo and luggage are stored in holds under the passenger compartments

Air conditioning duct Air is circulated through ducts fitted inside the fuselage

Fuselage main frame The skin of the fuselage is supported by strong ribs

Turbofan engine Large blades are used in the wide-diameter turbofan engines, which are relatively quiet and produce up to 98,000 lbs (444,500 N) of thrust

Center section fuel tank

Generator Electricity for the onboard systems is supplied by generators fitted to the engines

Tailfin Ribs and struts in the tailfin give it strength while minimizing weight

Static discharger

Rudder Made of a lightweight composite, the rudder controls the left-to-right motion (yaw) of the aircraft

Elevator The upward and downward pitch of the aircraft is controlled by raising and lowering the elevators

Touchdown

The main landing gear of the Boeing 777, located under the wings, consists of two six-wheeled trucks, rather than the conventional four-wheeled units. The extra wheels give better weight distribution on the runway and are fitted with an advanced brake design. The nose landing gear uses a standard two-wheel design.

Pressure bulkhead The bulkhead contains the pressurized atmosphere of the cabin

Tailplane

Cabin door/emergency exit A total of ten exits are provided for passengers and crew

Wing fuel tank 31,000 gallons (117,000 liters) of fuel are carried in tanks located in the wings and in the central section

Skin The wings are covered with a light aluminum alloy skin

Glass cockpit

In the cockpit of a Boeing 777, the most important flight, navigation, and engine information is clearly displayed on six large liquid-crystal screens that remain clearly visible, even in direct sunlight.

Undercarriage

Boeing 777-300

Introduced in 1995, the Boeing 777 series incorporates many of the latest features in aircraft design. The world's largest twin-engine aircraft, the 777-300, carries up to 394 passengers with a range of nearly 6,000 miles (9,500 km). Its advanced wing design, with a long span and added thickness, provides the capacity to climb quickly and cruise at high altitudes. A "fly-by-wire" flight-control relays instructions to all systems via a central electronic data bus—simpler and more reliable than traditional multiple-cable systems.

Port navigation lights Red and white lights are positioned at the leading and trailing edges of the wingtip, respectively

Static discharger Spikes on the trailing edge of the wings and tail help to prevent a buildup of static electricity

SEE ALSO: LIGHT AIRCRAFT *p.124* | JET ENGINES *p.128* | AIRCRAFT GUIDANCE *p.134*

Jet engines

Jet engines are a type of internal combustion engine that have been used to propel power boats and cars to world speed records. They are most familiar, however, as aircraft engines. Jet engines produce power by accelerating a mass of air through the engine. Burning fuel in the combustion chamber produces hot, expanding exhaust gases that are blasted out of the rear of the engine. Some jet engines also drive a propeller, while others, called turbofans, gain most of their power by pushing a "bypass" jet of cold air around the engine.

A jet engine produces thrust—a force that drives the engine forward. This force is explained by the principle of action and reaction. Hot, expanding gases are pushed out of the engine. This force, the action force, produces an equal and opposite reaction force from the exhaust gases to the engine. This reaction force, called thrust, propels the engine and the attached aircraft forward. Exactly the same principle operates when air escapes from an inflated balloon, propelling the balloon through the air.

British engineer Sir Frank Whittle (1907–1996) developed the jet engine in the 1930s and his design, the turbojet, is still used today, mainly in military aircraft. Modifications of this basic design include turboprop and turbofan engines. Turboprops couple a set of spinning blades called a turbine to a propeller, usually at the front of the engine. The turbine uses the exhaust jet's energy to drive the propeller, which provides most of the engine's thrust. Turboprops are economical and suitable for small, low-speed passenger aircraft. Turbofans use a turbine inside the engine to turn a large fan at the front that sucks in a huge volume of air. This air is bypassed around the hot engine core and expelled at the rear, producing most of the thrust. Turbofans are quieter but far more powerful than other jet engines and are used in large airliners.

Turbojet engine

Turbojet engines burn a mixture of air and liquid fuel, in the form of kerosene, to produce a stream of fast-moving, hot exhaust gases that drives the engine forward. Although turbojets consume large amounts of fuel and are very noisy, they are extremely powerful and relatively lightweight, which makes them suitable for use on supersonic aircraft such as Concorde.

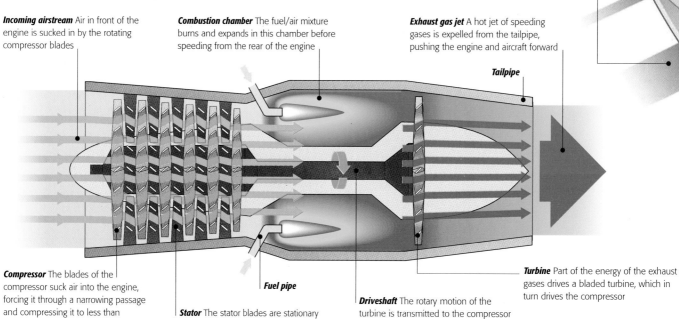

Compressor air intake Air entering the compressor is compressed to around 30 times atmospheric pressure

Bypass air intake The fan sucks a large volume of air into the bypass duct

Incoming airstream Air in front of the engine is sucked in by the rotating compressor blades

Combustion chamber The fuel/air mixture burns and expands in this chamber before speeding from the rear of the engine

Exhaust gas jet A hot jet of speeding gases is expelled from the tailpipe, pushing the engine and aircraft forward

Tailpipe

Compressor The blades of the compressor suck air into the engine, forcing it through a narrowing passage and compressing it to less than one-tenth its original volume

Fuel pipe

Stator The stator blades are stationary and angled to maintain a smooth airflow

Driveshaft The rotary motion of the turbine is transmitted to the compressor by this shaft

Turbine Part of the energy of the exhaust gases drives a bladed turbine, which in turn drives the compressor

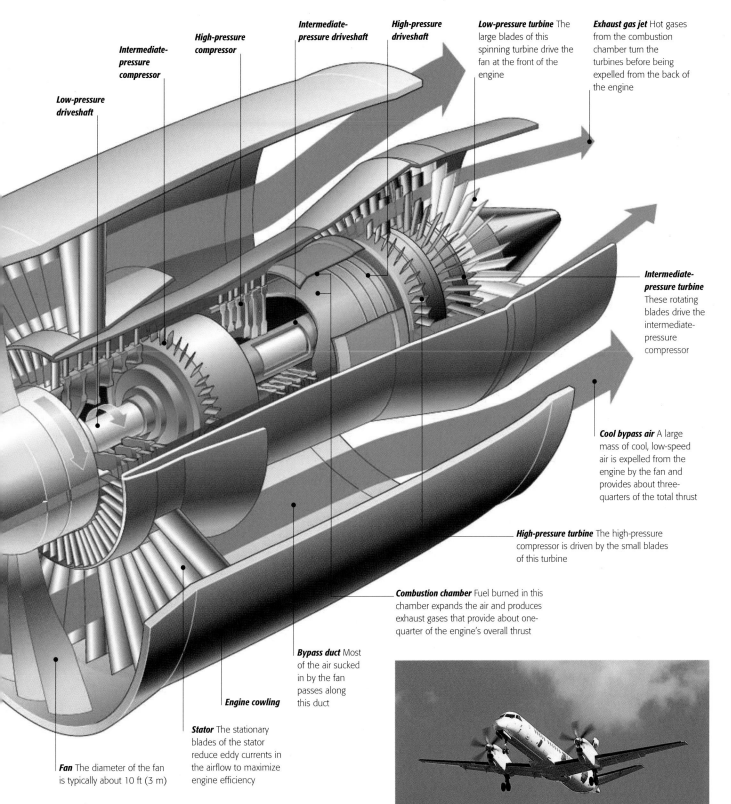

Low-pressure driveshaft

Intermediate-pressure compressor

High-pressure compressor

Intermediate-pressure driveshaft

High-pressure driveshaft

Low-pressure turbine The large blades of this spinning turbine drive the fan at the front of the engine

Exhaust gas jet Hot gases from the combustion chamber turn the turbines before being expelled from the back of the engine

Intermediate-pressure turbine These rotating blades drive the intermediate-pressure compressor

Cool bypass air A large mass of cool, low-speed air is expelled from the engine by the fan and provides about three-quarters of the total thrust

High-pressure turbine The high-pressure compressor is driven by the small blades of this turbine

Combustion chamber Fuel burned in this chamber expands the air and produces exhaust gases that provide about one-quarter of the engine's overall thrust

Bypass duct Most of the air sucked in by the fan passes along this duct

Engine cowling

Stator The stationary blades of the stator reduce eddy currents in the airflow to maximize engine efficiency

Fan The diameter of the fan is typically about 10 ft (3 m)

Turbofan engine

A turbofan engine has a large, front-mounted fan that sucks in air. Some air burns with fuel in the combustion chamber; the rest flows around the outside of the chamber to join the jet of hot gases exiting the rear of the engine. Turbofans are large, powerful engines that are used to power long-haul airliners because they are also fuel efficient and relatively quiet.

Turboprop engine
The exhaust jet leaving the rear of a turboprop engine drives a propeller at the front via a shaft. Turboprops are fuel efficient but unsuitable for high-speed flight, so their use is restricted to slower passenger aircraft.

SEE ALSO: PISTON ENGINES *p.110* | AIRLINERS *p.126* | SPACE ROCKETS *p.238* | MECHANICS *p.266*

Helicopters

Helicopters are highly maneuverable aircraft that generate lift using rapidly-turning wing-shaped rotor blades. In contrast, conventional aircraft fly by moving forward at speed to force air over fixed wings. Since helicopters do not require forward motion to produce lift, they can take off and land vertically, hover, and fly in any direction—even backward. The first practical helicopter was developed in 1939 by the Russian-born engineer Igor Sikorsky (1889–1972).

Most helicopters are powered by turboshaft engines—a type of jet engine in which a turbine turns a driveshaft that powers the rotors. In order to produce lift, each rotor blade must be pitched, or angled, upward relative to the oncoming air. The pilot controls the blade pitch using two controls, the collective and cyclic pitch sticks, which are connected to swash plates on the main rotor assembly. During takeoff, the pilot increases the pitch of all the rotor blades by the same amount with the collective pitch stick. The throttle is opened to speed up the rotor until the amount of lift produced exceeds the weight of the helicopter and it rises

Rotors and torque

As a helicopter's main rotor blade rotates, they produce a reactive torque (turning force) that tends to rotate the engine and fuselage (body) of the helicopter in the opposite direction. Most single-rotor helicopters use the thrust produced by a tail rotor to counteract the torque created by the main rotor. Twin-rotor helicopters balance out the effects of torque on the fuselage by having two main rotors that rotate in opposite directions.

vertically. To hover, the pilot reduces the collective pitch so that the lift becomes equal to the weight of the helicopter. To descend, the collective pitch is reduced until the lift is less than the weight.

Directional flight is achieved by tilting the swash plates with the cyclic pitch control, which alters the pitch of each blade as it rotates, so that every blade produces greatest lift at a particular point. Although vertical lift is still produced overall, thrust is also generated in the direction that the swash plates are tilted, which causes the helicopter to lean and fly in that direction. Further directional control is obtained by increasing or decreasing the thrust of the tail rotor that is fitted to most helicopters. One significant advance in helicopter design in recent years has been the development of no-tail-rotor (NOTAR) helicopters, which use fan-driven air circulation systems in place of tail rotors.

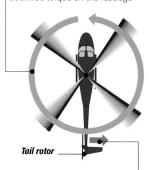

Rotation The counterclockwise rotation of the main rotors produces a reactive clockwise torque on the fuselage

Tail rotor

Thrust The torque is counteracted by thrust produced by the tail rotor

SINGLE-ROTOR HELICOPTER

Rotors The torque produced by the rotation of the front rotor is offset by that produced by a counter-rotating rear rotor

TWIN-ROTOR HELICOPTER

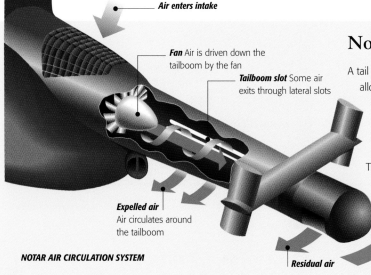

Air enters intake

Fan Air is driven down the tailboom by the fan

Tailboom slot Some air exits through lateral slots

Expelled air
Air circulates around the tailboom

NOTAR AIR CIRCULATION SYSTEM

Residual air

Thrust

No-tail-rotor helicopters

A tail rotor not only prevents a helicopter spinning out of control, it also allows the helicopter to be maneuvered in tight circles. It is, however, noisy and easily damaged, as well as being hazardous to people on the ground when it is in use. No-tail-rotor (NOTAR) helicopters provide a quieter, safer alternative. They allow a downwash of air from the main rotors to enter the fuselage of the helicopter. This air is driven down the tailboom by a large fan. Some of this air is forced out of slots along the side of the tailboom and circulates around it. This creates a force that counteracts the effects of torque. Directional control is provided by varying the amount of residual air that is expelled from a slot at the end of the tailboom.

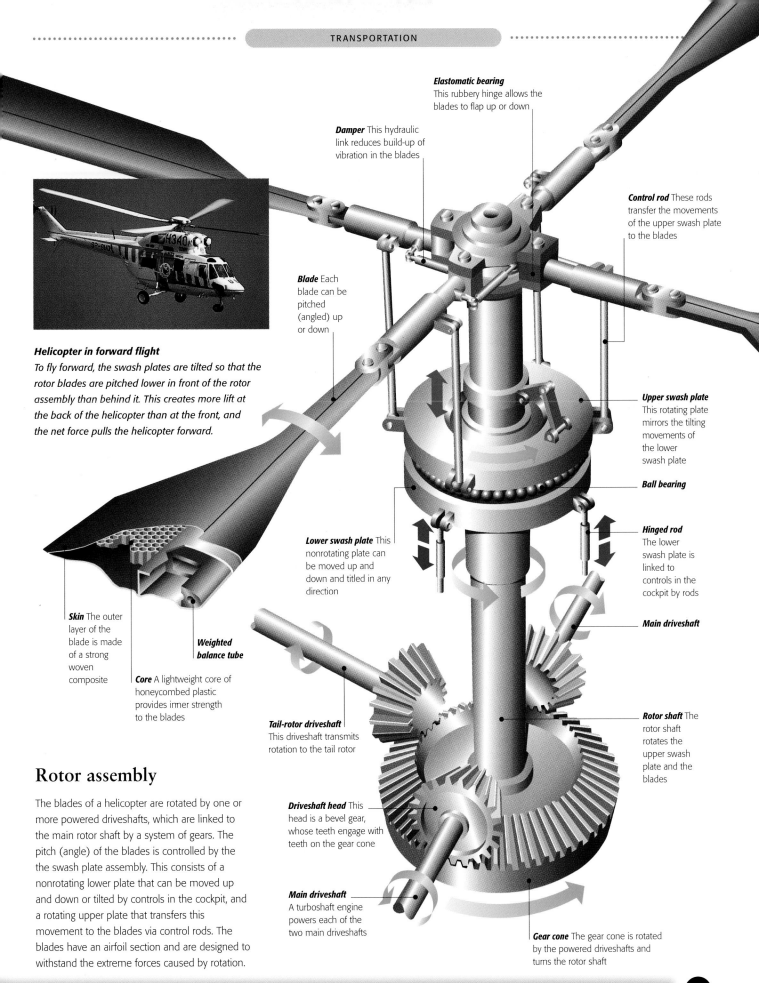

Elastomatic bearing This rubbery hinge allows the blades to flap up or down

Damper This hydraulic link reduces build-up of vibration in the blades

Control rod These rods transfer the movements of the upper swash plate to the blades

Blade Each blade can be pitched (angled) up or down

Helicopter in forward flight
To fly forward, the swash plates are tilted so that the rotor blades are pitched lower in front of the rotor assembly than behind it. This creates more lift at the back of the helicopter than at the front, and the net force pulls the helicopter forward.

Upper swash plate This rotating plate mirrors the tilting movements of the lower swash plate

Ball bearing

Lower swash plate This nonrotating plate can be moved up and down and titled in any direction

Hinged rod The lower swash plate is linked to controls in the cockpit by rods

Main driveshaft

Skin The outer layer of the blade is made of a strong woven composite

Weighted balance tube

Core A lightweight core of honeycombed plastic provides inner strength to the blades

Tail-rotor driveshaft This driveshaft transmits rotation to the tail rotor

Rotor shaft The rotor shaft rotates the upper swash plate and the blades

Rotor assembly

The blades of a helicopter are rotated by one or more powered driveshafts, which are linked to the main rotor shaft by a system of gears. The pitch (angle) of the blades is controlled by the the swash plate assembly. This consists of a nonrotating lower plate that can be moved up and down or tilted by controls in the cockpit, and a rotating upper plate that transfers this movement to the blades via control rods. The blades have an airfoil section and are designed to withstand the extreme forces caused by rotation.

Driveshaft head This head is a bevel gear, whose teeth engage with teeth on the gear cone

Main driveshaft A turboshaft engine powers each of the two main driveshafts

Gear cone The gear cone is rotated by the powered driveshafts and turns the rotor shaft

SEE ALSO: TRAFFIC GUIDANCE *p.26* | LIGHT AIRCRAFT *p.124* | COMPOSITE MATERIALS *p.144* | MECHANICS *p.266*

Navigation

Navigation is the science of fixing a vehicle's position and charting an efficient route to a destination. Until the advent of radio, navigation was once based on maps, charts, compasses, clocks, and the position of the Sun or stars. Ships and aircraft still use these techniques, but today a range of ground- and space-based radio navigation aids enables accurate navigation, even in conditions of zero visibility.

As traffic has increased, navigation has become crucially linked with traffic control and the maximization of the number of vehicles that can travel safely at the same time. Most modern air and sea navigation systems rely on various types of radio signals transmitted by fixed ground stations. The main exceptions are the GPS (Global Positioning System), which is a very precise satellite-based system, and IGS (Inertial Guidance System), which is a self-contained onboard system that works by recording every movement of a vehicle after it leaves a known starting point. Ships and aircraft both rely on radar (radio detection and ranging), in which a transmitter sends out radio waves and listens for echoes from other vehicles, surrounding terrain, and clouds. The delay and intensity of the echoes enable a computer to image the surrounding area.

Radio navigation

Two major radio navigation aids are LORAN and VOR. LORAN (long-range navigation) is widely used by ships and aircraft over sea. VOR (Very High Frequency Omnidirectional Radio Range) beacons mark waypoints on the air corridors (lanes) along which aircraft fly. When landing, aircraft may use an ILS (Instrument Landing System), which enables landing in zero visibility, or more advanced MLS (Microwave Landing System) and GPS-based systems.

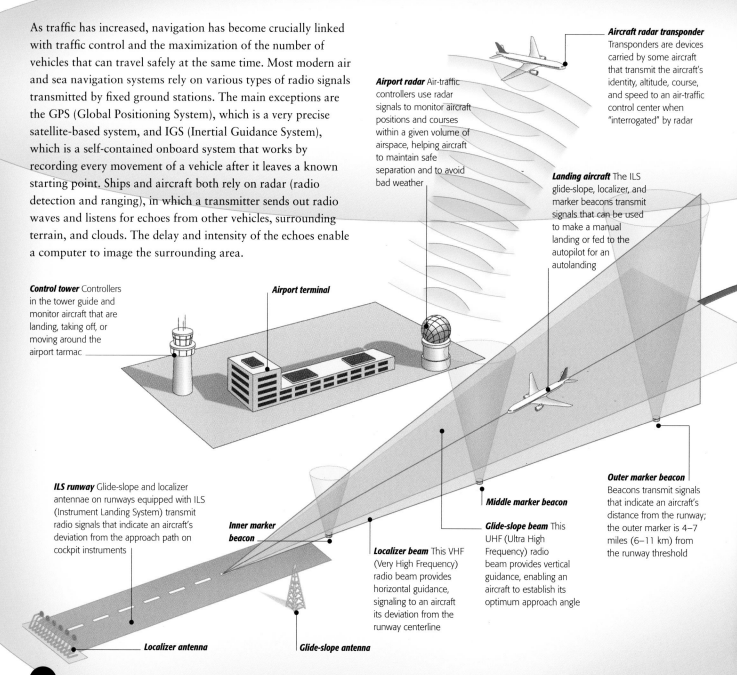

Aircraft radar transponder Transponders are devices carried by some aircraft that transmit the aircraft's identity, altitude, course, and speed to an air-traffic control center when "interrogated" by radar

Airport radar Air-traffic controllers use radar signals to monitor aircraft positions and courses within a given volume of airspace, helping aircraft to maintain safe separation and to avoid bad weather

Landing aircraft The ILS glide-slope, localizer, and marker beacons transmit signals that can be used to make a manual landing or fed to the autopilot for an autolanding

Control tower Controllers in the tower guide and monitor aircraft that are landing, taking off, or moving around the airport tarmac

Airport terminal

ILS runway Glide-slope and localizer antennae on runways equipped with ILS (Instrument Landing System) transmit radio signals that indicate an aircraft's deviation from the approach path on cockpit instruments

Inner marker beacon

Localizer beam This VHF (Very High Frequency) radio beam provides horizontal guidance, signaling to an aircraft its deviation from the runway centerline

Middle marker beacon

Glide-slope beam This UHF (Ultra High Frequency) radio beam provides vertical guidance, enabling an aircraft to establish its optimum approach angle

Outer marker beacon Beacons transmit signals that indicate an aircraft's distance from the runway; the outer marker is 4–7 miles (6–11 km) from the runway threshold

Localizer antenna

Glide-slope antenna

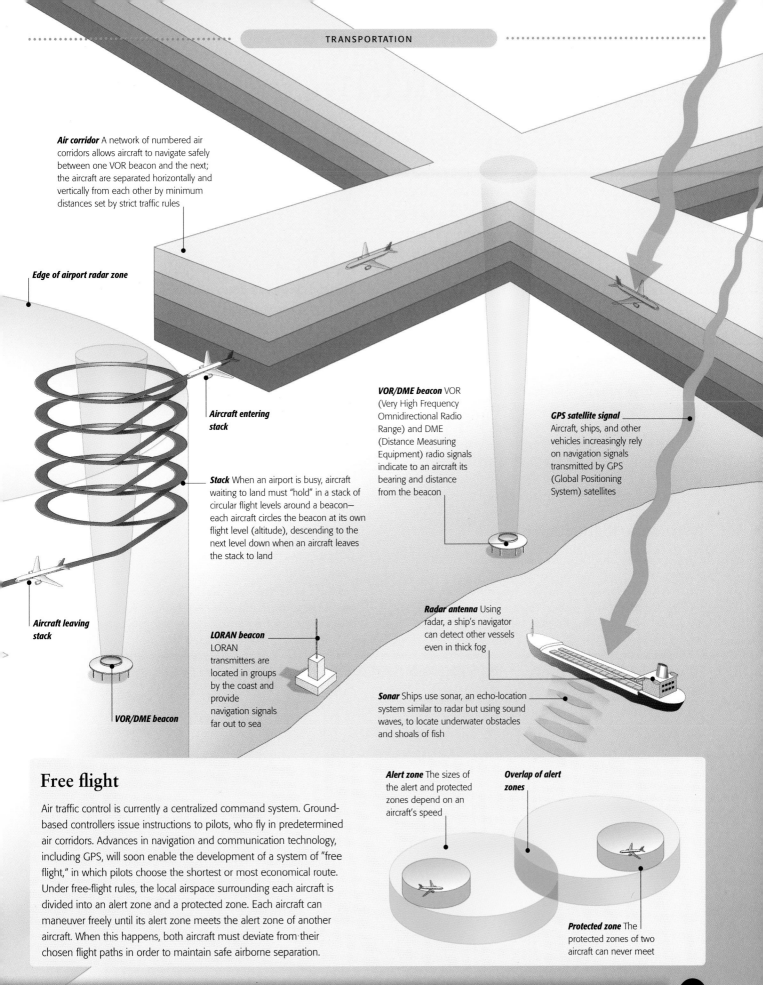

Air corridor A network of numbered air corridors allows aircraft to navigate safely between one VOR beacon and the next; the aircraft are separated horizontally and vertically from each other by minimum distances set by strict traffic rules

Edge of airport radar zone

Aircraft entering stack

Stack When an airport is busy, aircraft waiting to land must "hold" in a stack of circular flight levels around a beacon—each aircraft circles the beacon at its own flight level (altitude), descending to the next level down when an aircraft leaves the stack to land

VOR/DME beacon VOR (Very High Frequency Omnidirectional Radio Range) and DME (Distance Measuring Equipment) radio signals indicate to an aircraft its bearing and distance from the beacon

GPS satellite signal Aircraft, ships, and other vehicles increasingly rely on navigation signals transmitted by GPS (Global Positioning System) satellites

Aircraft leaving stack

VOR/DME beacon

LORAN beacon LORAN transmitters are located in groups by the coast and provide navigation signals far out to sea

Radar antenna Using radar, a ship's navigator can detect other vessels even in thick fog

Sonar Ships use sonar, an echo-location system similar to radar but using sound waves, to locate underwater obstacles and shoals of fish

Free flight

Air traffic control is currently a centralized command system. Ground-based controllers issue instructions to pilots, who fly in predetermined air corridors. Advances in navigation and communication technology, including GPS, will soon enable the development of a system of "free flight," in which pilots choose the shortest or most economical route. Under free-flight rules, the local airspace surrounding each aircraft is divided into an alert zone and a protected zone. Each aircraft can maneuver freely until its alert zone meets the alert zone of another aircraft. When this happens, both aircraft must deviate from their chosen flight paths in order to maintain safe airborne separation.

Alert zone The sizes of the alert and protected zones depend on an aircraft's speed

Overlap of alert zones

Protected zone The protected zones of two aircraft can never meet

SEE ALSO: AIRCRAFT GUIDANCE *p.134* | GLOBAL POSITIONING SYSTEM *p.136* | SHIPS *p.140* | LIGHT *p.262*

Aircraft guidance

Autopilots assist human pilots by guiding aircraft during tedious or difficult phases of flight, such as steady cruise and landing. Motion sensors feed data about changes in aircraft attitude (orientation) to a computer, which outputs signals to control surfaces on the wings and tail to maintain the desired course. Motion sensors are also used by the inertial guidance system, a self-contained navigation system.

Autopilots respond to changes in aircraft attitude and motion much faster than a human can, so they enhance aircraft stability and give a smoother ride. Some new aircraft are designed to be inherently unstable in order to give better maneuverability and cannot be flown without continuous autopilot assistance. The autopilot computer may be fed data from navigation systems, such as an inertial guidance system (IGS). Having initially been programmed with the coordinates of the starting position, the IGS uses data from motion sensors to track the aircraft's subsequent motion and pinpoint its position. Unlike other navigation systems, the IGS does not depend on external radio signals.

Automatic cruise

Basic autopilots relieve human pilots during tedious stages of flight, such as high-altitude cruise. Advanced autopilot and autothrust systems can carry out highly precise maneuvers that may be very demanding to perform manually, such as landing an aircraft in conditions of zero visibility.

Automatic flight control

A single-axis autopilot is a simple automatic flight-control system that augments aircraft stability and can maintain level flight. A gyroscope senses rotation around the roll axis and sends corrective signals to the ailerons. Such "wing-leveling" devices are widely used on light aircraft. The pilot can also dial commands into the autopilot control unit to make the aircraft turn without using the control stick, or can configure the autopilot to respond to compass signals in order to hold a heading automatically. Two-axis systems control roll and pitch (using the ailerons and elevators), while three-axis systems control roll, pitch, and yaw. Autothrust computers control engine thrust and work together with an autopilot to perform complex maneuvers.

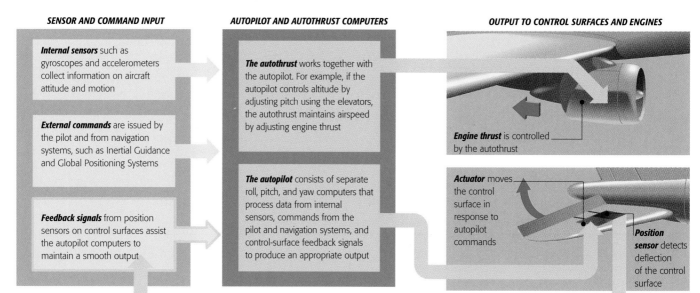

SENSOR AND COMMAND INPUT

Internal sensors such as gyroscopes and accelerometers collect information on aircraft attitude and motion

External commands are issued by the pilot and from navigation systems, such as Inertial Guidance and Global Positioning Systems

Feedback signals from position sensors on control surfaces assist the autopilot computers to maintain a smooth output

AUTOPILOT AND AUTOTHRUST COMPUTERS

The autothrust works together with the autopilot. For example, if the autopilot controls altitude by adjusting pitch using the elevators, the autothrust maintains airspeed by adjusting engine thrust

The autopilot consists of separate roll, pitch, and yaw computers that process data from internal sensors, commands from the pilot and navigation systems, and control-surface feedback signals to produce an appropriate output

OUTPUT TO CONTROL SURFACES AND ENGINES

Engine thrust is controlled by the autothrust

Actuator moves the control surface in response to autopilot commands

Position sensor detects deflection of the control surface

Gyroscopes and accelerometers

Aircraft guidance systems use gyroscopes and accelerometers to sense motion. A mechanical gyroscope consists of a heavy wheel spinning at high speed, mounted in a pivoted frame. The inertia of the spinning wheel means that it resists any change in its orientation. Rotation of the gyroscope frame around a particular axis produces a proportional reaction force that can be measured. Accelerometers use the inertia of a metal bar to detect acceleration along an axis (in a line).

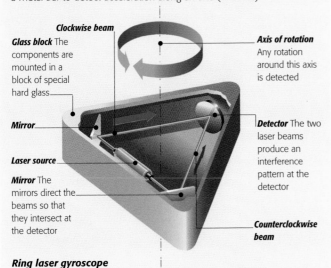

Glass block The components are mounted in a block of special hard glass

Clockwise beam

Mirror

Laser source

Mirror The mirrors direct the beams so that they intersect at the detector

Axis of rotation Any rotation around this axis is detected

Detector The two laser beams produce an interference pattern at the detector

Counterclockwise beam

Ring laser gyroscope
Unlike a mechanical gyroscope, a ring laser gyroscope has no moving parts. Two beams from a laser source are sent in opposite directions around a closed loop to meet at a detector, where they interfere. If the unit is rotated around an axis perpendicular to the loop, the travel times of the two beams become slightly different. This changes the interference pattern and gives a measure of the amount of rotation.

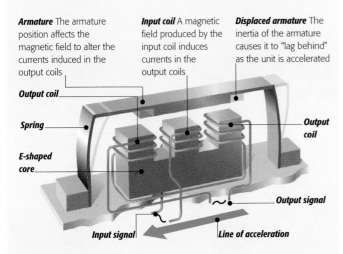

Armature The armature position affects the magnetic field to alter the currents induced in the output coils

Input coil A magnetic field produced by the input coil induces currents in the output coils

Displaced armature The inertia of the armature causes it to "lag behind" as the unit is accelerated

Output coil

Spring

E-shaped core

Output coil

Output signal

Input signal

Line of acceleration

Accelerometer
One input and two output coils are wound on an E-shaped core in this accelerometer. The coils are coupled electromagnetically by a spring-mounted armature, so that the input coil's alternating current appears across the output coils. Normally, the output coils have equal currents, but under acceleration, the armature "lags" due to its inertia, staying closer to one output coil and producing an imbalanced output signal.

Inertial Guidance System

An inertial guidance system (IGS) uses three accelerometers to measure aircraft acceleration. The accelerometers are mounted on an inertial platform stabilized by gyroscopes so that they always measure accelerations in the same directions relative to the Earth, no matter how the aircraft turns. Prior to departure, the platform is aligned with the aircraft and its latitude, longitude, and altitude data is programmed into the IGS computer. Speed and distance traveled can be calculated from acceleration.

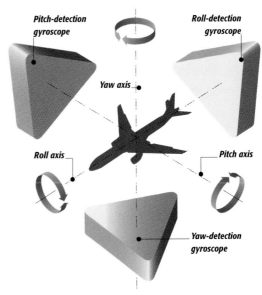

Pitch-detection gyroscope

Roll-detection gyroscope

Yaw axis

Roll axis

Pitch axis

Yaw-detection gyroscope

Gyroscopic stabilization
Three gyroscopes mounted at right angles to each other measure rotations around the pitch, roll, and yaw axes. This data is used to realign the inertial platform continuously so it always points in the same direction relative to its starting point. With ring laser gyroscopes, the stabilization is carried out in software by a computer, rather than by physically moving the platform.

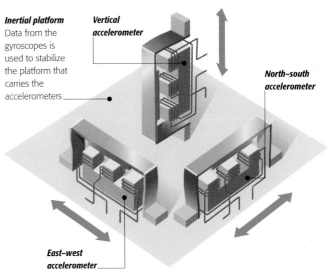

Inertial platform Data from the gyroscopes is used to stabilize the platform that carries the accelerometers

Vertical accelerometer

North–south accelerometer

East–west accelerometer

Inertial platform
Three accelerometers mounted on the inertial platform measure aircraft acceleration in three dimensions. The IGS computer calculates the aircraft speed and distance traveled and, using the aircraft's initial latitude, longitude, and altitude, is able to determine its position at any time.

SEE ALSO: LIGHT AIRCRAFT *p.124* | AIRLINERS *p.126* | NAVIGATION *p.132* | LASERS *p.220* | PRINCIPLES *p.000*

Global Positioning System

The global positioning system (GPS) is a satellite-based system that provides anybody using a GPS receiver anywhere in the world, in any weather, with accurate information on their position, speed, and the time. Twenty-four satellites in precise orbits, monitored by ground stations, broadcast signals to receivers on land, sea, in the air, and in space. GPS is the ultimate global reference system for position fixing, navigation, mapping, tracking, timing, and synchronizing events.

At the heart of the GPS are 24 Navstar satellites that continuously broadcast signals. A GPS receiver uses these signals to measure its range (distance) from each of the four or more satellites that are in view from its location at any one time. An almanac (astronomical calendar) in the receiver, which is updated with corrective signals from the satellites, specifies where the satellites are at any time. Given the locations of four satellites, and its range from each of them, the receiver can calculate its position. The GPS is coordinated by atomic clocks aboard the satellites, so the receiver also registers the exact time. By checking its position at successive moments, the receiver can calculate the speed of its movement. Standard receivers can fix location to within a few yards in any direction and the time to within one-millionth of a second. Advanced receivers are accurate to within a few inches.

The GPS provides a global standard for measuring space and time. Its great precision enables aircraft to fly closer together, with increased safety, and to follow more direct routes. Fishing fleets use the GPS to locate the best fishing areas. Navigation systems in cars combine GPS technology with moving map displays, making it impossible to "get lost." Watches with built-in GPS receivers are used by hikers and explorers. GPS tracking systems direct emergency vehicles, coordinate public transportion, and dispatch taxis and delivery trucks with great efficiency. The GPS also synchronizes a range of commercial activities from financial transactions to electricity distribution.

Safe orienteering

Wherever a person is in the world—in the woods or on a mountain—it should be possible to find the way back home with a portable GPS receiver. Many receivers feature a digital map that scrolls to show current position. If the receiver is programmed with the destination coordinates, it can issue directions using the digital display or a synthesized voice output.

Navstar orbits

Twenty-four Navstar satellites orbit the Earth at an altitude of approximately 12,500 miles (20,000 km). There are four satellites in each of six orbital planes. Though the orbits are precise, errors do occur and the satellites transmit navigational corrections to amend the almanacs in GPS receivers. Navigational corrections are uploaded to the satellites by ground stations that continuously monitor their position and speed.

Navstar satellite Each satellite continuously broadcasts ranging signals and navigational corrections as it orbits the Earth once every 12 hours

Orbit Navstar satellites move in precise, high orbits, designed so that at least four satellites are in view from any point on or above the Earth any given time

Earth

Rangefinding and triangulation

A GPS receiver fixes its position by calculating its range (distance) from each of at least four satellites, whose exact positions are known. Each satellite broadcasts signals, which take time to reach the receiver. The receiver's onboard clock is synchronized with the satellites' atomic clocks, enabling it to calculate the signals' travel times. The range of each satellite is calculated using the travel time of its signal and the signal speed (the speed of radio waves). Using a method called triangulation, the range measurements are combined with the satellite position data to locate the receiver.

Range from Satellite One The range from Satellite One is calculated to be 13,500 miles (21,600 km)

Satellite One

Sphere One The receiver must be located somewhere on the surface of this sphere. The sphere has a radius of 13,500 miles (21,600 km), centered on Satellite One

Ranging from one satellite

Suppose the receiver calculates the range to Satellite One as 13,500 miles (21,600 km). This means that it can be at any point in space this distance from the satellite—in other words, it can be anywhere on the surface of Sphere One.

Ranging from two satellites

A second range measurement locates the receiver on the surface of Sphere Two. Since the receiver must also be on the surface of Sphere One, it must be located where the two surfaces intersect—in other words, anywhere on the circumference of the pink circle.

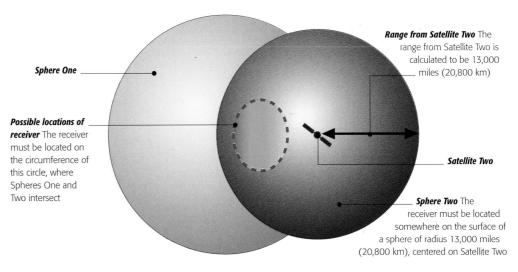

Sphere One

Possible locations of receiver The receiver must be located on the circumference of this circle, where Spheres One and Two intersect

Range from Satellite Two The range from Satellite Two is calculated to be 13,000 miles (20,800 km)

Satellite Two

Sphere Two The receiver must be located somewhere on the surface of a sphere of radius 13,000 miles (20,800 km), centered on Satellite Two

Triangulation

A third range measurement locates the receiver on the surface of Sphere Three. Since the receiver must also be on the surfaces of Sphere One and Sphere Two, it must be located where all three surfaces intersect—that is, where Sphere Three meets the pink circle. So the receiver must be located at one of only two points. One of these points is usually an unlikely position, so the receiver's position can be fixed with just three ranges. However, a fourth range is needed in order to synchronize the receiver and satellite clocks.

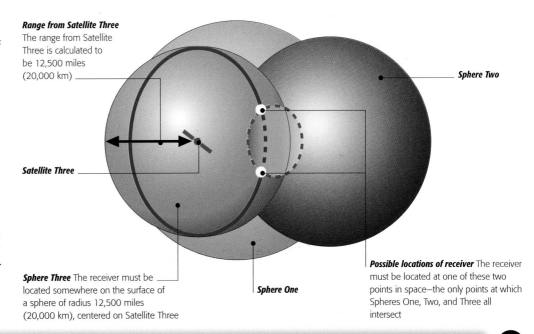

Range from Satellite Three The range from Satellite Three is calculated to be 12,500 miles (20,000 km)

Satellite Three

Sphere Two

Sphere Three The receiver must be located somewhere on the surface of a sphere of radius 12,500 miles (20,000 km), centered on Satellite Three

Sphere One

Possible locations of receiver The receiver must be located at one of these two points in space—the only points at which Spheres One, Two, and Three all intersect

SEE ALSO: TRAFFIC GUIDANCE *p.26* | SATELLITE COMMUNICATIONS *p.50* | AIRLINERS *p.126* | NAVIGATION *p.132*

Hovercraft and hydrofoils

Few ships travel at more than 35 mph (55 km/h) because of the drag on their hulls as they forge their way through water, which is about 800 times more dense than air. By traveling on a cushion of air, hovercraft overcome the slowing effects of water. Hydrofoils minimize their contact with water by using wing-shaped surfaces to raise their hulls out of the water when traveling at speed.

First devised early in the 20th century, hovercraft became a commercial and military reality in the 1960s. Hovercraft are among the world's most versatile boats because they are amphibious and can ride up onto the shore. A U.S. military hovercraft called the Landing Craft Air Cushion can carry a 70-ton (64-tonne) tank at speeds of up to 45 mph (75 km/h). It is also able to land on about three-quarters of the world's coastlines. Hydrofoils are even faster, with top speeds of up to 70 mph (115 km/h). The most powerful hydrofoils use diesel engines for low-speed travel and turbine-powered waterjets when traveling at high-speed. Hydrofoils, which were also developed early in the 20th century, are now widely used as high-speed ferries and fast patrol boats.

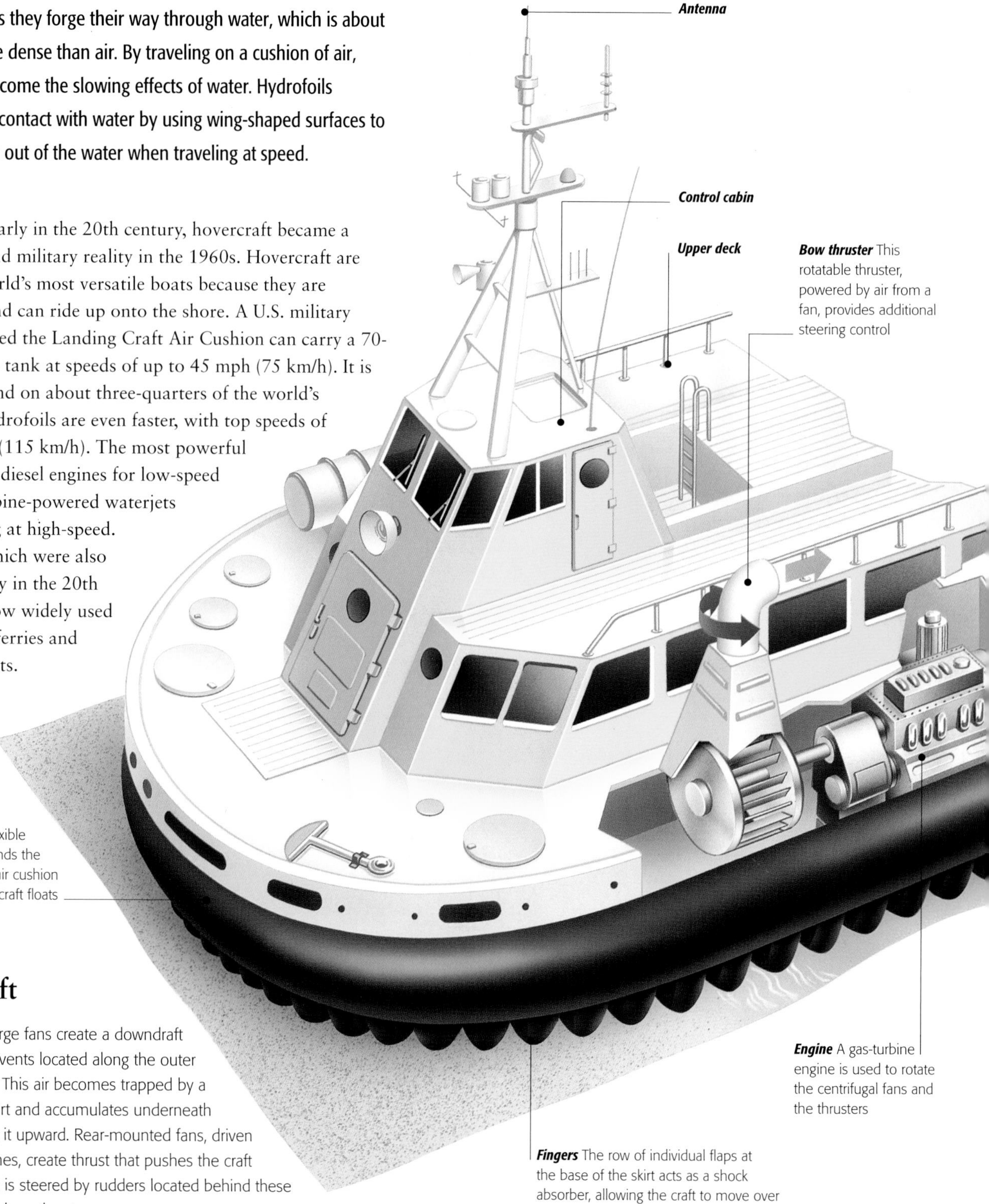

Antenna

Control cabin

Upper deck

Bow thruster This rotatable thruster, powered by air from a fan, provides additional steering control

Rubber skirt This flexible inflated skirt surrounds the craft and traps the air cushion on which the hovercraft floats

Hovercraft

In a hovercraft, large fans create a downdraft of air through air vents located along the outer edge of the craft. This air becomes trapped by a flexible rubber skirt and accumulates underneath the craft, pushing it upward. Rear-mounted fans, driven by separate engines, create thrust that pushes the craft forward. The craft is steered by rudders located behind these propellers and by bow thrusters.

Engine A gas-turbine engine is used to rotate the centrifugal fans and the thrusters

Fingers The row of individual flaps at the base of the skirt acts as a shock absorber, allowing the craft to move over waves and rough ground

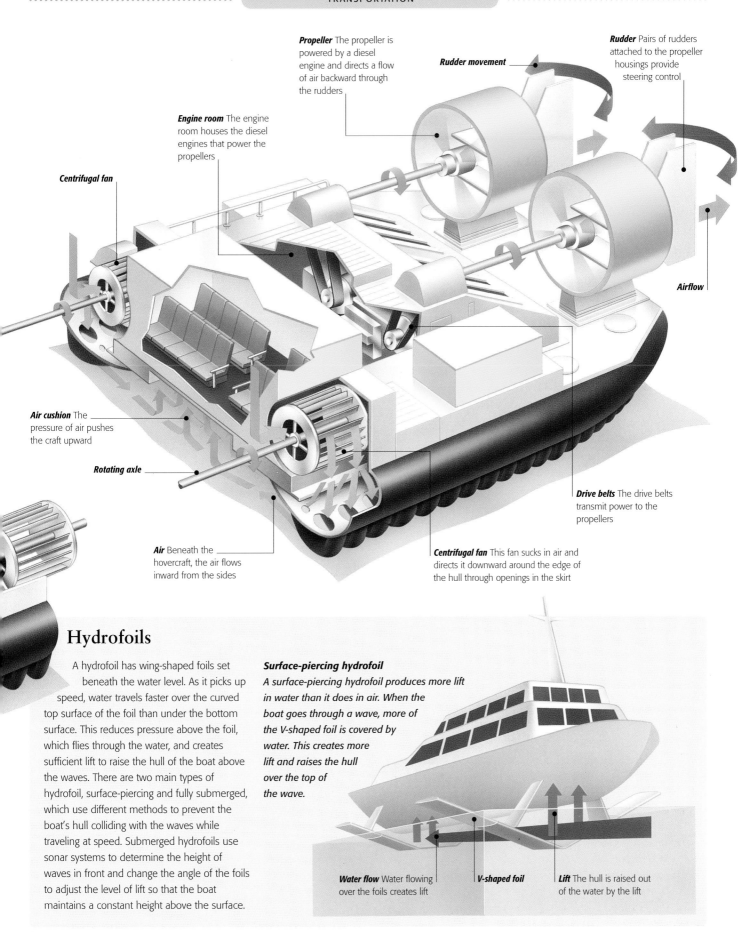

Propeller The propeller is powered by a diesel engine and directs a flow of air backward through the rudders

Rudder movement

Rudder Pairs of rudders attached to the propeller housings provide steering control

Engine room The engine room houses the diesel engines that power the propellers

Centrifugal fan

Airflow

Air cushion The pressure of air pushes the craft upward

Rotating axle

Drive belts The drive belts transmit power to the propellers

Air Beneath the hovercraft, the air flows inward from the sides

Centrifugal fan This fan sucks in air and directs it downward around the edge of the hull through openings in the skirt

Hydrofoils

A hydrofoil has wing-shaped foils set beneath the water level. As it picks up speed, water travels faster over the curved top surface of the foil than under the bottom surface. This reduces pressure above the foil, which flies through the water, and creates sufficient lift to raise the hull of the boat above the waves. There are two main types of hydrofoil, surface-piercing and fully submerged, which use different methods to prevent the boat's hull colliding with the waves while traveling at speed. Submerged hydrofoils use sonar systems to determine the height of waves in front and change the angle of the foils to adjust the level of lift so that the boat maintains a constant height above the surface.

Surface-piercing hydrofoil
A surface-piercing hydrofoil produces more lift in water than it does in air. When the boat goes through a wave, more of the V-shaped foil is covered by water. This creates more lift and raises the hull over the top of the wave.

Water flow Water flowing over the foils creates lift

V-shaped foil

Lift The hull is raised out of the water by the lift

Ships

Traditionally, boats and ships were built from timber planks, nailed together, and painted with a waterproof coating of pitch or tar, while sails or oars were used for propulsion. Traditional materials and relatively feeble power sources placed limits on the capabilities and performances of vessels that were not overcome until metal hulls and steam power were introduced in the 1800s.

As shipbuilding materials and propulsion systems have improved, ship designs have become larger and increasingly specialized. Today, there is a large variety ships for different functions, ranging from oil supertankers and aircraft carriers to refrigerated factory ships (used for transporting food and other perishable goods) and icebreakers. There are even plans to build a vast cruise ship called the *City of Freedom*. This floating city will be almost one mile (1.6 km) long and will house up to 40,000 people. The ship will be equipped with its own airport, an internal marina, and 200 acres (80 hectares) of recreational space.

Hull construction

The shape of a hull is a compromise between the role of the ship—which dictate the volume of space it needs to enclose, the arrangement of engines and equipment inside it, and practical concerns about its buoyancy—and its speed, which is diminished by the forces of drag that act on it as it cuts through the water. In order for a ship to be buoyant, the part of a ship below the water

Icebreaker
With their raised bows and reinforced hulls, icebreakers power through frozen seas, rising up on the pack ice and smashing a path through using their sheer weight. Icebreakers have been known to forge paths through 23 feet (7 m) of ice, although most are designed to tackle ice that is 8 feet (2.5 m) thick.

level must have a lower density than the water it displaces. Although slender hulls are best at cutting through the waves, they occupy relatively small volumes. If such a ship carries a heavy cargo, it may end up sinking into the seawater around it.

Metal hulls, whether of iron, steel, or aluminum, are thinner, stronger, and lighter than wooden hulls. Modern materials such as fiberglass composites are even lighter, and both metals and composites can be more easily shaped to an ideal design than wooden planks. In addition, they can keep their shape and support a superstructure without the need for heavy structural beams running across the vessel. Composite hulls are mainly used for yachts, while metal hulls remain more popular for large ships—as well as being more economical and easier to repair, they can carry much heavier cargos. Modern ships use welding rather than riveting to join metal plates together, and are often built up using prefabricated units.

Hull designs range from narrow, light racing yachts built for speed, to huge oil

Oil tanker
A supertanker is narrow at the waterline, to cut down drag, but bulges at the bottom, to increase capacity and buoyancy. Some supertankers are over 1,500 feet (455 m) long and 200 feet (60 m) wide. The largest oil-carrying tankers are also known as "ULCCs" (ultralarge crude carriers).

High-speed ferry
The Stena Line's HSS ferry is a catamaran. It is propelled by water jets that are powered by gas turbines and can reach a top speed of 40 knots (46 mph). Almost 420-ft (130-m) long, the ferry has a carrying capacity of 1,500 passengers and 375 cars. Its two narrow hulls, which are made of aluminum, spread the ship's weight and cut through the water easily.

tankers designed to carry the maximum load as cheaply as possible. Hulls often incorporate bulkheads that divide them into smaller volumes. This helps keep the cargo evenly distributed in the vessel, and also protects against the entire vessel flooding if the hull is breached at one place. Recently, multihulled designs developed for yachts such as catamarans and trimarans have been utilized by powered passenger ships. These designs reduce drag in the water, allowing these ships to reach higher speeds.

Means of propulsion
The arrival of steam power in the late 1700s created a revolution in marine transportation. The first oceangoing steamships used paddle wheels fitted to their sides. From the 1830s onward these were replaced by propellers that were more efficient because they remained permanently underwater. A propeller works because its shape pulls in water from ahead of the boat, and pushes it out behind at higher speed, forcing the boat in the other direction.

Propellers are still the most widespread propulsion device in shipping, but the engines used to drive them have changed considerably from the original steam engines. Diesel engines, first invented in the late 1800s, are still a popular sources of power, but many high-speed ships now use gas turbine engines. Gas turbines are modified jet engines that spin at very high speeds, and have to be stepped down through complex gearboxes to run a slowly turning propeller. Diesel engines are more efficient and run more slowly, but are much larger. Diesel-electric ships use diesel engines to power generators that provide electricity to electric motors.

The water jet is another recent form of propulsion that is spreading in popularity. A water jet is basically a long tube with a rapidly spinning propeller, called an impeller, in the middle. Seawater is sucked in through the front intake, accelerated by the impeller, and forced out of the back at high speeds, pushing the ship forward. The impeller itself can be powered by a gas turbine, diesel, or gasoline engine. Water jets were first used on small craft such as jetskis, but are now widely used on hydrofoils and other high-speed passenger ships.

Stability and self-righting

All ships will roll in rough seas, but there are several ways to increase stability. Interior compartments help ensure that cargo and ballast stay distributed evenly, and many ships now have stabilizers—small "wings" extending from the side of the hull which rotate to counteract the ship's roll. Even if a vessel does capsize, it may be able to right itself.

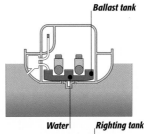

Ballast tank

1 Self-righting vessels
Many lifeboats have a water-filled ballast tank that adds stability in heavy seas. If the vessel capsizes, the ballast shifts, allowing the vessel to right itself.

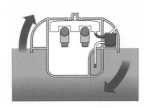

Water   *Righting tank*

2 The vessel rolls
As the vessel starts to roll over, a righting tank on one side of the boat starts to fill with water from the ballast tank.

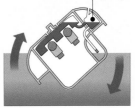

3 The vessel overturns
When the vessel is completely upside down, all the water from the ballast tank fills the righting tank. This creates additional weight on one side only, which pulls the vessel upright again.

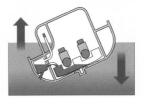

4 The vessel rights itself
As the vessel regains its upright position, the water from the righting tank returns to the ballast tank.

SEE ALSO: PISTON ENGINES *p.110* | COMPOSITE MATERIALS *p.144* | HOVERCRAFT & HYDROFOILS *p.138*

Submarine technology

The deep oceans are the least-explored environments on Earth and also the most extreme. Unprotected divers cannot penetrate these depths because of the immense pressure that builds up from the weight of water above. The only way humans can reach the deep seas is inside sealed vessels that can withstand the pressure of the water and maintain a warm, breathable atmosphere inside.

Most crewed deep-sea vehicles are submersibles—small submarines built around a spherical pressurized cabin for the crew. A spherically shaped hull is best suited to resist compression because the compression force is distributed evenly over its entire surface. Submersibles can reach far greater depths and withstand greater pressures than large military submarines.

Submersibles have a wide variety of uses, ranging from tourism to underwater engineering and from ecological research to archeology. They usually carry an array of equipment including cameras, powerful searchlights, and sophisticated manipulator arms for collecting samples and handling tools. Submersibles are usually propeller driven. They are designed to have an overall density close to that of water—this gives them neutral buoyancy, so that they remain suspended in the water when unpowered.

Deep-sea exploration is a dangerous and expensive activity, so remotely operated vehicles (ROVs) are often used in place of crewed submersibles. ROVs can be controlled by an operator at a console on the surface or in a nearby submersible. Commands are either sent along an electric cable called a tether, or are broadcast through the water in the form of sound waves, allowing the ROV to float free.

AUVs (autonomous underwater vehicles) are neither crewed nor under continuous remote control. Instead, they are fitted with computers that have a degree of "artificial intelligence" and are capable of making decisions according to preprogrammed rules in order to complete their mission. AUVs frequently spend days or weeks away from human control, before surfacing for retrieval. Some oceanographers predict that hundreds of AUVs might eventually be deployed in underwater networks to map and monitor environmental changes in the depths of the oceans.

Manipulator arm Each arm has six different types of movement controlled by a joystick in the pressure sphere

Manipulator This versatile tool collects samples from the water or the ocean floor and carries stereo video cameras and touch sensors that provide feedback to the operator

Deep-sea diving

The main problem faced by divers is water pressure, which prevents them from expanding their lungs when breathing from an air supply. Shallow-diving equipment (scuba gear) raises the pressure of gas that the diver breathes in order to resist the inward pressure, but works only to depths of 100 ft (30 m). Lower down, divers use long tubes called umbilicals, linked to a support ship. The umbilical supplies compressed air and also warm water that circulates inside the diving suit and insulates against cold. At even greater depths, down to 2,300 ft (700 m), divers use rigid atmospheric diving systems that allow them to breathe at normal pressure.

Atmospheric diving system (ADS)
An ADS is an all-encasing suit of rigid armor that may weigh up to a ton out of the water but that floats with neutral buoyancy once submerged. The diver maneuvers using rear-mounted propellers controlled by foot pedals and operates manipulators at the end of the suit's arms using hand controls. The suit has internal heating and an air supply so that the diver is completely isolated from the hostile environment.

How submersibles work

The Deep Rover 1002, a modern submersible, is designed to carry a crew of two to depths of 3,300 ft (1,000 m), but is safe to greater depths. The entire craft is designed around a spherical pressure hull, the only part of the vehicle that is kept at normal atmospheric pressure when underwater. The submersible is steered using four propeller thrusters—two on each side—and carries an array of television cameras and other equipment.

Body of submersible The main hull contains life support, fire suppression, navigation, maneuvering, buoyancy, and communications equipment

Acrylic pressure hull The pressure hull is a sphere—the best shape to resist water pressure—and will remain intact to depths of 13,000 ft (4,000 m)

Access hatch Crew enter the pressure hull through the bottom of the submersible

Lamp Powerful lamps illuminate the space around the submersible

Cockpit control Control joysticks feed back information from the manipulator arm sensors

"Vertrans" thruster A second pair of propellers helps maneuver the submersible in the water

Cockpit Instruments are displayed electronically on a computer screen, maximizing space inside the spherical pressure hull

Storage A locker stores samples and tools used by the manipulator arms

Horizontal thruster The submersible is pushed through the water by propellers. The relative speeds of the left and right thrusters are varied to change direction

Batteries Batteries provide power to the submersible

Outer hull The outer hull of the submersible can be detached, allowing the pressure sphere to rise safely to the surface

Trim and ballast tank A pair of tanks on either side of the submersible adjust its buoyancy so that it sinks or rises. Air-filled tanks would collapse under high pressure, so submersibles use kerosene-filled tanks to aid buoyancy

Remotely operated vehicles (ROVs)

ROVs are relatively small robotic vessels that carry out various underwater tasks under the command of an operator at a distance. They have a wide range of uses, from oil-rig and pipeline repairs to seabed mapping and exploration, often using sonar—an equivalent of radar that uses sound waves rather than radio waves. The ROV Argo, pictured here, was used to discover the wreck of the Titanic in the Atlantic Ocean in 1985.

SEE ALSO: ROBOTICS *p.210* | ASTRONAUTS' EQUIPMENT *p.242* | MECHANICS *p.270*

Composite materials

Composites are made by combining two or more materials in order to maximize their useful properties and minimize their weaknesses. One of the earliest composites was developed by the Ancient Greeks, who inserted iron rods into marble to strengthen it. Steel rods are now used in this way to reinforce concrete. Modern composites were developed as light and strong materials for the aerospace industry in the mid-20th century and have now found their way into a wide range of products, from stealth bombers to hockey sticks.

Graphite tennis racket
The strength and lightness of composites make them popular in the design of sports equipment. Tennis rackets are typically made from graphite-based composites reinforced with carbon fibers. The fibers are aligned specifically to reduce bending and twisting and improve stability.

What is a composite?

Most modern composites consist of fibers of one material tightly bound into another material called a matrix. The matrix binds the fibers together like an adhesive and makes them more resistant to damage, while the fibers make the matrix stronger and stiffer and help it resist cracks and fractures. The fibers are typically made of glass, carbon, silicon carbide, or asbestos, while the matrix is usually composed of a plastic polymer, a metal, or a ceramic material.

These three types of matrices produce three common types of composite. Polymer-matrix composites (PMCs), which are the most widely used, include glass-reinforced plastic (GRP). GRP combines glass fibers, which are strong but brittle, with plastic, which is flexible, to make a strong, flexible composite. Metal-matrix composites (MMCs) typically use silicon-carbide fibers embedded in a matrix made from an alloy of aluminum and mag-

nesium, but other matrix materials, such as titanium, copper, and iron, are increasingly being used. Typical applications of MMCs include bicycles, golf clubs, and missile-guidance systems. An MMC made from silicon-carbide fibers in a titanium matrix is currently being developed for use as the fuselage material for the U.S. National Aerospace Plane, which is planned to replace the Space Shuttle. Ceramic-matrix composites (CMCs) are the third major type, and examples include silicon-carbide fibers fixed in a matrix made from a borosilicate glass. The ceramic matrix makes them particularly suitable for use in components that must operate at high temperatures, such as parts for jet engines.

Making composites

Objects made from glass-reinforced plastic are often made using composite tapes. A long fiber is drawn through a tank containing the resin that will act as the matrix. The coated fiber is then pressed onto backing tape and formed into long sheets or continuous rolls called laminates. These tapes are particularly strong along one axis. Another kind of laminate can be used in which the fibers have been chopped up before they are stuck to the backing tape. Although this type of laminate is weaker than a continuous layer composite along its strong axis, it has the advantage of being equally strong in all directions.

Other composites can be made by heating two materials together so that they partially melt and bind together. A new type of composite that combines ceramic and metal, called cermet, is made by heating an aluminum block with a block of boron-carbide ceramic to around 1,470°F (800°C), so that the aluminum melts and penetrates the pores in the ceramic.

The world of composites

The strength and lightness of composites makes them particularly useful for transportation. From the $42 million B-2 "Stealth" Bomber to low-cost, homemade airplane kits, composites have

Continuous fiber laminates

In continuous-fiber laminates, long strands of fiber are laid parallel to each other in a resin-like matrix and pressed onto backing sheets. These sheets have great strength in the direction of the fibers but are weak in the other direction. In order to achieve multidirectional strength, each sheet is laid at 45 degrees to the adjacent layer.

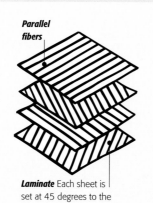

Parallel fibers

Laminate Each sheet is set at 45 degrees to the adjacent layer

made airplanes lighter, more economical, and more affordable, and solved problems such as cracking and metal fatigue. Composites have also made possible new craft called tilt-rotors—airplanes with a swiveling propeller at the end of each wing that can hover or take off vertically like a helicopter. With traditional materials such as aluminum, craft of this sort would be too heavy to carry their cargo. In some unusual ways, space rockets and satellites are also benefiting from composites. Instead of having fuel tanks that must be jettisoned partway through a mission, the next generation of spacecraft may have tanks made from composites that can themselves be burned up as fuel.

Composites are not only useful for aircraft. Automobiles of the future must be safer, more economical, and more environmentally friendly, and composites could help achieve all three aims. Although composites such as GRP have been used to make spare parts since the 1950s, most automobiles are still made from steel. Carefully designed composites could cut the weight of a typical automobile by 40 percent, increasing fuel economy by as much as a quarter, yet maintaining body strength and crashworthiness. High-temperature ceramic-matrix composites are also helping to make cleaner-burning, more fuel-efficient engines.

Composites are so versatile that they are now even used to build large-scale structures. Plans are being drawn up to build a 460-ft (140-m) long bridge in San Diego, California, with a deck made of fiber-reinforced plastic composites that is estimated to be one-fifth the weight of a concrete deck. NASA scientists and industry engineers are currently developing composites that could replace metals in offshore oil platforms and pipelines. Once developed, this technology is expected to yield extra-durable pipes that could be used for everyday applications, such as sewage disposal.

Bicycle wheels

Mountain bike wheels may be made from aerospace-grade carbon fibers in a nylon polymer matrix for toughness and lightness. Even more advanced Kevlar composites are used to make light and superrigid wheels with three or four spokes used in Olympic racing cycles.

Racing yacht

Laminates made from composites such as GRP have their fibers aligned in a single direction. To ensure a boat hull is equally strong in all directions, many layers of laminate tape are laid into a mold so that their fibers align in different directions. The layers are then covered in hard-setting resin.

SEE ALSO: STEEL & CONCRETE *p.24* | AIRLINERS *p.126* | PLASTICS *p.184* | CERAMICS *p.206*

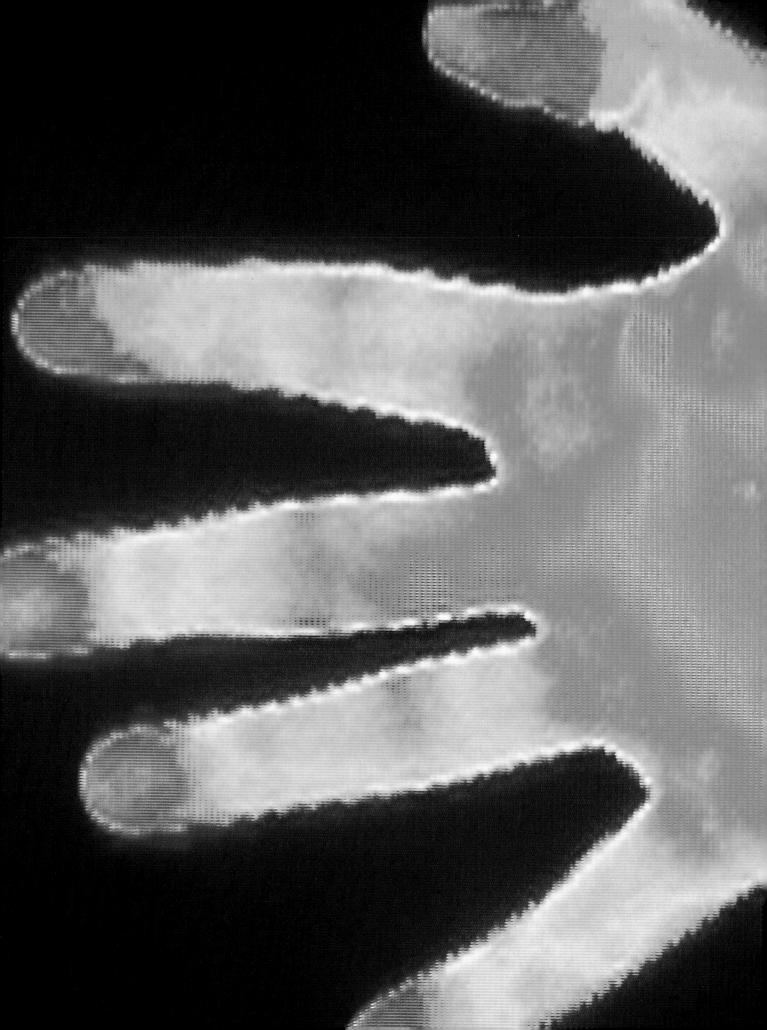

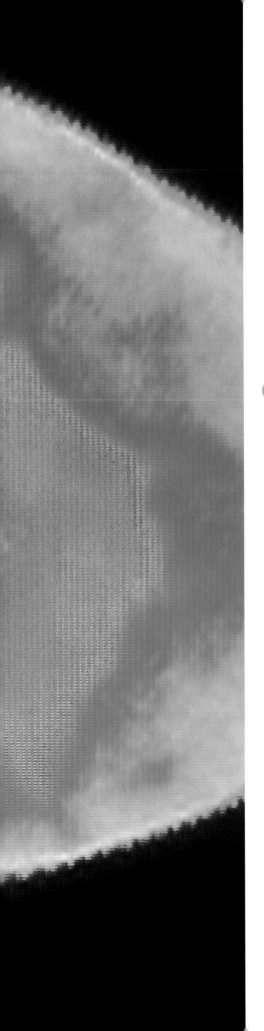

Crime and Security

One problem with the development of our increasingly technological society is that the opportunities for crime and, in particular, fraud have greatly increased. The main concerns of previous generations were burglary and counterfeiting, but the arrival of global telecommunications and increasing reliance on computers rather than people have given rise to whole new fields of crime, ranging from credit-card fraud to computer hacking. However, as this chapter shows, science and invention can also help combat the problem, increasing both our personal safety and the security of the systems we rely on every day, as well as helping to trace and identify suspects after a crime has been committed.

Locks

A lock of any kind is basically a smart bolt—one which, unlike a simple manual bolt or bar, can be locked or unlocked only with the correct key. Mechanical locks, which rely solely on ingenious arrangements of pins, levers, cylinders, and springs, remain the standard type in most homes, schools, and places of work; they are only gradually being replaced by electronic locks.

Many familiar types of mechanical locks in use today have their origins going back several hundreds or even thousands of years. The pin-tumbler mechanism seen in cylindrical locks—today the most widespread lock in the world—was first used in the Middle East some 4,000 years ago. The first lever lock was patented in 1778 by an English locksmith, Robert Barron, to improve on the very moderate security provided by the existing locks at that time. The Bramah lock was also invented in principle during the late 18th century and was claimed to be unpickable. This remained the case for 50 years. It is still in use today where high levels of security are required and for items that are left in the open and vulnerable to thieves, such as bicycles.

Triple lever lock

All lever locks are secured by a series of differently shaped levers, or tumblers. This example has three tumblers, each containing a rectangular slot. A bar attached to the bolt (called the bolt pin) passes through the slots in the tumblers. In the resting position (locked or unlocked), the bolt pin is held within notches adjacent to stumps that project downward into the slots. This prevents any movement of the bolt.

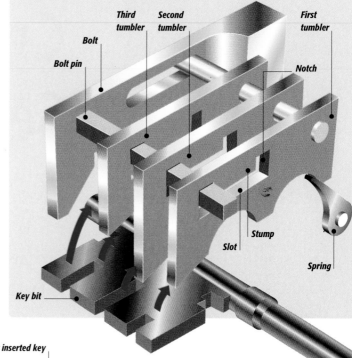

Third tumbler · Second tumbler · First tumbler · Bolt · Bolt pin · Notch · Stump · Slot · Spring · Key bit

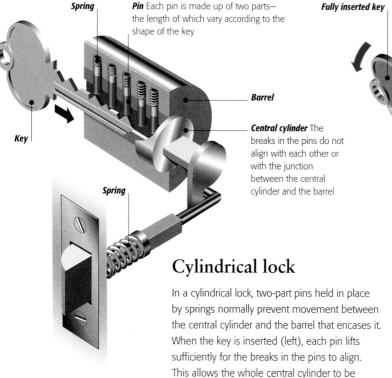

Spring

Pin Each pin is made up of two parts—the length of which vary according to the shape of the key

Barrel

Central cylinder The breaks in the pins do not align with each other or with the junction between the central cylinder and the barrel

Key

Spring

Fully inserted key

Aligned pins The breaks in the pins are all aligned with the junction between the central cylinder and the barrel, allowing the key to rotate the cylinder

Tailpiece

Cam The cam turns with the cylinder, engaging the bolt

Spring

Bolt The bolt is pulled back against the resistance of a spring

Cylindrical lock

In a cylindrical lock, two-part pins held in place by springs normally prevent movement between the central cylinder and the barrel that encases it. When the key is inserted (left), each pin lifts sufficiently for the breaks in the pins to align. This allows the whole central cylinder to be turned (right), pulling back the bolt against the resistance of a spring.

KEY INSERTION

UNLOCKING

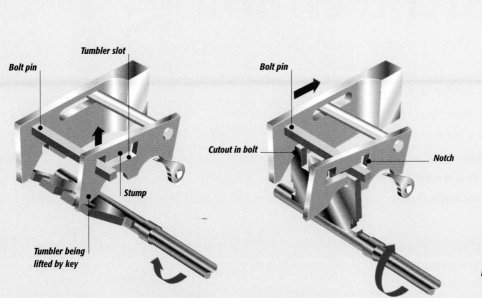

Bolt pin

Tumbler slot

Bolt pin

Bolt fully engaged in locked position

Cutout in bolt

Notch

Bolt pin secured in notch

Stump

Tumbler being lifted by key

Tumbler in resting position

Locking–stage 1
For the key to be turned, it must lift each of the tumblers sufficiently to permit the bolt pin to clear the stumps in the tumbler slots (only one of the tumblers is shown here). Because each tumbler is shaped differently, the key bit must have a unique shape to achieve this.

Locking–stage 2
Once the key has turned a little way, it engages a cutout in the bolt. As the key continues to turn, the bolt shifts toward the locked position (to the right in this example). The bolt pin moves through the slots in the tumblers toward the notches at their far ends.

Locking–stage 3
When the key has been fully turned, the bolt finishes fully engaged in the locked position (to the right in this example). The bolt pin now fits into the notches at the end of the tumbler slots as the tumblers fall back under gravity into their resting positions.

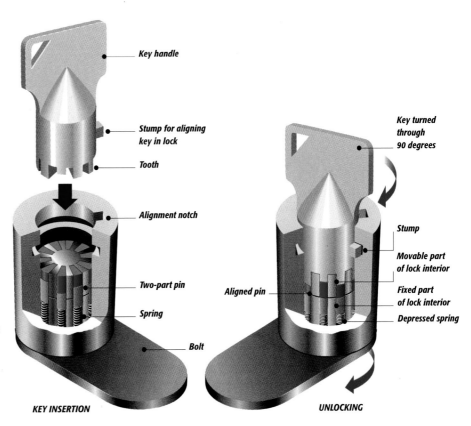

Key handle

Stump for aligning key in lock

Tooth

Alignment notch

Two-part pin

Spring

Bolt

KEY INSERTION

Key turned through 90 degrees

Stump

Movable part of lock interior

Aligned pin

Fixed part of lock interior

Depressed spring

UNLOCKING

Bramah-type lock

In this lock, the key is a hollow cylindrical tube, open at one end. Teeth are machined around the open edge. When the key is inserted into the lock, the teeth push down two-part pins against the resistance of springs. When all the pins are precisely depressed, the breaks in the pins align and the key can be turned. This action rotates a central cylinder and opens the attached bolt.

SEE ALSO: HISTORY OF CODES & ENCRYPTION *p.154* | COMPUTER SECURITY *p.156* | MECHANICS *p.266*

Plastic card technology

Until the 1960s, most financial transactions were based on paper (cash and checks). Today, the majority are electronic, and of these, a large proportion involve plastic cards. There are two main types of cards—cards with a magnetic strip, which include ATM (automated teller machine), credit, and swipe cards—and the more sophisticated "smart" cards. Both types are also used for security purposes.

The first step in any card transaction is validation of its use by the card holder. With magnetic strip cards, this process is performed by a remote computer once it receives the card number from the ATM or other card reader. For a swipe card giving entry to a gym, the only necessary checks may be that the card has not expired or been reported stolen. For a credit card, an additional check may be made that the user has not exceeded a credit limit. The user must also supply a signature. For an ATM card, the user enters a PIN (personal identification number) before transactions can occur. Smart cards differ in that the process of approving transactions is performed by a program on the card itself—the card knows the user's PIN. Smart cards are designed to hold confidential data in a secure form. In contrast, the data on a magnetic strip card is usually not secret at all.

Magnetic strip cards

Credit, ATM, and swipe cards all operate by means of the magnetic strip on the back. It holds about 200 bytes of data—usually the card number, owner's name, and expiration date. Each data bit is held in the strip as a pair of magnetic domains. A pair of domains pointing in the same direction represent a "0," those pointing in opposite directions, a "1."

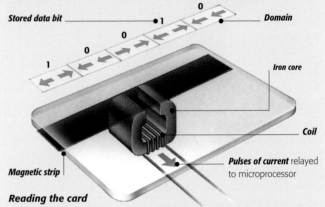

Reading the card
As the magnetic strip passes through the reader, the domains in the strip induce pulses of electric current in the coil wound around an iron core. The pulses are turned into binary code by a microprocessor.

Smart cards

A smart card is a credit card-sized piece of plastic containing an embedded integrated circuit chip. The chip provides the card with both computational capacity and memory. Data on the card is protected by advanced security features. A contact smart card is used by being inserted into a reader where surface contacts on the card interface with electrical connectors, causing data to be transferred. Contactless cards are passed near an antenna to carry out a transaction and data is transmitted by a weak electromagnetic signal. They are ideal for applications that require rapid processing, such as payment of bus or subway fares.

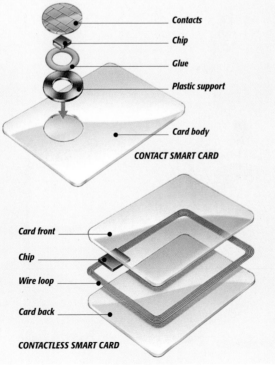

CONTACT SMART CARD

Card front
Chip
Wire loop
Card back

CONTACTLESS SMART CARD

Smart card applications
Although smart cards are most often used today as "electronic wallets," envisaged growth areas include security applications and storage of personal data.

Money Money units can be downloaded to the card from a bank account and spent at retail outlets or on buses/trains.

Medical and personal data Information such as blood group, driver's license, and passport number may be stored.

Biometrics To prove identity and so verify other stored data, a card may hold codes based on biometrics such as fingerprints.

Telephony Telephone credits, phone numbers, and passwords for Internet access via mobile phones may be stored.

Retailer loyalty schemes Programs to operate complex pricing systems based on retailer loyalty can be put onto a card.

Encryption/decryption A card may contain programs that can encrypt/decrypt data for secure Internet transmission.

Automated teller machine

A fully featured ATM can dispense bills, take deposits, answer balance inquiries, and handle fund transfers. The ATM's operation is controlled by its microprocessor, but tasks such as customer identification, account balance verification, and the issuing of transaction instructions are handled over a cable link by a computer belonging to the ATM owner (host).

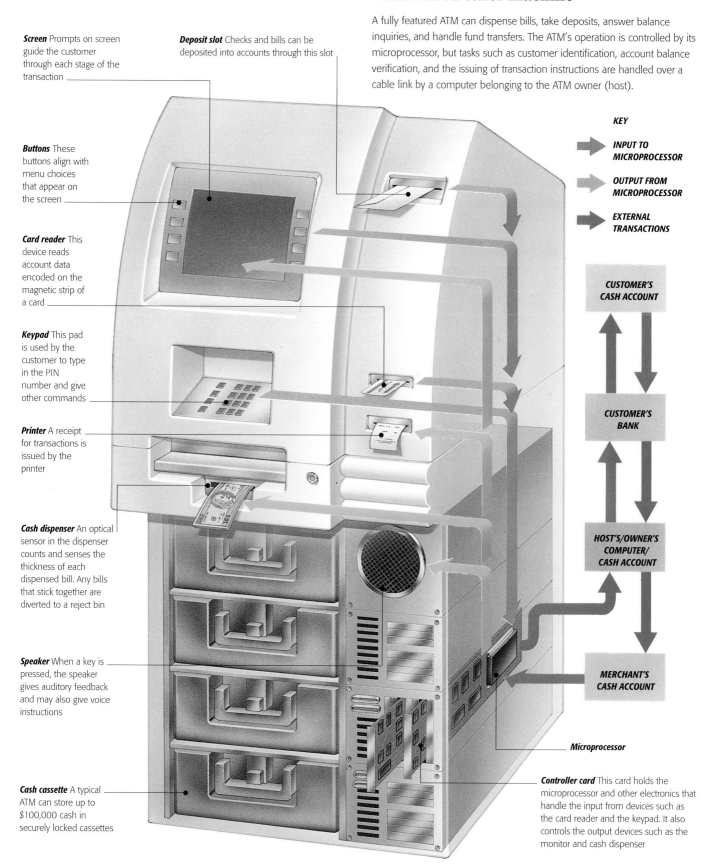

Screen Prompts on screen guide the customer through each stage of the transaction

Deposit slot Checks and bills can be deposited into accounts through this slot

Buttons These buttons align with menu choices that appear on the screen

Card reader This device reads account data encoded on the magnetic strip of a card

Keypad This pad is used by the customer to type in the PIN number and give other commands

Printer A receipt for transactions is issued by the printer

Cash dispenser An optical sensor in the dispenser counts and senses the thickness of each dispensed bill. Any bills that stick together are diverted to a reject bin

Speaker When a key is pressed, the speaker gives auditory feedback and may also give voice instructions

Cash cassette A typical ATM can store up to $100,000 cash in securely locked cassettes

KEY

➤ **INPUT TO MICROPROCESSOR**

➤ **OUTPUT FROM MICROPROCESSOR**

➤ **EXTERNAL TRANSACTIONS**

CUSTOMER'S CASH ACCOUNT

CUSTOMER'S BANK

HOST'S/OWNER'S COMPUTER/ CASH ACCOUNT

MERCHANT'S CASH ACCOUNT

Microprocessor

Controller card This card holds the microprocessor and other electronics that handle the input from devices such as the card reader and the keypad. It also controls the output devices such as the monitor and cash dispenser

SEE ALSO: COMPUTER MEMORY p.66 | COUNTERFEIT PROTECTION p.152 | COMPUTER SECURITY p.156

Counterfeit protection

Manufacturing paper money is a complex and secretive business, reflecting the fact that currency bills have always been an obvious target for thieves and counterfeiters. Most aspects of paper money design are aimed at rendering the bills as difficult as possible to forge. This means using special paper and inks, a range of different printing techniques, and innovative devices such as holograms.

Fully printed currency bills have only been in circulation since about 1850. Before then, part-printed notes, which had to be signed like checks, circulated in some major financial centers. Bill printing involves several stages, using different techniques. The first of these, used to print most features of each bill, is offset printing—the same process used to print books and magazines. Next is intaglio printing, which involves pressing an engraved metal plate against the paper and is used to add the more finely detailed parts of the design, giving them a distinctive "raised" feel. Special features such as holograms may then be added, and finally the serial numbers are stamped onto the bills. In Australia, plastic bills, which last longer and are more difficult to copy, are issued.

Despite the sophistication of modern bill protection technology, counterfeiting remains a problem—and it has become more of a threat in some countries since the introduction of high-quality color photocopiers. Aside from trying to stay "one step ahead" of counterfeiters in technical terms, the principal weapon used by governments is to punish offenders with fines and heavy prison sentences.

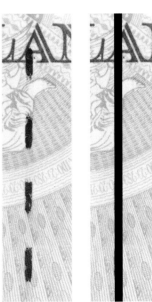

$100 BILL, FRONT

English £20 note

The £20 note shown below was first issued in 1999. As well as the security features highlighted, the note contains a watermark; a holographic foil on the front that shows the figure "20" or an image of Britannia according to the angle of view; raised (intaglio) printing for words such as "Bank of England," and serial numbers that vary in size and color.

U.S. $100 bill

The $100 bill shown right was first issued in 1996. In addition to the security devices depicted, the bill has fine-line printing in the background to the portrait and special "never-dry" ink is utilized. Viewed from the front, a watermark is located at the far right of the bill.

£20 NOTE, FRONT

£20 NOTE, BACK

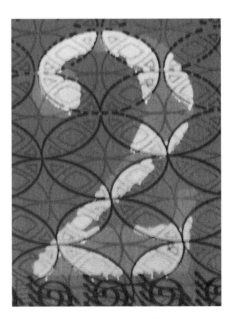

Fluorescent feature When the front of the note is viewed under ultraviolet light, a fluorescent "20" can be seen in an area toward the lower left corner. This feature is not apparent in normal light

Metallic thread A thread is embedded in the paper, showing as a series of silver dashes on the reverse of the note (left). When held up to the light, the thread shows as a solid dark line (right)

Thread A polymer thread is embedded vertically in the bill and indicates, by its position, the bill's denomination. The words "USA 100" can be seen on the thread when held up to a bright light

Microprinting The words "THE UNITED STATES OF AMERICA" are printed in tiny capital letters along Benjamin Franklin's coat lapel, and "USA 100" is printed repeatedly within the "100" at bottom left

Paper The paper for this bill is made of cotton and linen and has a strong but pliable feel. It has hundreds of tiny red and blue fibers embedded within it

Fine detail Most currency bills include portraits or natural objects as part of the design, providing ample opportunity for incorporating fine details, such as hair or mustaches, that are difficult to counterfeit

Color shifting ink The "100" in the lower right corner appears green when viewed straight on but looks black when viewed from an angle

Detecting counterfeit bills

A small proportion of the currency bills circulating in most countries are counterfeit. Any counterfeit bill presented to a bank is liable to be confiscated without compensation, so it pays to be on guard. Spotting a counterfeit in broad daylight is usually quite easy, but detection in an environment such as a darkened bar can be more difficult. Good-quality color photocopiers can produce quite convincing counterfeits. Such bills used to reflect a violet sheen under ultraviolet light, but counterfeiters have now switched to using UV-dull paper. The best solution is to be aware of all the security features on the bills you handle regularly, and check for those features in any bills you are handed. Also pay particular attention to the feel of the bills.

History of codes and encryption

Encryption is the use of codes and ciphers to scramble messages so that they can be read only by the intended recipients (who are made aware of the decoding method in advance). Encryption for military and espionage purposes has a long history. Today, it is also used for guarding commercial secrets and for providing secure data transmission over the Internet.

Simple substitution ciphers

One of the earliest code makers was Julius Caesar. He is said to have written to friends using a simple substitution cipher, in which each letter was replaced by a letter three places further down the alphabet. So if he'd lived in the modern era, he might have encoded his name as MXOLXV FDHVDU. Any cipher in which there is a fixed alphabetic distance between the normal letter and the cipher letter is now known as a Caesar cipher. Ciphers of this type may be "cracked" in a few minutes because there are only 26 possible coding permutations to investigate. Slightly more difficult

Cipher disk

This disk, made during the Civil War, was used to encode messages based on a Caesar-type cipher. The outer ring was rotated to a position that aligned each letter in the outer ring to one in the inner ring. Each letter in the message was then found in the outer ring and encoded by the adjoining letter in the inner ring.

to break are ciphers in which the cipher letters are not a fixed distance from the plain letters but are alphabetically scrambled. These can still be cracked by seasoned code breakers within a few hours, by analysis of relative letter frequencies in the enciphered message.

Polyalphabetic ciphers

Once it was realized how easily simple substitution ciphers could be broken, cryptographers devised stronger ciphers called polyalphabetic substitution ciphers, in which a letter of the alphabet in the plain text could be represented by different letters in the cipher text, depending on its position within the message. These ciphers use a "key" (often a number) that both the encrypter and recipient need to know. The key is used to translate the letters of the plain message into the cipher message and to decode the cipher message. With this cipher, a particular letter in the plain message may be represented by any of several different letters in the cipher message. Although ciphers of this sort are more difficult to crack than simple substitutions, there are methods for breaking them. For example, the first step is to find the length of the key. To do this, the cipher message is shifted against itself and the number of matches between letters is counted. A shift that is a multiple of the key length will give a high number of matches.

The Enigma code

Perhaps the most famous cipher-making apparatus of all time is the Enigma machine used by Germany during World War II. The machine was basically a means of encoding a text message using a polyalphabetic substitution cipher and a scrambling key of extreme complexity. The Enigma "scrambler" consisted of three rotors at the heart of the machine. Each rotor was a disk with a ring of electrical connectors on each side. A mass of wiring within

Use of a key in a polyalphabetic cipher

In this example, the key is the number 46215. The digits of this number are written repeatedly beneath the letters of the plain message. Each letter in the plain message is then shifted alphabetically by the number underneath it to give a cipher letter. So, VENI VIDI VICI becomes ZKPJ AMJK WNGO.

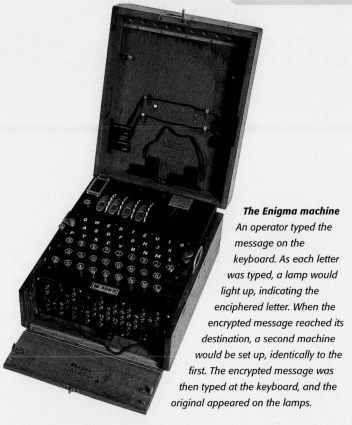

The Enigma machine
An operator typed the message on the keyboard. As each letter was typed, a lamp would light up, indicating the enciphered letter. When the encrypted message reached its destination, a second machine would be set up, identically to the first. The encrypted message was then typed at the keyboard, and the original appeared on the lamps.

each rotor carried an electrical signal arriving at a connector on one side a rotor to another connector on the other side of that rotor, and then on to the next rotor.

When a key on the Enigma machine was pressed, a signal passed through the three rotors and back again by a circuitous route to one of 26 lamps. After each key press, one or more of the rotors progressed a "notch" and the wiring connections changed. To add to the complexity of the machine, each rotor could have 26 starting positions, and a part of the machine called the plugboard could be used to introduce further permutations. Although there were one thousand billion billion different combinations of settings, the Allies managed to crack the Enigma code in 1943 with help from an early computer called Colossus.

Encryption for the Internet

Today, encryption continues to be used in many areas of life, but a difficulty arises when the people who are passing information to each other cannot meet to agree the "key" to their system. This is a particular problem for transmission of data over the Internet—for example, credit card numbers in E-commerce transactions.

One method that has been devised for solving this difficulty is public key encryption. Under this system, anyone requiring data to be sent to them in encrypted form requires two "keys," called the public key and the private key. These keys are strings of digits based on very large prime numbers and are generated using special software. They can be used to encrypt and decrypt data in conjunction with some math-based software. An individual's public key is published widely and can be used by anyone wishing to send confidential information to that person. However, the public key cannot be used to decrypt data. The person must use a private key—known only to him- or herself—to perform the decoding.

Public key cryptosystem

This diagram shows the basic operation of a public key cryptosystem. When Susan wants to send a secret message to Frank, she uses his public key (which he's sent to her or advertised on a web site) to encrypt the data. When Frank receives the message, he uses his private key to decrypt it. For additional security, Susan may insist on a trustworthy authority verifying that Frank's public key really is his before she uses it. This protects against spies and fraudsters who may attempt to issue public keys in other people's names and then intercept their mail.

1 Susan composes her secret e-mail message to Frank in the normal way.

2 To encrypt the message, Susan uses Frank's public key together with readily available encryption software.

3 Susan addresses the coded message to Frank and then sends it in the normal way.

4 The message travels to Frank via the Internet. If anyone intercepts it, it can't be deciphered, even if his public key is known.

5 Frank decrypts the message using a private key known only to himself and readily available software.

6 Frank can now read the secret message. To reply secretly to Susan, he first needs to obtain her public key.

SEE ALSO: HISTORY OF COMPUTERS *p.52* | INTERNET & E-MAIL *p.70* | WORLD WIDE WEB *p.72*

Computer security

Computer security systems exist to combat computer crime, which has been around as long as computers. The motives for computer crime are diverse and include: fraud, theft, blackmail, espionage, sabotage, and revenge, as well as self-aggrandizement. Three of the most common forms of computer crime are attacks by hackers, the creation of viruses, and the interception of private E-mail.

Hackers are people who illegally break into private, corporate, or government computer systems, usually via the Internet, in order to alter files, remove confidential data, or simply for the challenge. A computer virus is a program that runs automatically, causing disruption by using up memory, affecting stored data, and even infecting application programs. Viruses replicate themselves and spread rapidly through computer networks.

There are ways to make a network more resistant to such attacks. The first line of defense is the firewall, a software or hardware device that acts as a gatekeeper between the Internet and a company's internal network. A firewall is often run in conjunction with a proxy server—a program that helps match incoming data packets with user requests to the Internet. Within the firewall-protected zone, sensitive data can be passed around, and workers can feel reasonably free from hacker attack. But a firewall does not protect against viruses, which must be guarded against by running antivirus software on individual computers.

A third category of computer crime is the interception of E-mail and other Internet traffic. With the growth of e-commerce, which can involve transmitting credit-card data over the Internet, this is of increasing concern. Sensitive data can now be encrypted before it is transmitted over the Internet, using public and private key cryptographic systems, for which special software exists.

Battling the hacker

In the scenario shown here, firewalls protect the more sensitive areas of a company's computer network. Unfortunately, a hacker has accessed the network by a "backdoor route," because an employee has left a PC plugged into a telephone socket. This allows the hacker to implant hacking tools onto the network and start downloading confidential data.

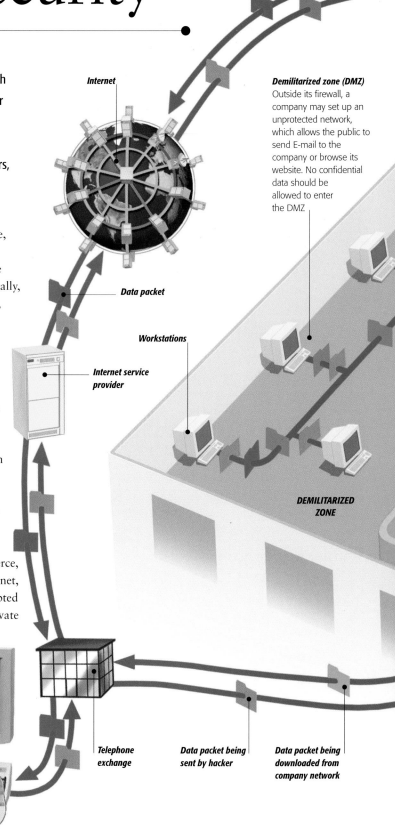

Internet

Demilitarized zone (DMZ)
Outside its firewall, a company may set up an unprotected network, which allows the public to send E-mail to the company or browse its website. No confidential data should be allowed to enter the DMZ

Data packet

Workstations

Internet service provider

DEMILITARIZED ZONE

Telephone exchange

Data packet being sent by hacker

Data packet being downloaded from company network

Hacker

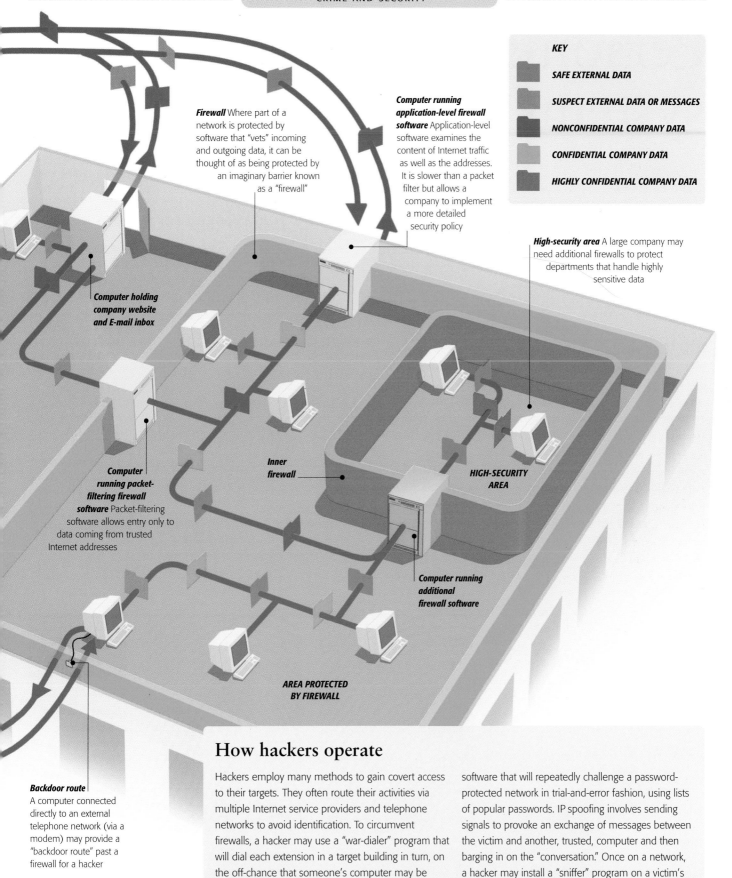

KEY

SAFE EXTERNAL DATA

SUSPECT EXTERNAL DATA OR MESSAGES

NONCONFIDENTIAL COMPANY DATA

CONFIDENTIAL COMPANY DATA

HIGHLY CONFIDENTIAL COMPANY DATA

Firewall Where part of a network is protected by software that "vets" incoming and outgoing data, it can be thought of as being protected by an imaginary barrier known as a "firewall"

Computer running application-level firewall software Application-level software examines the content of Internet traffic as well as the addresses. It is slower than a packet filter but allows a company to implement a more detailed security policy

High-security area A large company may need additional firewalls to protect departments that handle highly sensitive data

Computer holding company website and E-mail inbox

Computer running packet-filtering firewall software Packet-filtering software allows entry only to data coming from trusted Internet addresses

Inner firewall

HIGH-SECURITY AREA

Computer running additional firewall software

Backdoor route A computer connected directly to an external telephone network (via a modem) may provide a "backdoor route" past a firewall for a hacker

AREA PROTECTED BY FIREWALL

How hackers operate

Hackers employ many methods to gain covert access to their targets. They often route their activities via multiple Internet service providers and telephone networks to avoid identification. To circumvent firewalls, a hacker may use a "war-dialer" program that will dial each extension in a target building in turn, on the off-chance that someone's computer may be connected to a telephone socket. To access individual PCs or networks, they may use password-guessers—software that will repeatedly challenge a password-protected network in trial-and-error fashion, using lists of popular passwords. IP spoofing involves sending signals to provoke an exchange of messages between the victim and another, trusted, computer and then barging in on the "conversation." Once on a network, a hacker may install a "sniffer" program on a victim's hard disk that will record all keyboard and network activity, helping the hacker discover more passwords.

SEE ALSO: INTERNET & E-MAIL *p.70* | WORLD WIDE WEB *p.72* | HISTORY OF CODES AND ENCRYPTION *p.154*

Person-recognition systems

Person-recognition systems are a major spin-off from improvements in computing power, together with research into biometrics (the measurement of biological characteristics) and the development of software to exploit this research, during the 1980s and 1990s. These systems work by measuring features that are unique to an individual and then using the data for security purposes.

In broad terms, all person-recognition systems work in the same way. A user must first be enrolled onto the system by either speaking into a microphone or by having a feature, such as an eye or a hand, scanned, analyzed, and turned into a digital code. The code is stored on a database, sometimes with a supplementary identifier, such as a personal identification number (PIN). When the user wishes to gain access to a protected environment, such as a building or a computer network, the scan is repeated and a second code generated. This code may then be compared either with a particular stored code (by reference to the additional identifier supplied by the user) or with all the codes on the database, which takes a little longer but requires no other identifier. Modern systems have very low false-accept and false-reject rates, which are the main indicators of their effectiveness.

Hand-based recognition systems

Hand-recognition systems involve the user putting a hand, normally palm down, into a reader for a second or two. A light or infrared source within the reader projects images of the hand that are captured as silhouettes by a high-resolution digital camera. Sophisticated software analyzes the images to produce an output, which is compared with a mathematical template recorded during an enrollment session.

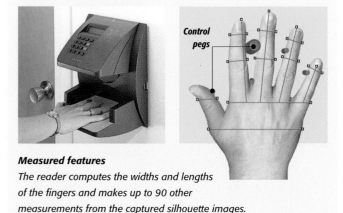

Control pegs

Measured features
The reader computes the widths and lengths of the fingers and makes up to 90 other measurements from the captured silhouette images.

Fingerprint system

Several different fingerprint-recognition systems have been devised. The system shown here produces an image by detecting heat variations. Other systems are based on straightforward optical imaging or on measuring small electrical variations across the finger's surface.

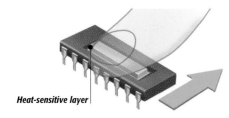

Heat-sensitive layer

1 *The user sweeps his or her finger across a sensor, which is a single microchip topped by a heat-sensitive layer that contains 14,000 imaging elements.*

2 *The chip converts the tiny heat variations it detects into a series of image slices, each showing a fingerprint ridge pattern. About 50–100 slices are generated.*

3 *Special software reconstructs the slices into a full image in 0.1 seconds. The image may be displayed on a monitor and saved into a database.*

```
0001111101010000 1
1100000011101010 0
1110111101010000 1
0000010001010010
```

4 *For security systems, a computer further processes the image using complex algorithms and generates a digital identification code.*

Iris-based recognition systems

The iris is the colored ring of tissue surrounding the pupil of the eye. Every iris has a unique pattern of features—striations, freckles, and so on—that remains constant throughout life and cannot be faked or mimicked. The iris is thus a highly effective personal identifier.

Enrollment

During enrollment, the unique features of a person's iris are analyzed. The data is then stored on a database as a digital code.

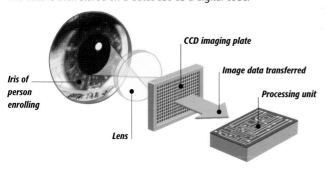

CCD imaging plate

Image data transferred

Processing unit

Iris of person enrolling

Lens

1 *A series of images of the eye is captured over several seconds by a device similar to a camcorder. The device contains a lens that projects the images onto a CCD (charge-coupled device). Analog image data from the CCD is digitized and sent to a processing unit for analysis.*

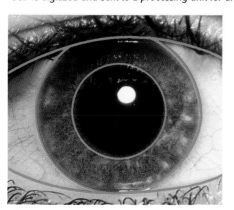

2 *Special software processes the images. First it identifies and removes any eyelid data, then it locates the borders between iris, pupil, and white of the eye, to isolate the data relating solely to the iris.*

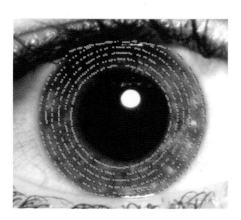

3 *The program maps the data to points on the iris, using a coordinate system that takes into account the degree of constriction of the pupil. Brightness data is gathered for each point and analyzed.*

4 *The generated code is a 512-byte binary number (several thousand 1s and 0s).*

`00111110101000011100000011101010011`
`10101000011100000011101010001111010`

Recognition

During recognition, the same scanning procedure is conducted as in enrollment and a digital code is generated. This code is then compared with the digital codes stored on the database.

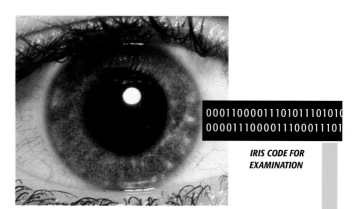

`0001100001110101110101`
`0000111000011100011101`

IRIS CODE FOR EXAMINATION

1 *A person requiring security clearance has his or her iris scanned in the same way as during the enrollment process.*

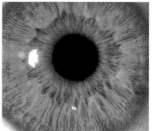

`000111110101000`
`1110000001110101`

NONMATCHING CODE

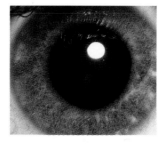

`0001100001110101`
`0000111000011100`

MATCHING ENROLLED IRIS CODE

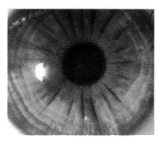

`001010100001101`
`0001110101001110`

NONMATCHING CODE

2 *The generated code is now compared with the database codes, to seek a match. An exact match is not needed. Iris codes are so different that even a 75 percent degree of matching between two codes provides compelling evidence that they represent the same iris. The chances that two such codes are from different people has been calculated to be 1 in 10 billion billion.*

SEE ALSO: DIGITAL DATA *p.56* | SCANNING DEVICES *p.64* | CAMERAS *p.92* | VIDEO TECHNOLOGY *p.104*

Suspect identification

A number of technologies can assist in identifying the perpetrators of crime. Ordinary fingerprinting is the best known, but since its invention in the 1980s, DNA fingerprinting has proven even more valuable, especially for more serious crimes. In many cases it produces near-conclusive evidence of guilt or innocence. Computers are also increasingly used to help match crime evidence against vast databases of information on known criminals and their activities.

All techniques of suspect identification rely on finding a link or match between something left at the crime scene and the suspect. If the suspect was seen, then an attempt may be made to build a photocomposite of his or her face based on the witness's recollection. Physical evidence can often prove conclusive, and material left at the crime scene that can be used for identification ranges from palmprints and fingerprints to fibers, which can often be matched for color and other properties to the suspect's clothing. Bite marks found on a victim may also be matched to an assailant's dental record. But the most valuable evidence is biological material, such as blood, saliva, semen, or even a single hair, because it can be subjected to DNA analysis. DNA (deoxyribonucleic acid) is the hereditary material found in all body cells. In the 1980s, it was shown that certain sequences of bases, the subunits that make up DNA, varied significantly across the population, and that the overall pattern of these sequences was unique for each person (identical twins excepted). A single cell from a biological sample is sufficient for laboratory analysis of the pattern of base sequences in the cell's DNA, producing a DNA "fingerprint" of the person who left the sample.

How DNA fingerprinting is done

At the crime scene, samples are taken of any biological material that may have come from a criminal so that the sequence of bases in the sample's DNA can be analyzed. A DNA molecule consists of two intertwined strands that complement each other chemically. The presence of specific sequences in the test DNA can be detected by cutting it up, separating the strands, and adding complementary fragments of single-stranded DNA that will bind only to the sought-after sequences.

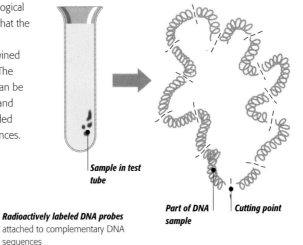

Sample in test tube

Part of DNA sample | **Cutting point**

1 DNA is extracted from the sample and cut at selected points, using substances called restriction enzymes. The DNA fragments are placed on a gel and separated using an electric field. This causes the fragments to align according to length. Thus, specific DNA sequences end up at specific locations on the gel.

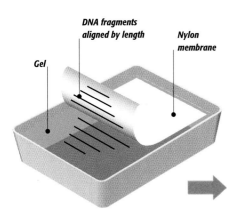

DNA fragments aligned by length   **Nylon membrane**

Gel

2 The DNA fragments are transferred to a nylon membrane and made single stranded. DNA probes are then added. These are radioactively labeled fragments of single-stranded DNA that bind to specific complementary DNA sequences if those sequences are present on the membrane.

Radioactively labeled DNA probes attached to complementary DNA sequences

3 The sheet is exposed to X-ray film which, when developed, provides a visual "barcode" pattern of the DNA fragments to which the probes have attached. This is the DNA "fingerprint" and it is compared with DNA fingerprints from suspects to see if there is a match. In the example shown here, the DNA fingerprint from one suspect closely matches the DNA fingerprint from the crime scene.

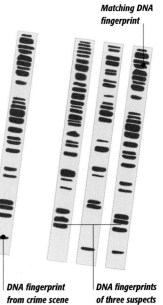

Matching DNA fingerprint

DNA fingerprint from crime scene   **DNA fingerprints of three suspects**

Creating a composite

Originally performed by overlaying sheets of clear film printed with different facial features, composites of crime suspects are now most often created on a computer. A number of different computer systems are used by police forces throughout the world. More sophisticated systems under development include features that model how a suspect might look after aging or that compare computer-created composites with databases of the faces of known criminals.

Building a picture
The sequence of images below provides an example of how a composite might be built up. The person that the witness was trying to recall (not a genuine criminal) is shown left.

Facial feature menus (hair, eyes, mouth, etc.)

Feature manipulation menu (lighten/ darken, rotate, distort, delete, etc.)

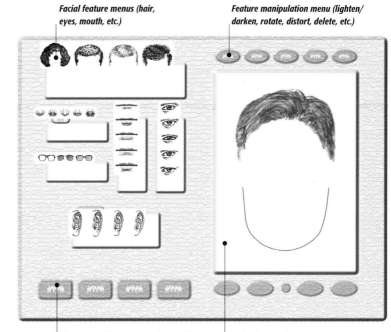

Main menu (Save, Load, Clear, Exit)

Composition area

Software interface
Programs typically allow the artist to select facial features from a database, combine them, and manipulate them.

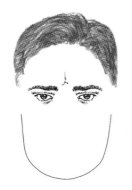

ADD EYES, EYEBROWS, AND HAIR

ALTER EYES AND CHIN

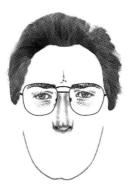

ADD NOSE AND GLASSES, ALTER HAIR

ADD EARS AND MOUTH, ALTER GLASSES

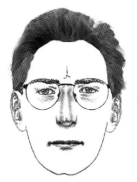

ENLARGE MOUTH

ADD FACIAL HAIR, ALTER EARS AND NOSE

REVISE FACIAL HAIR

GLASSES TINTED TO COMPLETE COMPOSITE

SEE ALSO: PERSON-RECOGNITION SYSTEMS *p.158* | FACIAL RECONSTRUCTION *p.162* | HUMAN GENOME PROJECT *p.214*

Facial reconstruction

Facial reconstruction has a tradition going back to the Victorian era, when the faces of individuals such as Bach were reconstructed from their skulls. Today, it is performed either forensically, in an attempt to identify otherwise unidentifiable victims of a crime or accident, or by archaeologists in an effort to show how long-dead people, such as Neanderthals, might have looked in life.

Manual facial reconstruction relies on applying average tissue-depth values at 21 specific landmark sites on the skull and building the face up by hand from clay—tissue depth is the distance between the skin and the underlying bone. During Victorian times scientists began collecting tissue-depth data from cadavers. This method is only now being gradually replaced with measurements taken from living people using techniques such as ultrasound scanning and MRI. In manual reconstruction, the face is physically constructed over the skull. New digital methods use digital models of actual skulls built up from 3-D laser scans.

It is a common misconception that facial reconstruction can produce an exact likeness. In fact, only a resemblance to the living person can be hoped for because variables such as obesity, age, ethnic group, and even gender can only be estimated from a skeleton. Forensic facial reconstructions are used to stimulate public interest as a last resort in identifying a body and so are left deliberately unembellished to allow viewers to impose their own interpretations. The reconstruction is presented in the media, and many names may be put forward. The investigating authorities work through the list of names, eliminating some. Identity is ultimately confirmed using dental records or DNA analysis.

Digital reconstruction

Researchers at a number of centers in the U.S. and Europe are working on the development of computerized systems for facial reconstruction. Methods generally rely on obtaining a digital model of the skull by 3-D laser scanning, morphing this model into a face using known data about tissue depths, and then adding a surface texture. The approach outlined here is based on work at the University of Sheffield in the U.K. To improve reconstructions, more accurate tissue depth data is being collected by magnetic resonance imaging (MRI). Computerized 3-D animated models showing obesity, aging, and ethnic ancestry have also been constructed in order to broaden the range of possible outcomes of reconstructions.

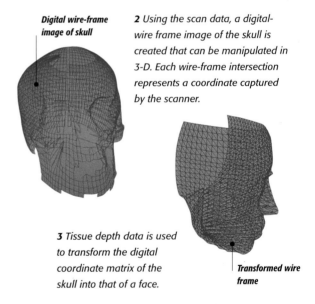

Digital wire-frame image of skull

2 Using the scan data, a digital-wire frame image of the skull is created that can be manipulated in 3-D. Each wire-frame intersection represents a coordinate captured by the scanner.

3 Tissue depth data is used to transform the digital coordinate matrix of the skull into that of a face.

Transformed wire frame

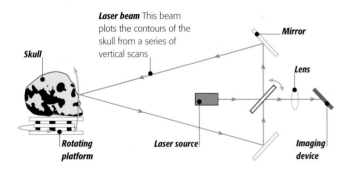

Laser beam This beam plots the contours of the skull from a series of vertical scans

Mirror

Skull

Lens

Rotating platform

Laser source

Imaging device

1 For 3-D laser scanning, the skull is placed on a rotating platform. Using a system of mirrors, a laser beam is reflected off the skull and onto an imaging device. Signals are then sent to a powerful graphics computer, which calculates the skull's shape from the pattern of reflected laser light.

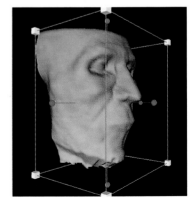

4 Graphical editing software can be used to add color, texture, lighting, and 3-D features such as hair or earrings onto the reconstructed face. Once this has been done, the completed reconstruction can be placed on a website for viewing and downloaded to a videotape or printer.

Manual reconstruction

Traditional methods of facial reconstruction rely on building up a face on the skull physically by hand, using clay. Before the process begins, the skull is examined carefully for any clues to the age, gender, build, and ethnic origin of the subject that could assist the reconstruction. The reconstruction process may take a few days.

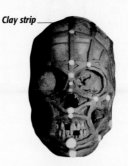

Clay strip

1 Dowels are placed on the skull at the 21 landmark sites, using average tissue-depth data. Tissue-depth measurements from MRI (magnetic resonance imaging) or ultrasound are gradually superseding data collected from cadavers.

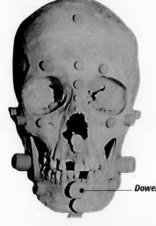

Clay fill-in

Dowel

2 Facial shape between the landmarks is reconstructed using clay. In the Russian method, the facial musculature is built up layer by layer. In the American method, shown here, strips of clay are used to model the shape of the face from the contours of the underlying bone.

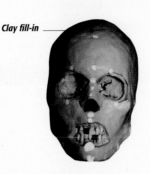

3 Additional clay is used to fill in the gaps between the strips. The approximate positions of the eyelids, lips, and base and tip of the nose can be determined from the skull. The shapes of these features cannot be predicted, however, and so adding them relies on educated guesswork.

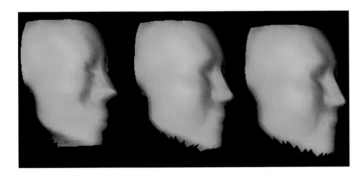

5 Although build can be determined from a skeleton, bodyweight cannot. Facial reconstructions produced using emaciated and obese tissue-depth values are being used as the basis of computerized 3-D interpolation models, which morph between the thin and fat extremes of an individual facial reconstruction.

Tissue-depth data collection

Tissue-depth data collection, comprising thousands of measurements per head, is now being undertaken on living persons using magnetic resonance imaging (MRI). Eventually, the MRI-generated data will be used to transform digitally constructed skull wire frames into 3-D facial models. This innovation will make facial reconstructions quicker to perform and more accurate.

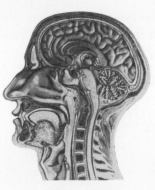

MRI SCAN OF HEAD

6 In adults, it is possible to estimate age from the skeleton to within a range of only five years, at best. Often much broader age range estimates have to be accepted. Facial reconstructions produced for the youngest and oldest extremes of the age range are being used as the basis of computerized 3-D models for aging, which it is hoped may assist in identifications.

30-YEAR-OLD VARIANT

50-YEAR-OLD VARIANT

80-YEAR-OLD VARIANT

SEE ALSO: DIGITAL WORLD *p.54* | SUSPECT IDENTIFICATION *p.160* | MRI & PET SCANNING *p.198* | LASERS *p.220*

Seeing in the dark

The human eye is naturally limited at seeing in the dark—it can detect shapes only faintly and see them only in shades of gray. Modern image enhancement technology allows us to overcome some of these limitations to improve night vision. Image intensifiers are used in a variety of situations, from security and military applications to wildlife monitoring and search-and-rescue operations.

Devices that allow the user to see in poor lighting conditions fall into one of two categories. Image intensifiers collect what little visible light may be available and then amplify it many thousands of times to produce a bright image on a screen, while thermal imagers transform the invisible infrared (heat) radiation that many objects emit even in the dark into visible light. Intensifiers are able to work only in clear conditions, when there is no obscuring fog, mist, or smoke. Thermal imagers, by contrast, can see warm objects, such as people, through other, cooler materials, such as solid walls or rubble. Image intensifiers are often fitted with an infrared floodlight to illuminate the scene—they can amplify the reflected infrared in the same way as normal light.

Intensifier tube

Hand grip

Eyepiece

Control button

BINOCULAR INTENSIFIER

Lens

Infrared lamp Lamps on either side of the lens emit invisible infrared light, improving visibility

Electron released from photocathode

Image intensification

In an image intensifier, light is focused by a lens onto a plate and individual photons ("packets" of light) hitting this plate trigger the release of electrons. In order to amplify the signal, the number of electrons is multiplied. This cascade of electrons is then used to trigger the release of photons from a phosphorescent screen.

Photons
Light enters the objective lens in the form of "packets" called photons

Obscured objects At night, objects reflect faint light from the moon or stars, but are too dim to be seen by the naked eye

Objective lens Light reflected from the surroundings is focused into parallel beams by the intensifier's objective lens

Photocathode This thin, negatively charged metal plate emits electrons when stuck by photons. The charge on the plate repels the electrons, causing them to move down the tube at high speed

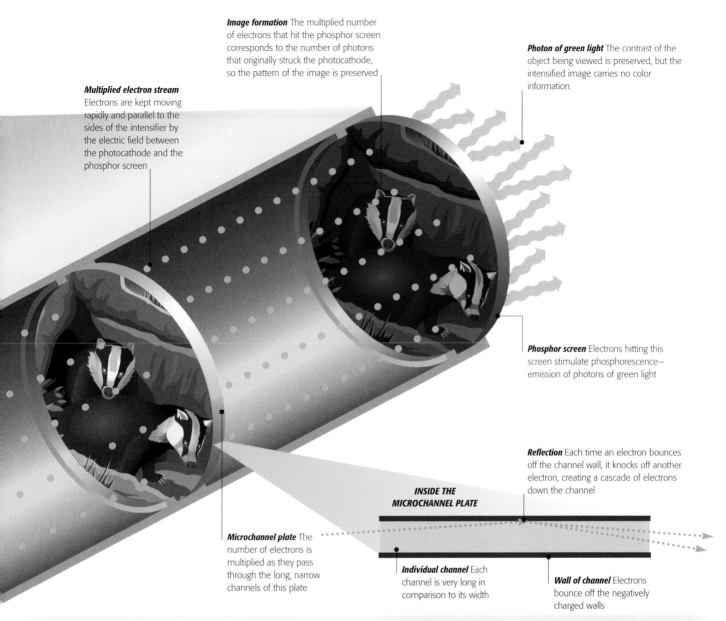

Image formation The multiplied number of electrons that hit the phosphor screen corresponds to the number of photons that originally struck the photocathode, so the pattern of the image is preserved

Photon of green light The contrast of the object being viewed is preserved, but the intensified image carries no color information

Multiplied electron stream Electrons are kept moving rapidly and parallel to the sides of the intensifier by the electric field between the photocathode and the phosphor screen

Phosphor screen Electrons hitting this screen stimulate phosphorescence—emission of photons of green light

Reflection Each time an electron bounces off the channel wall, it knocks off another electron, creating a cascade of electrons down the channel

INSIDE THE MICROCHANNEL PLATE

Microchannel plate The number of electrons is multiplied as they pass through the long, narrow channels of this plate

Individual channel Each channel is very long in comparison to its width

Wall of channel Electrons bounce off the negatively charged walls

Thermal imaging

Thermal imaging detects temperature differences in a scene. Infrared radiation from objects of different temperatures is focused using germanium lenses, and absorbed on an electronic detector array. The array is divided into pixels of a ceramic that releases electrons according to its temperature. The electrons are collected and read out as an electric current in the same way as they are on an optical charge-coupled device (CCD). This current can then be used to reconstruct images on a television monitor using false colors.

False-color infrared image (thermogram)
On a thermal imager, hot objects, such as the mug of coffee in this image, show up brightest. Cooler objects are much darker, and some allow heat from objects hidden behind them to pass straight through. Rescue teams use thermal imaging to search for disaster survivors underneath rubble, and the police use it to track criminals at night.

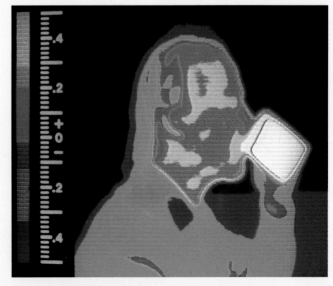

Protective materials

The human body's remarkable design can protect against heat and cold, cushion blows, and even repair its own injuries. But it has not evolved fast enough to keep pace with the threats posed by human inventions. Our bodies are not fireproof or bullet-proof, they cannot protect us against the extremes of temperature and pressure found deep in the ocean or out in space, and they are poorly suited to defend us against injury in high-speed accidents. Fortunately, technological evolution has developed a range of protective materials that offer better defense against these threats.

Fire and flames

Humans are believed to have used fire more than a million years ago, but it was not until the 20th century that people developed effective protection against its destructive effects. The widespread use of electricity and petroleum in the last century or so has made fire protection all the more important.

Fire means intense heat and temperatures, so fire-protective materials must be capable of resisting heat for some time without

Fire-fighting suits
People who work close to intense heat, such as firefighters, wear protective clothing made from lightweight, fireproof materials. Aluminum-based fabrics that reflect radiant heat are used in heavy-duty fire-fighting suits. Fire-fighting visors are made from Kapton coated with Teflon, two tough fireproof materials used in the outer layers of Apollo space suits.

Bulletproof vest
The grid of Kevlar fibers absorbs and dissipates the energy of a knife blow or bullet before it can damage the body beneath it. Kevlar fibers are also chemical and flame resistant. Kevlar is the material of choice in flak jackets, bulletproof vests, antimine boots, and chainsaw-proof clothing.

Bulletproof vest *The vest contains a protective plate made of Kevlar*

Damaged Kevlar plate *The plate inside the bulletproof vest reveals the damage a bullet causes as the Kevlar slows it to a halt*

themselves catching fire. Some plastics and composites are naturally heat- and fire-resistant. Polyvinyl chloride (PVC), for example, is used as an insulator in domestic electrical wiring, because it is difficult to set on fire and prevents flames from spreading. A composite of silica and alumina, two naturally non-flammable minerals, is used in industrial buildings such as power stations to offer heat protection up to 2,200°F (1,200°C).

Another promising use for the silica-alumina composite is in airplanes, whose combustible plastic interiors give off as much energy as their equivalent weight in petroleum when they catch fire. In 1998, Geopolymer, a lightweight, nonflammable fabric laminate based on the alumina-silica composite, became the first material to withstand arduous Federal Aviation Authority (FAA) flammability tests based on an aviation fuel fire. Totally fireproof materials such as this can be expensive, however, and an alternative approach is to wrap flammable materials in flame-retardant ones. For example, airplane seat cushions are typically made from flammable polyurethane, but can be made safer by coating them with fire-resistant Kevlar.

Motorcycle clothing
A motorcycle helmet reduces head injuries in a crash by spreading the impact with a hard outer casing of Injection-molded plastic or fiberglass, and absorbing the energy with a soft inner liner of styrofoam. Padded leathers also soften any impacts and protect the skin from abrasion, although today's "leathers" are more likely to be made from Kevlar woven with flexible Lycra and nylon.

Bullets and blows

If the human body is badly designed to withstand fire, it has even less chance where bullets, knives, and bombs are involved. But revolutionary materials that act like a protective outer skin can turn a potentially fatal gunshot into little more than a bruise.

Five times stronger than steel and much lighter, Kevlar has revolutionized body protection. It is a fibrous material made from long molecular chains of a polymer (plastic) called polyparaphenylene terephthalamide. Woven tightly together, the fibers knit into a tough protective grid that can withstand knife blows and cuts and some types of gunfire. An even tougher material called Spectra Fiber is made from a polyethylene (polythene) composite, in which the fibers are woven at right angles to one another, embedded in a flexible resin and coated with a laminate film. This material is ten times stronger than steel, yet extremely light, and provides better protection against automatic weapons than materials such as Kevlar.

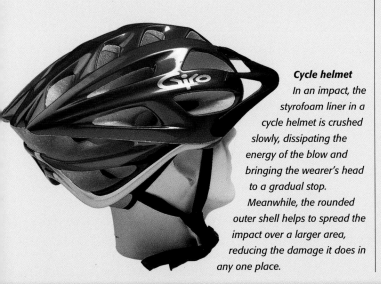

Cycle helmet
In an impact, the styrofoam liner in a cycle helmet is crushed slowly, dissipating the energy of the blow and bringing the wearer's head to a gradual stop. Meanwhile, the rounded outer shell helps to spread the impact over a larger area, reducing the damage it does in any one place.

It is not always practical for people to wear bulletproof clothing—sometimes it is more convenient to protect them by armoring the vehicles in which they travel. Armor-plated cars have long been used by heads of state and celebrities. Originally, they were built with heavy-duty steel reinforcement between the vehicle's bodyshell and chassis. Today, composite materials provide the same protection as antiballistic steel but add much less weight to the vehicle. Armored vehicles are also typically fitted with bulletproof "glass"—a tough sandwich of glass and polycarbonate (a transparent energy-absorbing plastic).

From bicycle helmets to UV protection

Not everyone needs a defense against petroleum fires or automatic weapons, but everyone is exposed to some risk every day. Urban traffic makes riding a bicycle an increasingly dangerous undertaking, but using a helmet can reduce the risk of sustaining a serious head injury by about 85 percent. Most helmets consist of a styrofoam (expanded polystyrene, or EPS, foam) liner and an outer shell made of a plastic or a composite.

Some people have to wear protective materials, all the time. Sufferers of a rare genetic condition called xeroderma pigmentosum have skin that lacks natural protection against ultraviolet (UV) radiation. Without protective clothing, UV in sunlight readily burns their skin and causes skin cancer. To reduce the risk, they typically wear hats and gloves impregnated with a UV-absorbing chemical called benzotriazole, plus sunglasses or masks with UV-resistant visors similar to those worn by astronauts. They also have to apply sunblock to their bodies regularly throughout the day. Sunblocks contain chemicals that absorb or reflect different wavelengths of UV light and prevent them from reaching the skin. The increasing damage to the Earth's ozone layer, which guards us from harmful ultraviolet rays, means that such protective measures will become more common for everyone in the future.

SEE ALSO: COMPOSITE MATERIALS *p.144* | PLASTICS *p.184* | ASTRONAUTS' EQUIPMENT *p.242*

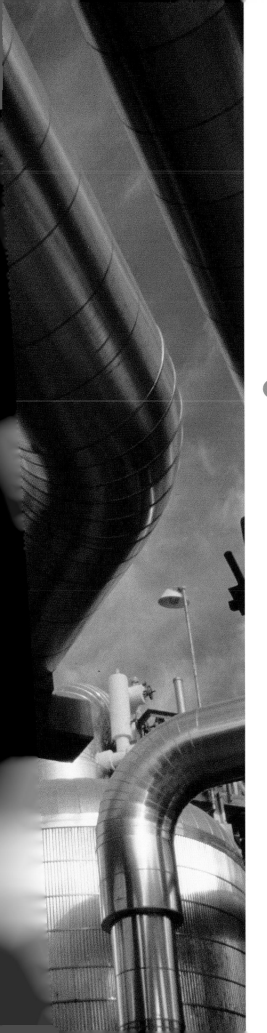

Power and Industry

The industrial revolution, which began in Britain during the 18th century, brought with it improved agricultural techniques, new power sources, new techniques of metal production, and the mechanization of the manufacturing process. Since that time, the ongoing application of science to industrial processes has continued to create new ways of detecting and extracting raw materials and manipulating them for industrial use. This chapter reveals just how far that process has come, showing how new technology has continued to revolutionize the way in which power is generated and products are manufactured in both the traditional heavy industries and in the fast-expanding field of biotechnology.

Power stations

When inventor Thomas Alva Edison (1847–1931) constructed one of the first power stations on Pearl Street, New York City in 1882, he revolutionized the way people used energy. Previously, energy had had to be produced in a useful form on the site where it was needed. By generating electricity at a power station, energy could be transmitted by cables in a clean, convenient form to other locations.

At the heart of a power station is a generator powered by a set of turbines. Inside the generator, a spinning electromagnet induces an electric current in surrounding copper coils. In a conventional thermal power station, the heat generated from the combustion of fossil fuels is used to turn water into steam. This steam is then directed at high pressure into a turbine, causing it to rotate and drive the generator. A typical coal-burning power station converts only 35 percent of the energy released by burning fuel into electrical energy; the rest is waste heat. New combined-cycle power stations improve efficiency by utilizing a second set of turbines to extract more of the energy released by combustion.

Hydroelectric power
In a hydroelectric power station, the kinetic energy of falling water spins a turbine that turns a generator to produce electricity. The water may be released from a large dammed reservoir to satisfy peak electricity demand during the daytime and pumped back up into the reservoir at night. This is known as pumped storage.

Combined-cycle power station

A combined-cycle power station has two sets of turbines and generators. The first set of turbines is spun by expanding combusted gases. Heat from the combusted gases is then used to produce steam to drive a second set of turbines. This design is up to 15 percent more efficient than a power station with a single set of turbines.

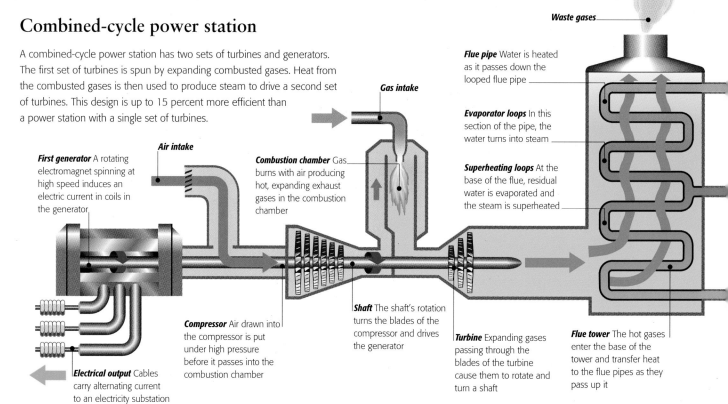

Gas intake

First generator A rotating electromagnet spinning at high speed induces an electric current in coils in the generator

Air intake

Combustion chamber Gas burns with air producing hot, expanding exhaust gases in the combustion chamber

Flue pipe Water is heated as it passes down the looped flue pipe

Evaporator loops In this section of the pipe, the water turns into steam

Superheating loops At the base of the flue, residual water is evaporated and the steam is superheated

Waste gases

Compressor Air drawn into the compressor is put under high pressure before it passes into the combustion chamber

Shaft The shaft's rotation turns the blades of the compressor and drives the generator

Turbine Expanding gases passing through the blades of the turbine cause them to rotate and turn a shaft

Flue tower The hot gases enter the base of the tower and transfer heat to the flue pipes as they pass up it

Electrical output Cables carry alternating current to an electricity substation for distribution

How a generator works

In 1831, British chemist Michael Faraday (1791–1867) found that when he rotated a copper disk between the poles of a magnet, an electric current was induced in the disk. Power station generators use this principle to convert kinetic energy into electrical energy.

In an alternating-current generator, a metal coil connected to two slip rings spins between the poles of a magnet. The two sides of the coil cut through the magnetic field, but move in opposite directions relative to the field. This causes a current to flow up one side of the coil and down the other. The sides of the coil swap position every half turn, reversing the induced current. This alternating current flows through carbon brushes that press against the rotating slip rings, into an external circuit.

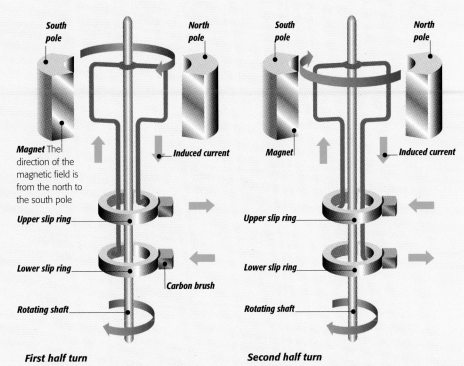

South pole

North pole

Magnet The direction of the magnetic field is from the north to the south pole

Induced current

Upper slip ring

Lower slip ring

Carbon brush

Rotating shaft

South pole

North pole

Magnet

Induced current

Upper slip ring

Lower slip ring

Rotating shaft

First half turn
When the green side of the coil cuts the magnetic field moving inward, a current flows up that side of the coil and down the other.

Second half turn
When the green side of the coil cuts the magnetic field moving outward, the induced current changes direction.

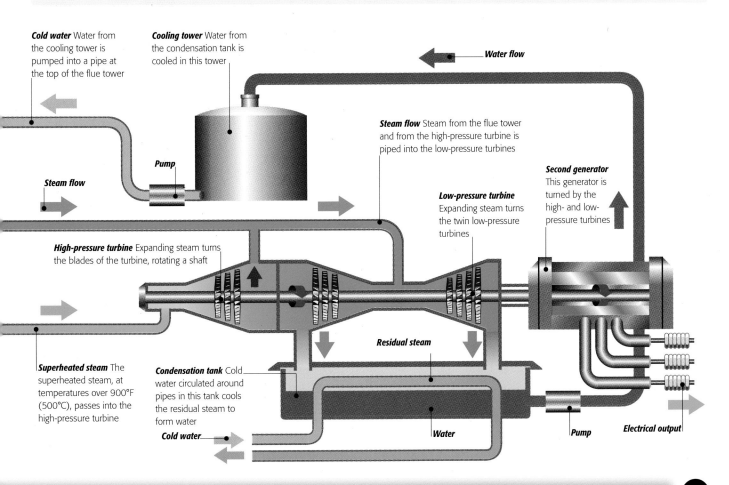

Cold water Water from the cooling tower is pumped into a pipe at the top of the flue tower

Cooling tower Water from the condensation tank is cooled in this tower

Water flow

Steam flow Steam from the flue tower and from the high-pressure turbine is piped into the low-pressure turbines

Second generator This generator is turned by the high- and low-pressure turbines

Low-pressure turbine Expanding steam turns the twin low-pressure turbines

Pump

Steam flow

High-pressure turbine Expanding steam turns the blades of the turbine, rotating a shaft

Residual steam

Superheated steam The superheated steam, at temperatures over 900°F (500°C), passes into the high-pressure turbine

Condensation tank Cold water circulated around pipes in this tank cools the residual steam to form water

Cold water

Water

Pump

Electrical output

SEE ALSO: NUCLEAR POWER *p.172* | ALTERNATIVE ENERGY *p.176* | ELECTRICITY & MAGNETISM *p.260*

Nuclear power

Nuclear power is potentially a huge source of cheap energy. Unlike conventional power stations, nuclear reactors do not rely on burning fossil fuels. Instead, they split apart individual atoms of the metal uranium, releasing huge amounts of energy in the process. In theory, nuclear power could generate far more energy than the world's remaining fossil fuel reserves, but the cost of building reactors, and the dangerous waste they produce, have held it back.

Radioactivity is widespread in nature—the atoms of substances such as uranium have naturally unstable nuclei that spontaneously split apart, decaying into two nuclei of two lighter elements and releasing energy (a process called fission). Nuclear reactors are designed to produce fission on demand in a controlled chain reaction in which the products of the reaction tend to make it continue. Heat released during the reaction is used to boil water and produce steam, which spins turbines to generate electricity, in the same way as a conventional power station.

To achieve fission, a subatomic particle called a neutron is fired at a uranium atom, causing the nucleus to split. This releases energy and more neutrons, which in turn hit the nuclei of nearby atoms and set off a chain reaction. Harmful radiation is released along with enormous heat, so nuclear power stations must contain this radiation and control the speed of the reaction. The smaller nuclei produced during fission are also often highly radioactive, so nuclear waste is dangerous and must also be handled carefully.

In the future, nuclear fission may be abandoned in favor of a cleaner and even more powerful source of energy—nuclear fusion, the process powers the Sun. In a fusion reaction, nuclei of hydrogen—the lightest element—are forced together to make heavier elements, releasing energy in the process. However, this energy is difficult to harness because of the enormous temperatures and pressures needed to produce fusion.

Harnessing nuclear power

More than two-thirds of the world's nuclear power stations contain the process of nuclear fission inside pressurized water reactors. In these reactors, water is used to moderate the speed of neutrons that are released from uranium nuclei. The heat energy released by fission heats the water to produce steam, which drives a turbine linked to electricity generators.

Nuclear fission

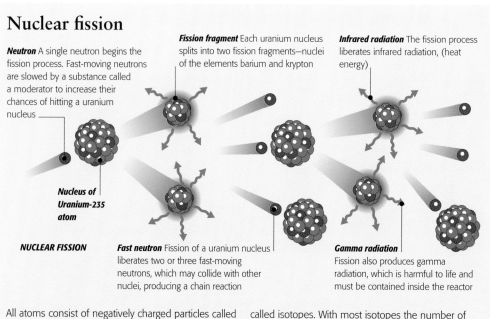

Neutron A single neutron begins the fission process. Fast-moving neutrons are slowed by a substance called a moderator to increase their chances of hitting a uranium nucleus

Nucleus of Uranium-235 atom

NUCLEAR FISSION

Fast neutron Fission of a uranium nucleus liberates two or three fast-moving neutrons, which may collide with other nuclei, producing a chain reaction

Fission fragment Each uranium nucleus splits into two fission fragments—nuclei of the elements barium and krypton

Infrared radiation The fission process liberates infrared radiation, (heat energy)

Gamma radiation Fission also produces gamma radiation, which is harmful to life and must be contained inside the reactor

Welded steel jacket This jacket resists the pressure of the water inside the reactor

Fuel assembly Each fuel assembly contains hundreds of fuel rods and control rods. Control rods determine the rate of nuclear fission by absorbing the fast neutrons produced by the chain reaction

Pressurized water Water acts as both a coolant and a moderator—a substance that slows fast neutrons down so that they are more likely to hit uranium nuclei and cause a chain reaction

All atoms consist of negatively charged particles called electrons, orbiting a central nucleus made of positive protons and uncharged neutrons. An element is defined by its number of protons. Atoms of the same element can contain different numbers of neutrons. These different varieties of the same element are called isotopes. With most isotopes the number of neutrons and protons is roughly equal, but some—heavy isotopes—have many more neutrons than protons. This imbalance makes them unstable. The Uranium-235 isotope is an example of an unstable radioactive isotope that undergoes radioactive decay.

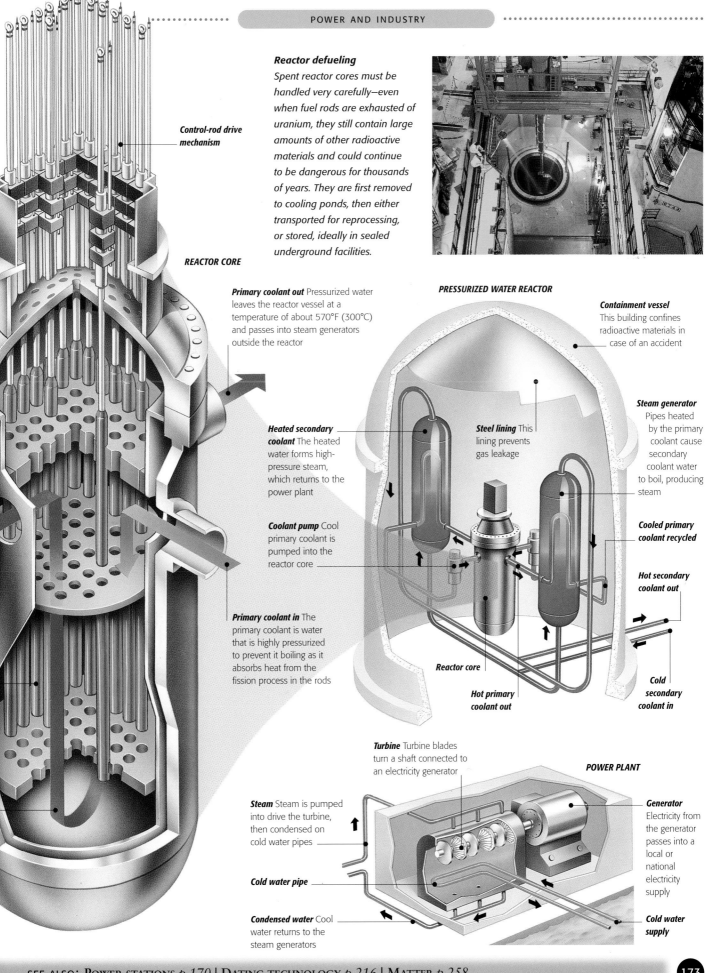

Control-rod drive mechanism

REACTOR CORE

Reactor defueling
Spent reactor cores must be handled very carefully—even when fuel rods are exhausted of uranium, they still contain large amounts of other radioactive materials and could continue to be dangerous for thousands of years. They are first removed to cooling ponds, then either transported for reprocessing, or stored, ideally in sealed underground facilities.

Primary coolant out Pressurized water leaves the reactor vessel at a temperature of about 570°F (300°C) and passes into steam generators outside the reactor

PRESSURIZED WATER REACTOR

Containment vessel This building confines radioactive materials in case of an accident

Steam generator Pipes heated by the primary coolant cause secondary coolant water to boil, producing steam

Heated secondary coolant The heated water forms high-pressure steam, which returns to the power plant

Steel lining This lining prevents gas leakage

Cooled primary coolant recycled

Coolant pump Cool primary coolant is pumped into the reactor core

Hot secondary coolant out

Primary coolant in The primary coolant is water that is highly pressurized to prevent it boiling as it absorbs heat from the fission process in the rods

Reactor core

Hot primary coolant out

Cold secondary coolant in

Turbine Turbine blades turn a shaft connected to an electricity generator

POWER PLANT

Steam Steam is pumped into drive the turbine, then condensed on cold water pipes

Generator Electricity from the generator passes into a local or national electricity supply

Cold water pipe

Condensed water Cool water returns to the steam generators

Cold water supply

SEE ALSO: POWER STATIONS *p.170* | DATING TECHNOLOGY *p.216* | MATTER *p.258*

Solar energy

As early as the third century BCE, people were aware of the power of focused sunlight. The Greek mathematician Archimedes (287–212 BCE) is famously said to have defended Syracuse (Sicily) from attack using giant mirrors that directed intense beams of sunlight onto the enemy ships, causing them to burst into flames. Today, solar power is used for more peaceful purposes, harnessed by a range of technologies to generate clean, renewable energy.

The potential of solar energy

A huge amount of sunlight strikes the Earth and is absorbed or reflected back into space during the day. The average intensity of this solar energy is rougly equivalent to 100 watts—the power of a light bulb—per square foot (1,100 per square meter). Most of the energy arrives in the form of heat rather than visible light. According to some estimates, the amount of solar energy arriving at the Earth's surface is enough to supply current power demands 20,000 times over. However, this vast resource is unevenly distributed and variable. Regions close to the equator receive much more sunshine than those at higher latitudes, and passing clouds can absorb or scatter most of the energy before it even reaches the ground. For this reason, many applications of solar energy are only practicable in areas where bright sunshine is the norm.

Probably the simplest solar-powered device is the solar oven. This is simply a box with a glass lid that allows sunlight to enter, a black metal interior that absorbs nearly all the solar energy striking it, and an insulated lining to prevent heat loss. In bright sunshine, the oven can reach very high temperatures (well above the boiling point of water), and can be used to cook food. It is particularly useful and effective in hot countries where wood or other fuel for cooking fires may be scarce.

Solar array
One of the world's largest operating solar power complexes is in California's Mojave Desert. Here, 1,000 acres (405 hectares) of computer-controlled mirrors superheat synthetic oil to 735°F (390°C), which generates steam that drives turbines producing up to 275 megawatts of power.

Solar panels

Solar panels are becoming increasingly widespread for domestic heating. They work by heating up a "working fluid" (often water, oil, or glycol), which then transfers its heat to a water system.

A solar panel is designed to transfer the maximum amount of heat to the working fluid and it is covered with glass, which transmits light and heat as well as protecting the unit. To increase the surface area exposed to the Sun, working fluid is pumped through a zigzagging pipe made of copper—an exceptionally good heat conductor. This pipe is sandwiched between a black heat-absorbing material, which sits on top and captures heat from the sunlight, and a layer of aluminum foil and thick insulation under the pipe, which reflects back any heat not absorbed in the pipes. Passing through this pipe, the fluid absorbs radiant energy and heats up. It then flows into a heat exchanger—another zigzag of pipes running through a water tank. In some panels, the heat is transferred to a bed of stones, which heat up and retain heat more easily than water.

Solar water heaters
Rooftop solar panels are most effective in hot climates with plenty of sunshine, such as here at Alanya, Turkey. A household's hot water needs can be completely provided for using such solar heaters.

Solar furnace

In this solar furnace in France, sunlight is directed onto the huge curved reflector shown here by an array of computer-controlled flat mirrors. This reflector, which is made up of 9,500 separate mirrors, focuses the Sun's energy into the collector tower, where temperatures can approach 6,900°F (3,800°C).

Electricity from the sun

There are currently two main ways of generating electricity from radiant solar energy. The most widespread is the solar or photovoltaic cell, which is a piece of semiconductor that converts light energy into electric energy. Photocells were first developed to power satellites and space probes, but are now found on everything from calculators to experimental automobiles.

However, the best prospect for large-scale power generation comes from solar furnaces. These use huge arrays of tilting mirrors called heliostats, which turn to follow the sun around the sky, and focus its rays onto a collector, typically located high above the ground in a tower. Water pumped through the collector boils rapidly, and the force of the expanding steam can be used to turn a turbine. This turbine drives a generator that produces electricity in the same way as many other types of power stations.

So far, solar power stations make a relatively small contribution to the total world generation of electricity—most are still experimental, although a few are now linked to national grids. Solar power will never completely replace more traditional sources of energy, because sunlight is absent for large parts of the day. But combined with other forms of alternative energy production, solar energy will play an important role.

Photovoltaic solar cells

A photovoltaic solar cell is a sandwich made of two layers of the semiconductor silicon that have been "doped" with other elements to alter their electric properties. One layer (n-type) has more negatively charged electrons than usual, and the other (p-type) has fewer. When p-type and n-type semiconductors are sandwiched together, electrons from the n-type layer drift into the p-type layer. This sets up a voltage across the junction between the two semiconductors. When sunlight strikes the photovoltaic cell, it has enough energy to free electrons. Electrons released close to the junction move across it, encouraged by the junction voltage, and cause a current flow.

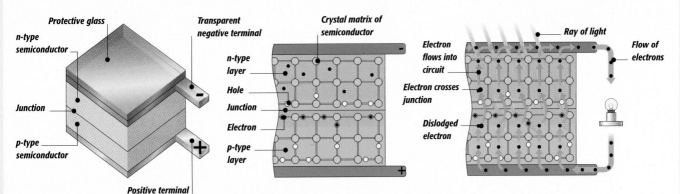

Structure of a solar cell
N-type and p-type semiconductors are joined between two terminals that conduct electricity. The whole cell is protected by a glass cover.

Junction potential
A voltage is created between the n- and p-type semiconductors, by the movement of a few electrons and holes across the junction.

Current flow
Sunlight strikes the cell, dislodging electrons. They move under the influence of the junction voltage, flowing around an external circuit.

Alternative energy

Electricity has traditionally been generated by the burning of fossil fuels, but fears about the environmental cost to the planet and the sustainability of continued fossil fuel consumption have prompted research into cleaner methods of generating electricity from renewable resources. Even multinational oil companies have begun to invest heavily in the development of such technologies, and predict that in 50 years' time half of the world's power will come from them.

Burning fossil fuels not only depletes resources but also releases carbon dioxide that had previously been locked in the Earth's crust. Excess atmospheric carbon dioxide has been linked with global warming and climate change, making the harnessing of energy sources that have a minimal environmental impact a must for the 21st century. Renewable resources are sources of energy cannot be significantly depleted by generating power from them. Such energy sources include radiation from the Sun, power from the wind, waves and tides, and water falling under gravity, and heat from the Earth itself. The Sun drives the weather systems, continually keeping huge masses of air circulating the globe. The gravitational attraction of the Moon pulls great volumes of water across the globe following the Moon's orbit, and residual heat from the formation of the Earth is conducted from the core as the Earth slowly cools. Generating electricity from these resources causes significantly less pollution than fossil fuel combustion.

Geothermal power
At a few locations around the world, known as "hotspots" such as the "Blue Lagoon" at Svartsengi, Iceland, upwelling within the Earth's mantle brings molten rock close to the surface. Water circulating underground at these sites is heated, giving rise to hot springs and geysers. This geothermal energy is captured by using the naturally-occurring steam to drive turbines that generate electricity.

Wind power
Wind farms, like the one in this picture, are becoming an ever more common sight. They are built on high ground, open plains, coastal ridges, or even offshore. Wind turbines have a few simple components—positioning gears that orient the turbine blades into the wind, and a generator that is coupled to the turbine's axle—but their design is often very advanced. Currently, wind farms provide small communities with power or supplement the main grid, but the cost of wind-generated power is falling and the technology could mature into a significant source of electrical power.

The Earth's largest and most constant source of energy is held in the enormous mass of the oceans as they are dragged across the surface of the Earth by the wind and the gravitational pull of the moon. The two modes of ocean movement are the continuous motion of the waves and the periodic changing of the tides. Different techniques are used to exploit these modes for power generation.

Wave power

Gully generators utilize the almost unceasing motion of the waves to generate power. A concrete chamber built on the shore is open at the sea end so that the water level inside the chamber moves up and down with each successive wave. The air above the water is alternately compressed and decompressed, driving a turbine that is connected to a generator.

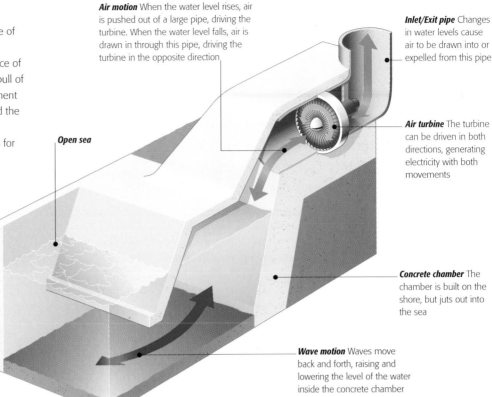

Air motion When the water level rises, air is pushed out of a large pipe, driving the turbine. When the water level falls, air is drawn in through this pipe, driving the turbine in the opposite direction

Inlet/Exit pipe Changes in water levels cause air to be drawn into or expelled from this pipe

Air turbine The turbine can be driven in both directions, generating electricity with both movements

Open sea

Concrete chamber The chamber is built on the shore, but juts out into the sea

Wave motion Waves move back and forth, raising and lowering the level of the water inside the concrete chamber

Tidal power

Tidal barrages use the difference between water levels at high and low tide to generate electricity. They are constructed across the mouths of tidal estuaries. When the tide is rising, water is allowed through the barrage, filling the estuary behind it. As the tide falls, floodgates are closed and a head of water builds up behind the barrage. Water is then allowed to flow back to the sea, driving turbines that are connected to generators as it does so.

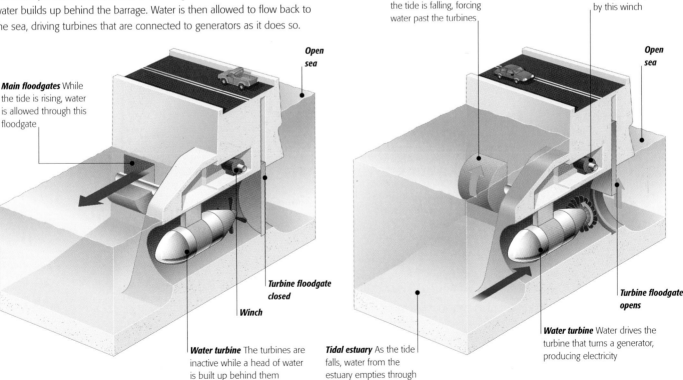

Main floodgates While the tide is rising, water is allowed through this floodgate

Open sea

Main floodgates The giant floodgates are shut when the tide is falling, forcing water past the turbines

Winch The turbine floodgate is opened by this winch

Open sea

Turbine floodgate closed

Winch

Water turbine The turbines are inactive while a head of water is built up behind them

Tidal estuary As the tide falls, water from the estuary empties through turbine floodgate

Turbine floodgate opens

Water turbine Water drives the turbine that turns a generator, producing electricity

RISING TIDE

FALLING TIDE

SEE ALSO: ALTERNATIVE CARS *p.114* | POWER STATIONS *p.170* | SOLAR ENERGY *p.174* | MECHANICS *p.268*

Batteries

Since electricity cannot always be generated where and when it is needed, batteries were developed to provide an easily transportable power supply. Although many different types of batteries exist, they all work by storing energy in chemical form and converting it into electric energy when it is required. Batteries power a wide range of machines, from watches and cameras to satellites in space.

All batteries contain one or more cells—electrochemical devices that convert chemical energy into electric energy. A cell has three main components: a positive electrode, a negative electrode, and an electrolyte, which is in contact with both electrodes. When the battery is connected in a circuit, chemical reactions occur that produce an excess of electrons in the negative electrode and a deficit in the positive electrode. The electrons flow around the external circuit. This movement of electric charge is balanced by a movement of ions (other charged particles) through the electrolyte.

Cells can be divided into primary and secondary. Primary cells include alkaline long-life batteries, typically used in flashlights, which generate power by slowly using up their own store of chemicals. Secondary cells include the lead-acid batteries used in automobiles and NiMH (nickel metal hydride) batteries used in cellular phones. Unlike primary cells, secondary cells can be recharged by passing a current through them in the reverse direction to that in which they normally cause a current to flow.

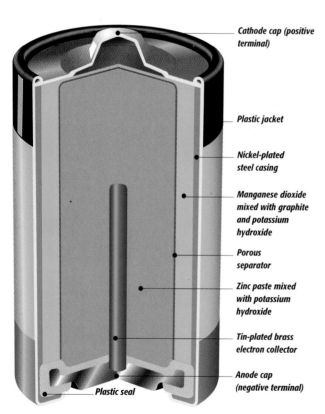

Cathode cap (positive terminal)

Plastic jacket

Nickel-plated steel casing

Manganese dioxide mixed with graphite and potassium hydroxide

Porous separator

Zinc paste mixed with potassium hydroxide

Tin-plated brass electron collector

Anode cap (negative terminal)

Plastic seal

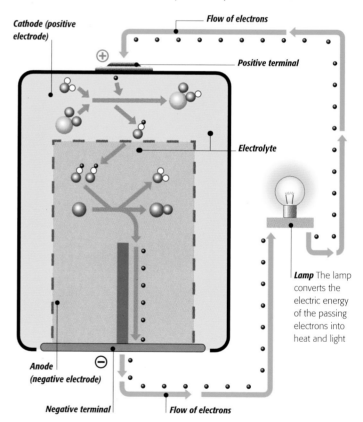

Cathode (positive electrode)

Flow of electrons

Positive terminal

Electrolyte

Anode (negative electrode)

Negative terminal

Flow of electrons

Lamp The lamp converts the electric energy of the passing electrons into heat and light

Long-life alkaline cell

In a zinc-manganese dioxide cell, the anode (negative electrode) consists of zinc paste, an electron collector. The cathode (positive electrode) consists of manganese-dioxide powder in a graphite matrix and the metal cell casing. Metal caps form the electrode terminals. The electrolyte, a solution of potassium hydroxide, bathes both electrodes. Ions in the electrolyte can pass through the porous separator between the electrodes.

Operation of an alkaline cell
At the anode, zinc turns into zinc oxide, releasing electrons that flow around an external circuit to the cathode, where they are collected by manganese dioxide. Inside the cell, negatively-charged hydroxyl ions move from cathode to anode to complete the flow of charge.

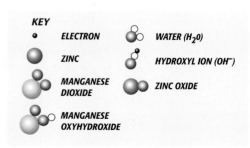

KEY

- ELECTRON
- ZINC
- MANGANESE DIOXIDE
- MANGANESE OXYHYDROXIDE
- WATER (H_2O)
- HYDROXYL ION (OH^-)
- ZINC OXIDE

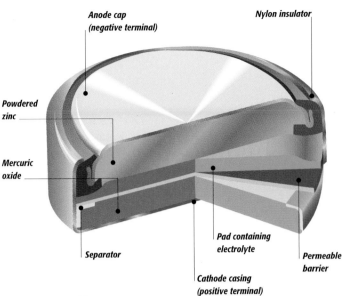

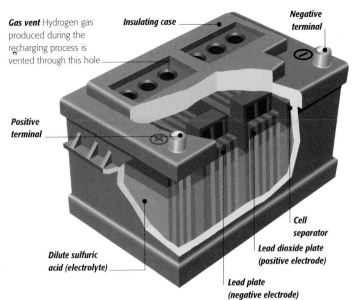

Button cell

Button cells are used inside calculators and watches. The anode (negative electrode) is in the top part of the cell and consists of powdered zinc. The cathode (positive electrode) at the bottom of the cell contains mercuric oxide. Between the two electrodes is an absorbent pad soaked in an electrolyte of potassium hydroxide. During operation, the zinc anode loses electrons and forms zinc oxide. These electrons flow in external circuit and return to the cathode, where the mercuric oxide gains electrons to form metallic mercury.

Automobile battery

A lead-acid accumulator contains three or six separate cells inside a tough plastic case. Each cell contains a series of lead and lead-dioxide plates, with separators between them, and an electrolyte of dilute sulfuric acid. During operation, the sulfuric acid gradually turns into water, both the lead dioxide and lead electrodes turn into lead sulfate, and the battery gradually loses power. But, unlike a primary cell, it can be recharged by passing a current through it in the opposite direction, which reverses the chemical reactions.

Fuel cells

Unlike a battery, which gradually loses its ability to generate electricity as the chemicals inside it get used up, a fuel cell generates electricity from a continuous supply of chemicals. Like a battery, a fuel cell has positive and negative electrodes and an electrolyte in between. Fuel cells are used in unmanned space probes and in the Space Shuttle; they are expected to be used in the cars of the future, because they are twice as efficient as internal combustion engines and produce nothing more polluting than water.

How an alkaline fuel cell works

In this type of fuel cell, hydrogen (H_2) and oxygen (O_2) combine to form water, with the release of electric energy and heat. At the negative electrode, hydrogen gas reacts with hydroxyl (OH^-) ions in the electrolyte to form water. The reaction releases electrons, which flow around the external circuit. Electrons are taken up at the positive electrode and combine with oxygen and water to form hydroxyl ions, completing the cycle of chemical reactions.

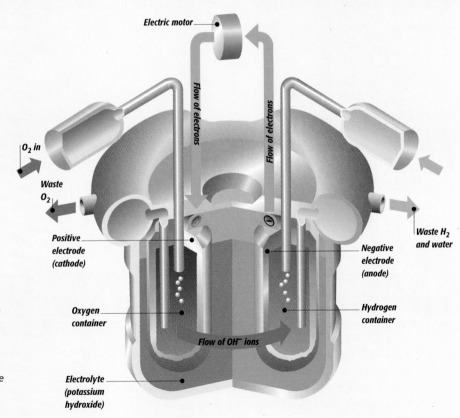

SEE ALSO: ALTERNATIVE CARS *p.114* | ELECTRICITY & MAGNETISM *p.260*

Oil production

The demand for oil has grown enormously over the last 150 years and it is now the world's most important source of energy as well as being the raw material for a diverse range of petrochemical products. As the more readily accessible onshore reserves are being depleted, oil companies have developed technology for extracting oil using offshore rigs and wells sunk deeper and deeper into the Earth.

The first oil wells were drilled using percussion, a technique that involved a steam engine repeatedly raising and dropping a heavy steel tool into soft rock. The rotary drill bit, introduced in 1908, allowed oil to be extracted from a wider range of locations because it enabled wells to be sunk to greater depths and through harder rock. The greatest advance since then has been directional drilling, which has made it possible to drill horizontally as well as vertically. Modern drill bits contain their own motors, sensors that can detect what material lies ahead, and computerized steering tools that allow drilling to be targeted precisely.

The discovery of vast reserves of oil under the seabed has prompted the development of a range of offshore rigs, ranging from massive gravity platforms to maneuverable drilling ships, which are able to extract both crude oil and natural gas—a valuable by-product of both on- and offshore drilling operations. In 1999, over 72 million barrels of crude oil were produced each day worldwide. A third of these came from offshore oil reserves.

Cement The bore hole is shored up by pumping cement down the sides of the casing

Drill string This is a series of rotating steel pipes whose combined weight drives the drill bit through the rock

Drill collar The drill collar adds weight and stability to the drill bit

Drill bit Diamond-studded teeth cut through even the hardest rock

Casing The casing for the drill string consists of a series of large steel pipes

Waste Rock fragments are forced out of the bore hole by the mud

Mud This synthetic mixture is pumped out of the end of the drill bit and forces rock fragments out of the bore hole

Rotary drill

Oil wells are drilled using special diamond-studded steel drill bits on the end of a strong steel pipe called a drill string. This is rotated by either a motor at the surface or a turbine down the bore hole. Rock fragments are carried upward and out of the bore hole by pumping a combination of chemicals and water known as "mud" down the drill string. This mud also prevents the drill bit from getting too hot. Additional drill-string lengths are added at the top of the bore hole to lengthen the drill.

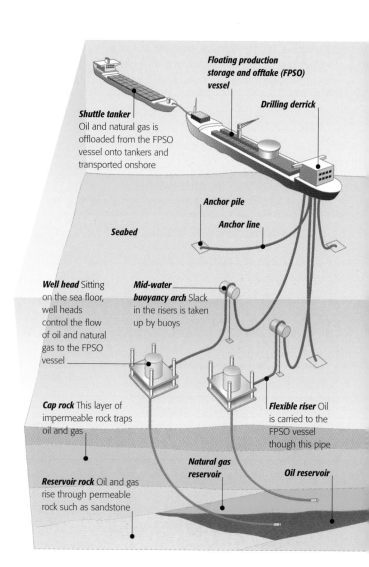

Shuttle tanker Oil and natural gas is offloaded from the FPSO vessel onto tankers and transported onshore

Floating production storage and offtake (FPSO) vessel

Drilling derrick

Anchor pile

Anchor line

Seabed

Well head Sitting on the sea floor, well heads control the flow of oil and natural gas to the FPSO vessel

Mid-water buoyancy arch Slack in the risers is taken up by buoys

Cap rock This layer of impermeable rock traps oil and gas

Reservoir rock Oil and gas rise through permeable rock such as sandstone

Natural gas reservoir

Flexible riser Oil is carried to the FPSO vessel though this pipe

Oil reservoir

Offshore production platform

Most offshore oil production platforms are assembled in quiet inland waterways and then towed into position by a fleet of tugs. The drilling platform shown above was assembled in Louisiana and is being towed down the Mississippi Delta to offshore oil fields in the Gulf of Mexico.

Onshore oil production

Beam pumps, sometimes known as "pump jacks," are used to extract oil from wells that do not have enough natural pressure to force the oil out at an economic rate. The rocking action of the beam pushes a plunger down the well, forcing oil upward past a series of nonreturn valves.

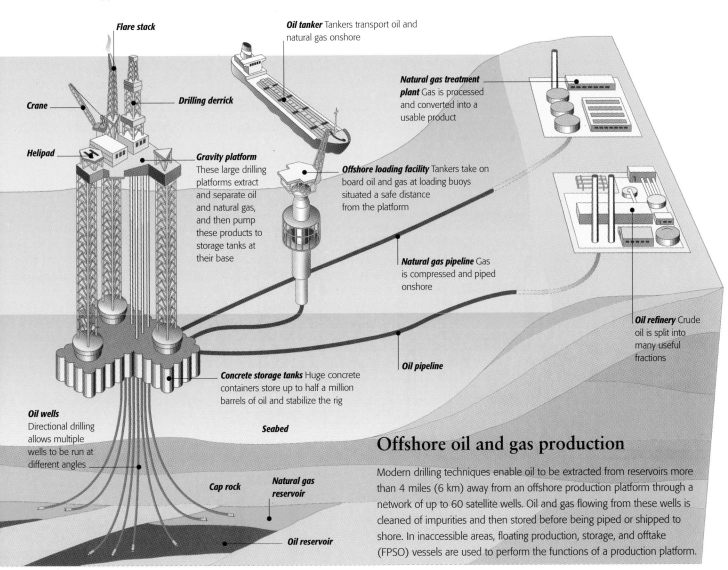

Flare stack

Oil tanker Tankers transport oil and natural gas onshore

Crane

Drilling derrick

Natural gas treatment plant Gas is processed and converted into a usable product

Helipad

Gravity platform
These large drilling platforms extract and separate oil and natural gas, and then pump these products to storage tanks at their base

Offshore loading facility Tankers take on board oil and gas at loading buoys situated a safe distance from the platform

Natural gas pipeline Gas is compressed and piped onshore

Oil refinery Crude oil is split into many useful fractions

Concrete storage tanks Huge concrete containers store up to half a million barrels of oil and stabilize the rig

Oil pipeline

Oil wells
Directional drilling allows multiple wells to be run at different angles

Seabed

Offshore oil and gas production

Modern drilling techniques enable oil to be extracted from reservoirs more than 4 miles (6 km) away from an offshore production platform through a network of up to 60 satellite wells. Oil and gas flowing from these wells is cleaned of impurities and then stored before being piped or shipped to shore. In inaccessible areas, floating production, storage, and offtake (FPSO) vessels are used to perform the functions of a production platform.

Cap rock

Natural gas reservoir

Oil reservoir

SEE ALSO: SHIPS *p.140* | OIL REFINING *p.182* | GEOLOGICAL EXPLORATION *p.218* | MECHANICS *p.266*

Oil refining

Crude oil, or petroleum, is a dark, sticky liquid mixture of compounds that was formed underground by the decay of ancient marine animals. This crude oil is refined and processed to make fuels, plastics, and other products for the chemical industry. During refining, the various components of the mixture, called fractions, are first separated, and then chemically altered to produce useful compounds.

In the refinery, crude oil is separated into different fractions, each of which boils at a different temperature, by a process called fractional distillation, which takes place in tall steel towers. When the fractions are sufficiently purified, they may be further processed to alter their chemical structure. These processes include cracking, in which the fraction's molecules are broken down into smaller molecules by heat and a catalyst (a substance that helps a reaction to occur), and polymerization, in which thousands of identical small molecules are chemically bonded together to form long chainlike molecules of plastics.

Bubbling vapors

Fractional distillation relies on different pure fractions having different boiling points. Distillation towers contain a series of perforated trays at various heights on which cooling vapors condense to form liquid oils. When hot vapors rise through a hole in a tray, a device called a bubble cap forces them to bubble through liquid on the tray. The hot vapor heats the liquid, causing any remaining volatile molecules of lighter fractions to evaporate and rise up the tower, thus purifying the liquid.

Rising vapor Hot vapor bubbles heat the liquid as they rise through it, before rising further up the tower

Condensed liquid As the liquid heats up, volatile molecules in it evaporate and join the vapors rising up the tower

Bubble cap

Siphoning pipe When the oil fraction in the tray is sufficiently purified, it is removed through this pipe

Rising vapors

Overflow channel

Falling liquid This liquid has overflowed from a tray further up the tower

Oil refinery

In a refinery, heated crude oil is fed into the bottom of a distillation tower, mainly as a mixture of vapors. As the different vapors rise up the tower, they condense out as liquids at different levels, according to their boiling points. The lighter and more easily vaporized the vapor fraction, the higher up the tower it condenses.

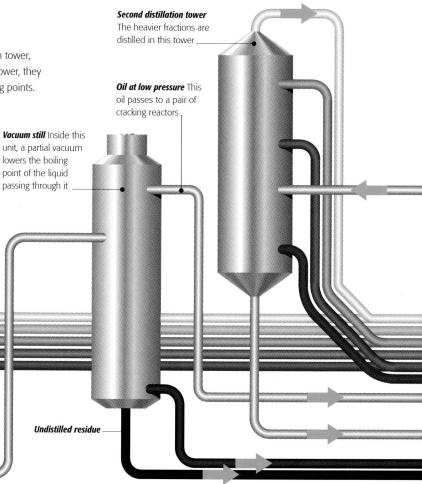

First distillation tower The temperature in the tower decreases from bottom to top, so that fractions with different boiling points condense at different levels in the tower

Condensation tray

Crude oil Liquid oil is heated to a temperature of 750°F (400°C) before entering the first distillation tower mainly in the form of a vapor

Liquid oil The heaviest fraction of the crude oil remains unvaporized in the first tower due to its high boiling point

Vacuum still Inside this unit, a partial vacuum lowers the boiling point of the liquid passing through it

Second distillation tower The heavier fractions are distilled in this tower

Oil at low pressure This oil passes to a pair of cracking reactors

Undistilled residue

Mossmorran oil refinery, Fife, Scotland
In a typical oil refinery, crude oil is separated into its various components in dozens of towers connected by up to 1,000 miles (1,600 km) of pipes. Oil and its products may be delivered to and collected from a refinery by road, rail, sea, or pipeline.

Petrochemical products

Oil refining splits crude oil into many different fractions, which have a wide range of applications in industry and currently provide fuel for most forms of transportation. Refineries are capable of controlling the temperature within the tower to alter the balance of fractions to meet demand—if diesel becomes more widely used, more of that fraction may be collected.

The range of petrochemical products shown below is not exhaustive, but covers the main types of fuels and products generated directly from the refining process. These products are arranged by density, ranging from bottled gas, the lightest fraction, down to bitumen, the heaviest fraction.

Bottled gas
Light fractions are gases used as fuel.

Gasoline
Automobiles use this fraction for fuel.

Kerosene
This product is used as fuel for jet engines.

Diesel oil
Diesel-powered vehicles use this fraction for fuel.

Heating oil
Some domestic heating systems use this oil for fuel.

Lubricating oil
This oil lubricates engines and other equipment.

Fuel oil
This heavy oil is used to fuel ships.

Bitumen
This heaviest fraction is used to surface roads.

Cracked oil vapors
Cracked vapors pass from the cracking reactors to the second distillation tower

Carbon dioxide
This waste gas is produced during the cracking process

First cracking reactor

Second cracking reactor

Choked molecules
While reacting with the oil molecules, the catalyst molecules become "choked," or covered with carbon

Regenerated catalyst molecules
In the second reactor, carbon molecules are burned off the choked catalyst molecules at a temperature of 1,340°F (725°C)

Catalyst molecules
The catalyst molecules react with the oil molecules to break them down into smaller units

Undistilled oil Oil that remains undistilled in the second tower returns to the cracking reactors, where its molecules are further broken down

SEE ALSO: OIL PRODUCTION *p.180* | PLASTICS *p.184* | GEOLOGICAL EXPLORATION *p.218* | MATTER *p.258*

Plastics

Plastics are the most versatile materials ever invented. Indeed, the word "plastic," which derives from the Greek word *plastikos*, meaning to mold or form, has come to be used as a general description for anything particularly adaptable or flexible. Since the first synthetic plastic was developed as a replacement for elephant ivory in the 1860s, many different types of plastic, including nylon, polyethylene, and Teflon (PTFE, or polytetrafluoroethylene) have revolutionized the manufacture of products as diverse as photographic film, stockings, and automobile-body parts. Although plastics are finding ever wider application, concerns have been raised about the environmental effects of their use and disposal because most plastics are not biodegradable.

What are plastics?

Plastics are polymers, long molecules made up of many copies of a basic molecular unit, called a monomer. Most plastics are synthesized from hydrocarbons (compounds of hydrogen and carbon), which are obtained by refining petroleum. The monomer molecules may be coupled together through a process called additive polymerization to form a long chain polymer molecule. Alternatively, in condensation polymerization, some atoms from each monomer are removed so that they can join together, again forming a long chain. There are basically two different kinds of plastic. Plastics that can be repeatedly softened by heating and which harden again upon cooling are called thermoplastics, and include polyethylene and polystyrene. Thermosetting plastics, such as Bakelite, cannot be repeatedly softened by heating. When a thermoset is first synthesized, crosslinks form between the polymer molecules, fixing their shape. Thermosets are particularly suitable for products that must withstand hot environments.

How plastics are made

Plastic products begin life as a resin produced by polymerization to which various other materials are added. Some of these additives provide color or texture, while others give the plastic particular physical properties, such as fire resistance, electrical conductivity, or additional strength. Additives called plasticizers make the plastic flow more easily. Other additives called stabilizers and antioxidants help to prevent the plastic breaking down over time due to agents such as ultraviolet radiation in sunlight.

The final product is made from the raw plastic by a range of different manufacturing processes. For example, in extrusion, plastic is squeezed through a shaping die (like squeezing toothpaste from a tube) to make products such as pipes. Injection molding, used to make bowls, involves heating resin pellets until they melt, then forcing them under pressure into a mold, where

Nylon stockings
Nylon stockings were an instant success when they first went on sale in May 1940, but became very hard to buy during World War II because all nylon supplies were allocated for parachutes and other war items. Huge lines formed outside shops selling "nylons" when they reappeared in 1945.

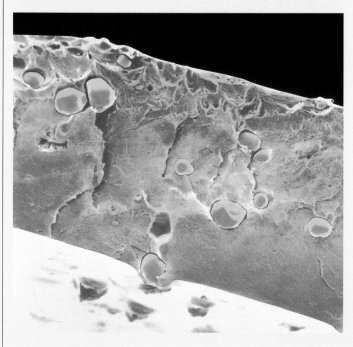

Biodegradable plastic
Granules of starch, shown in orange, can be seen clearly in this colored scanning electron micrograph of a slice of biodegradable plastic. Starch granules are added to plastic, because they absorb water and expand in the soil, breaking the plastic into small fragments. This gives bacteria that are able to digest the plastic a greater surface area of plastic to attack.

they cool and harden. Blow molding uses air pressure to push a bubble of plastic into a mold and is used to make bottles. Sheets are created by calendering, which involves squeezing plastic between rollers. Thermosetting plastics are poured into a mold, heated to complete the synthesis, then allowed to set in a process known as casting.

The world of plastic

Since the 1860s, chemists have synthesized dozens of different kinds of plastic for almost every conceivable use. Polyethylene is used for food wrapping, shopping bags, and plastic bottles. Polypropylene is easily drawn into strong fibers and woven into ropes and carpets. Polystyrene can be expanded with air bubbles to make Styrofoam, a light packaging material with particularly good heat insulation properties that is used in containers for hot food and drinks. Polyvinyl chloride (PVC) is a cheap and versatile plastic that can be formed into a wide range of items, including imitation leather, "vinyl" records, and plastic pipes. And Teflon (PTFE) is a very slippery heat- and chemical-resistant plastic used as the nonstick coating in frying pans.

Plastics are such a dominant feature of the modern world that it seems hard to imagine new uses for them. Yet chemists continue to pioneer improved methods of polymerization and continually produce revolutionary new plastic materials. Plastic-based composites have long been used to manufacture automobile components, but manufacturers such as Daimler-Chrysler are now looking to produce automobile bodies built purely from plastics such as PET (polyethylene terephthalate), a material commonly used to make plastic bottles. Daimler-Chrysler claims the plastic shells are as crash resistant as steel and composites, but could halve the cost of some conventional automobiles.

Other new plastics promise a range of different benefits. One of the latest developments, light-emitting polymers (LEPs), could replace cathode ray tubes and expensive flat-panel LCD displays, and provide a cheap alternative to the semiconductor lasers used in CD players. One of the most unusual new plastics being developed is a polymer called 3GT, which has a kind of "stretch memory," that could be used to make seats that remember the shapes of their occupants or clothes that mold to peoples' bodies. For all their benefits, plastics do present a serious problem: their

Polythene sheeting
This polythene (polyethylene) sheeting shown in this picture is made from a mixture of recycled polythene pellets that have been soaked in water. The recycled polythene is melted and then extruded into a shaping die. Hot air passing through the die blows the polythene into a tube shape and dries it. The polythene tube is then drawn upward by rollers and pinched into a sheet, which is shown undergoing inspection in the background.

sheer durability means they persist in the environment for many years, and have a damaging impact on wildlife if not properly disposed of—over a quarter of the world's seabirds are estimated to have injested plastic residue. More waste plastics are now being recycled into a range of useful goods, such as upholstery padding. Starch can be added to plastic to make it more biodegradable and new types of plastic are being developed that break down completely in soil. Most of the new biodegradable plastics, such as polyhydroxybutyrate (PHB), are produced from natural polymers rather than petroleum and are consequently more expensive than other plastics.

Teflon cooking pan
Invented by DuPont in 1938, Teflon is familiar as a nonstick coating for frying pans. It has also been used as a fire- and abrasion-resistant coating for space suits. Teflon is so slippery that it takes several sandblasting and baking processes and a primer chemical to make it stick to the pan.

SEE ALSO: COMPOSITE MATERIALS *p.144* | PROTECTIVE MATERIALS *p.166* | OIL REFINING *p.182*

Mining and extracting metals

Most of the 92 naturally occurring elements are metals. Metals are generally shiny, dense, and conduct heat and electricity well. Many are malleable and ductile—they can be hammered into sheets or drawn into wire. Rarely found in their pure state, most metals occur in compounds, often with oxygen, called ores. The ore is mined from surrounding rock, and the metal can then be extracted from the ore.

The simplest method of metal extraction is chemical reduction, in which the ore is heated with a more reactive substance that removes the oxygen, leaving the pure metal. This is how iron is extracted from its ore in a blast furnace—the ore (iron oxide) is heated with coke (carbon) in an atmosphere poor in oxygen. As the coke burns, it "robs" oxygen from the ore, leaving pure iron. Some other metals must be extracted by more complex methods such as electrolysis, in which electricity is used to split a compound apart. Whatever method of extraction is used, the waste products are often toxic and must be carefully disposed of to prevent damage to the environment.

Bauxite mining
Metal ore mines can be deep or open-cast. Deep mining uses shafts and tunnels to reach seams of metal ore that have formed underground. Open-cast mining involves blasting or scraping at the ground in areas where ores such as bauxite (aluminum ore) are close to the surface.

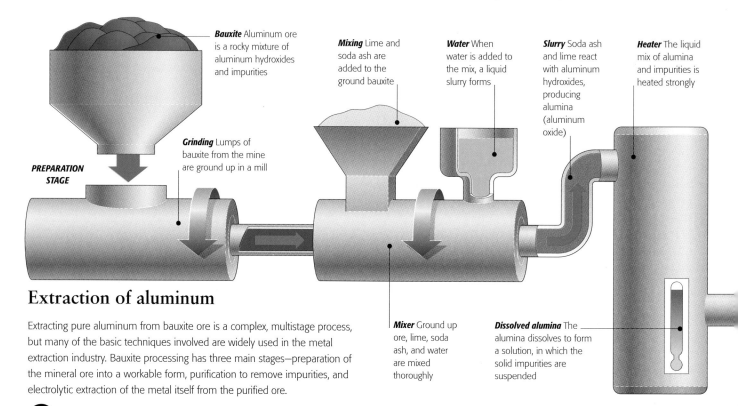

Bauxite Aluminum ore is a rocky mixture of aluminum hydroxides and impurities

Grinding Lumps of bauxite from the mine are ground up in a mill

PREPARATION STAGE

Mixing Lime and soda ash are added to the ground bauxite

Water When water is added to the mix, a liquid slurry forms

Slurry Soda ash and lime react with aluminum hydroxides, producing alumina (aluminum oxide)

Heater The liquid mix of alumina and impurities is heated strongly

Mixer Ground up ore, lime, soda ash, and water are mixed thoroughly

Dissolved alumina The alumina dissolves to form a solution, in which the solid impurities are suspended

Extraction of aluminum

Extracting pure aluminum from bauxite ore is a complex, multistage process, but many of the basic techniques involved are widely used in the metal extraction industry. Bauxite processing has three main stages—preparation of the mineral ore into a workable form, purification to remove impurities, and electrolytic extraction of the metal itself from the purified ore.

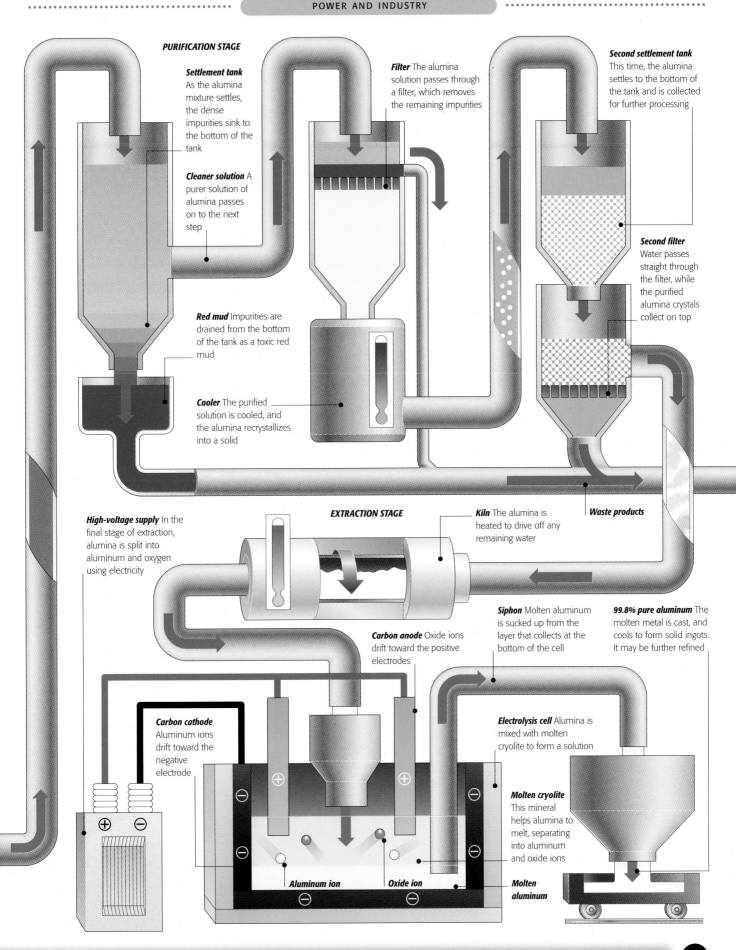

PURIFICATION STAGE

Settlement tank As the alumina mixture settles, the dense impurities sink to the bottom of the tank

Cleaner solution A purer solution of alumina passes on to the next step

Filter The alumina solution passes through a filter, which removes the remaining impurities

Second settlement tank This time, the alumina settles to the bottom of the tank and is collected for further processing

Second filter Water passes straight through the filter, while the purified alumina crystals collect on top

Red mud Impurities are drained from the bottom of the tank as a toxic red mud

Cooler The purified solution is cooled, and the alumina recrystallizes into a solid

High-voltage supply In the final stage of extraction, alumina is split into aluminum and oxygen using electricity

EXTRACTION STAGE

Kiln The alumina is heated to drive off any remaining water

Waste products

Siphon Molten aluminum is sucked up from the layer that collects at the bottom of the cell

99.8% pure aluminum The molten metal is cast, and cools to form solid ingots. It may be further refined

Carbon anode Oxide ions drift toward the positive electrodes

Carbon cathode Aluminum ions drift toward the negative electrode

Electrolysis cell Alumina is mixed with molten cryolite to form a solution

Molten cryolite This mineral helps alumina to melt, separating into aluminum and oxide ions

Aluminum ion

Oxide ion

Molten aluminum

SEE ALSO: STEEL AND CONCRETE *p.24* | GEOLOGICAL EXPLORATION *p.218* | ELECTRICITY & MAGNETISM *p.260*

Agricultural technology

Higher yields and improved quality of crops and foods such as cereals, vegetables, fruit, meat, milk, and eggs are the twin prizes farmers have strived for since the dawn of agriculture. Due to the rapid rise in the world's population, which almost tripled in size during the 20th century, this quest has become evermore pressing, demanding major innovations in farming practices.

Mechanization of farms

The seeds of the agricultural revolution were sown in the 18th century with the invention of a number of laborsaving devices. The first of these was a horse-drawn hoe-and-seed drill invented by the English farmer Jethro Tull. This reduced the waste that resulted from scattering seed on the ground by cutting a shallow trench and placing the seed in this trench. In North America, Eli Whitney's cotton gin, first used in 1793, mechanized the harvesting of cotton, which until then had been a labor-intensive task.

In the 19th century, increasingly large crops of grain were grown on the plains of the Midwest. Three inventions of the 1830s facilitated planting, harvesting, and processing these crops: John Deere's steel plow, which cut cleaner furrows than the existing iron plows; Cyrus McCormick's reaper machine, which used blades mounted on a rotating frame to cut the corn; and Hiram and John Pitt's portable threshing machine, which separated the corn seed mechanically. The widespread adoption of these inventions greatly reduced the amount of labor required for crop production and contributed to the dramatic shift of population

Combine harvesters
Machines that could both harvest cereal crops and thresh the grain only came into use at the beginning of the 20th century. Modern combine harvesters, like the ones shown here at work on North America's prairies, are large, self-propelled vehicles fitted with rotating blades.

from rural areas to the industrial cities. The next major transformation of arable farms came with the introduction of all-purpose, gasoline-fueled tractors in the 1920s. These tractors could haul a wide range of mechanical devices used to plow, sow, and harvest crops that until that time had been hauled by draft animals. The advent of four-wheel drive in the second half of the 20th century and the introduction of diesel engines made tractors more powerful and cheaper to run. There are currently over 16 million tractors in use worldwide, a quarter of which operate in the U.S. alone.

Milking machine
The systematic electrification of rural areas has led to milking machines being widely adopted on dairy farms. Powered by electric motors, the machine shown here is able to milk more than 20 cows at the same time.

The mechanization of livestock husbandry has seen significant advances since the electrification of rural areas, which was not undertaken on a wide scale in the U.S. until the 1930s. The arrival of this power source has enabled farmers to introduce a range of automated equipment, such as milking machines, and prompted the development of intensive indoor production methods, such as "battery farming" for chickens. Electric power is also used for sprinkler systems and water pumps.

Animal husbandry and crop production

Significant improvements in the quality of livestock began with intensive selective-breeding practices. In the late 1700s, an English farmer named Robert Bakewell showed that by continually breeding animals selected for their desirable traits, marked changes could be effected. He became best known for a breed of sheep that fattened quickly and so could be raised at low cost for slaughter as well as for wool.

In the 18th century, a four-year crop cycle, devised in England by Charles Townsend, made farmland more productive, enabling crops to be grown all-year-round without depleting nutrients in the soil. Cereal crops were grown in rotation with turnips and crops such as alfalfa, which restore nitrogen and other nutrients to the soil. Artificial fertilizers were not used until the 1920s, when an economical way of obtaining nitrogen from the atmosphere was found. Since then, artificial fertilizers have been used extensively. The introduction of the insecticide DDT and the herbicide 2,4-D in the 1940s further increased the use of chemicals on the land. While the use of artificial fertilizers, insecticides, herbicides, and fungicides has greatly boosted farm output, it has had its costs. Their use has caused damage to ecosystems, and concerns about the levels of toxic chemicals entering food chains and the water system has prompted research into safer alternatives.

The 1960s saw the introduction of high-yield hybrid varieties of wheat and rice, produced by laboratory plant-breeding techniques. The resultant increase in yields was so dramatic that the introduction was dubbed the "green revolution." Modern genetically modified crops, whose genetic code has been altered to produce desirable traits such as pest-resistance, are promising to offer even greater yields.

Crop-dusting plane
In order to reduce wastage, crop-dusting planes spray pesticides through a electrostatic nozzle that imparts a positive charge to the droplets of spray. When the droplets land on the crops, a negative charge is induced on their surface, creating an attractive force that binds the spray to the crop.

Precision farming and hydroponics

Advances in satellite and computer technology have led to the introduction of systems that allow farmers to measure the precise distribution of crops in a field. Electronic devices fitted in a combine harvester measure the flow of grain entering its tank, while the Global Positioning System is used to track the position of the machine. From this data, a computer produces a yield map. This map enables farmers to target the correct amount of chemicals to the areas they are needed, which cuts down wastage and reduces environmental damage. Other methods of reducing the impact of chemicals on the land include organic agriculture and hydroponics. The latter is a method of growing fruit, vegetables, and herbs without soil, under carefully controlled conditions, in places where arable land is scarce. The roots of the plants grow in water enriched with nutrients and are anchored by a porous material, such as gravel, fiber, or sand. Since the nutrient-enriched water is contained, it does not pose the same threat as runoff water from fertilized soil. After use, hydroponic water is piped into beds where it is purified by organic methods similar to those used in sewage works, before being pumped back into a river.

Hydroponically grown lettuce
The lettuces in this greenhouse are being grown in nutrient-enriched water rather than soil. Unsuitable land can thus be used for growing crops. Sodium lights take the place of sunlight so that the lettuces can grow in any season.

SEE ALSO: THE WATER SYSTEM *p.30* | GLOBAL POSITIONING SYSTEM *p.136* | GENETICALLY MODIFIED FOODS *p.190*

Genetically modified foods

Genetically modified (GM) foods are foods that have been altered in some way as a result of manipulation of their genes by genetic engineering. In many cases, this involves introducing a gene from one species into a totally unrelated species. For example, an animal gene may be inserted into a plant in order to produce a desired characteristic, such as resistance to insects or tolerance to cold.

Most GM foods are of plant origin, the main reason being that it is usually easier to produce a GM plant than a GM animal. Producing GM animals usually involves adding a foreign gene to a fertilized egg so that a copy of the gene will be present in every cell as the egg divides. This manipulation of the sex cells and subsequent implantation of the fertilized egg back into the uterus of a female is a complex procedure. In contrast, to produce a genetically modified plant, a cell from any part of the plant can be used. After modification, it can then easily be cultured to produce a whole new plant.

The creation of most GM crops involves using a vector—a piece of DNA, usually obtained from a bacterium, that is used to carry genetic material from one species to another. The bacterium *Agrobacterium tumefaciens* naturally infects many different plant species and inserts genes that it carries in a loop of DNA into the plant chromosomes. Genetic engineering uses this loop of DNA, called a plasmid, as a vector to insert carefully selected genes into plant cells. These genes may be either modified forms of a plant's own genes or completely foreign genes. Other techniques for the incorporation of new genes into plant cells have been investigated. For example, genes and plant cells may be mixed together. An electric current is then applied to make them combine. Generally, these methods are not as successful as those that use DNA vectors.

Examples of beneficial GM crops include a strain of rice that contains a new gene for producing vitamin A. Deficiency of this vitamin is a common cause of blindness in the developing world. Consumption of foods made from these crops increases dietary vitamin A intake and has reduced the incidence of vitamin A-related blindness. Other genetic modifications create sturdier plants, thus benefiting crop producers. Such modification may make crops poisonous to common pests, or allow crops to be blanket-sprayed with herbicide, or make crops more tolerant to arid or cold conditions. An example of a GM food from animals is the leaner pork obtained from pigs that carry a growth hormone gene derived from cows.

Genetic invaders

The bacterium *Agrobacterium tumefaciens* causes the disease "crown gall" in many types of plant. The bacterial cells attack a plant cell, breaking down its tough outer wall. The disease is caused by the bacterium transferring some of its genes into the chromosomes of the plant. A genetically modified form of *A. tumefaciens,* in which the disease-causing genes have been removed, is commonly used by plant geneticists as a means to transfer foreign or modified useful genes into plants. *A. tumefaciens* does not infect all plants; many important food crops, such as certain bananas and cereals, are naturally resistant to the bacterium. However, ways have now been found to infect even these plants with the bacterium to create GM crops.

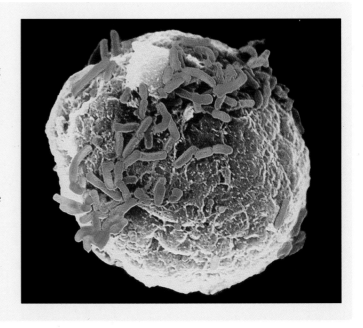

Bacteria invading a plant cell
This image shows bacteria of the species A. tumefaciens *invading a tobacco plant cell.* A. tumefaciens *is capable of penetrating the tough walls of plant cells.*

Genetically modified plants

Genes are lengths of DNA that are found along chromosomes within the cell nuclei of plants and animals. They control the synthesis of proteins that directly affect physiological processes. To make a protein, the gene makes a copy of itself, forming a molecule called messenger RNA (mRNA). The mRNA provides the cell with a template for the manufacture of the protein. Genetic engineering in plants introduces modified versions of the plant's own genes or foreign genes into the plant-cell chromosome. This causes the plant to either increase or inhibit the production of an existing protein, or to manufacture a new protein. Some plants make this new gene product—typically, an insecticidal protein or a cold-tolerant protein—throughout their lives, while others do so only at a certain stage in the plant's life cycle.

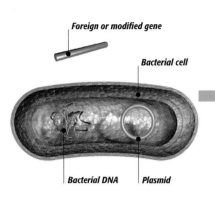

Foreign or modified gene

Bacterial cell

Bacterial DNA **Plasmid**

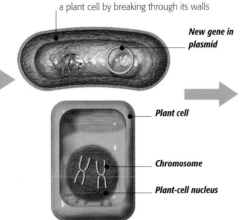

GM bacterial cell This bacterium infects a plant cell by breaking through its walls

New gene in plasmid

Plant cell

Chromosome

Plant-cell nucleus

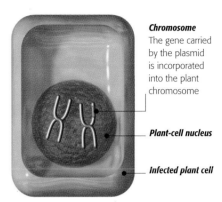

Chromosome The gene carried by the plasmid is incorporated into the plant chromosome

Plant-cell nucleus

Infected plant cell

Bacterial vectors
Some bacterial cells such as A. tumefaciens *contain plasmids—rings of DNA outside the chromosome that are able to replicate themselves independently of the cell's chromosomal DNA. Foreign or modified genes can be inserted into this genetic element.*

Gene splicing
The foreign or modified gene with the desired effect or product is spliced into the plasmid using restriction enzymes, which act like "biological scissors" to cut open the plasmid at specific points for gene insertion.

Infected plant cell
The GM bacterium invades the plant cell and transfers its DNA into the plant cell's own chromosomes.

New plant

Expression of genes in an engineered plant
Genes introduced into a plant's cells influence its physiology in various ways. If the inserted gene is identical to one of the plant's own genes, it will increase protein production. If it is a modified gene, a modified protein will be produced. If the inserted gene is arranged to produce "antisense" mRNA (complementary to the plant's natural mRNA), the two pieces of mRNA bind, inhibiting protein production. Insertion of a foreign gene produces a protein that the plant could not naturally make.

Adult plant

Culturing
The plant cell with the modified gene is then cultured and induced to multiply, each new cell carrying the new gene. The plant cells become new GM plants, each with the desired characteristic.

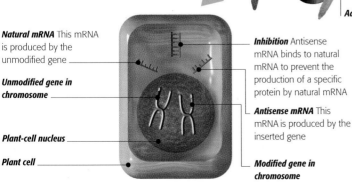

Natural mRNA This mRNA is produced by the unmodified gene

Unmodified gene in chromosome

Plant-cell nucleus

Plant cell

Inhibition Antisense mRNA binds to natural mRNA to prevent the production of a specific protein by natural mRNA

Antisense mRNA This mRNA is produced by the inserted gene

Modified gene in chromosome

INHIBITING PROTEIN PRODUCTION

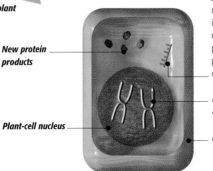

New protein products

Plant-cell nucleus

Modified mRNA The mRNA produced by the inserted gene increases natural protein production, or produces proteins that the plant cell could not ordinarily make

Inserted gene in chromosome

Plant cell

ENHANCING PROTEIN PRODUCTION

SEE ALSO: AGRICULTURAL TECHNOLOGY p.188 | GENETIC ENGINEERING p.212 | HUMAN GENOME PROJECT p.214

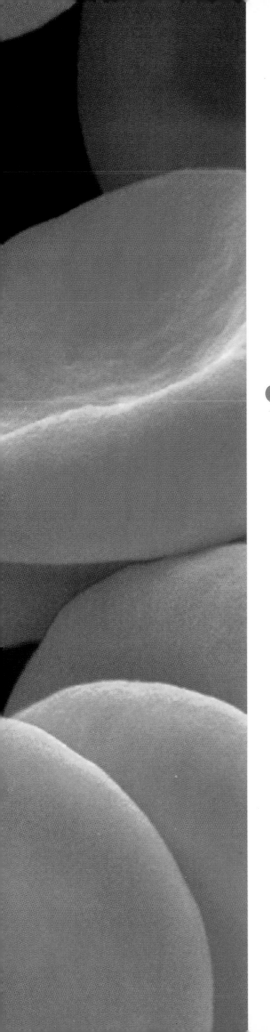

Medicine
and Research

Invention and innovation are vital processes at the cutting
edge of scientific research. With new processes, techniques,
and instruments, researchers can push forward the frontiers
of knowledge, ultimately resulting in new forms of applied
science that affect us all. As this chapter shows, many of the
most obvious benefits of pure research are seen in the field
of medicine, where new diagnostic techniques, drugs, and
other treatments develop only from years of study—as
shown by the enormous effort put into decoding the human
genome. However, research into every aspect of our
environment—ranging from the subatomic to the cosmic—is
also having a profound impact on how we view our world.

Ultrasound

Sound at frequencies above the upper limit of human hearing (by convention, 20 kHz) is called ultrasound. Ultrasonic waves transport energy and can be reflected, refracted (bent), and focused. These properties are exploited in numerous applications, such as sonar. In medicine, ultrasound is used in a range of diagnostic and treatment procedures, from prenatal imaging to destruction of kidney stones.

Medical ultrasound imaging uses very high frequency sound waves, typically 5 to 10 MHz, to produce images of internal organs and tissues. The most common imaging procedure is prenatal scanning to check fetal development; ultrasound is also used to image the kidneys, liver, heart, and blood vessels. The

principle of ultrasound imaging is similar to that of echo location as used by bats and in sonar. A probe outside the body focuses ultrasonic pulses into the region to be examined. The pulses penetrate the body and are reflected at the boundaries between different tissue types. Reflected pulses are detected and processed by a computer to produce an image. Doppler ultrasound—a technique that uses the Doppler effect (the apparent variation in frequency caused by movement of the source of waves or echoes) —enables motion such as blood flow to be imaged.

High-intensity ultrasound may have destructive effects on body tissue. A carefully tuned and focused burst of intense ultrasound may be used to destroy abnormalities such as kidney stones from outside the body, without the need for an operation.

Shock wave lithotripsy

Some people are prone to developing hard masses called stones in their urinary system. Small stones can pass out of the body, but large stones traditionally needed surgical removal. Today, a noninvasive procedure called lithotripsy makes such surgery unnecessary. During lithotripsy, the patient lies still on a treatment table and the kidney stone is located using X-rays, or, less commonly, low-intensity ultrasound. High-intensity ultrasonic shock waves from an external source are then focused onto the stone, crumbling it into fine particles that can pass out of the body with urine.

X-ray detector Data from the X-ray detector is used to focus the ultrasonic shock waves onto the kidney stone

Beam of X-rays

Kidney

Ultrasonic shock waves

Shock wave The shock wave is focused on the kidney stone

Kidney

Ureter Fine particles of the kidney stone pass through the ureter with urine for excretion

Kidney stone The hard kidney stone is crumbled by the shock waves

Water-filled balloon The balloon and a jelly layer on the patient's skin ensure that shock waves pass into the body without disruption

Lithotripsy table

Shock wave source One type of source uses hundreds of piezoelectric elements arranged in a concave dish. When a voltage is applied across them, these elements produce ultrasonic shock waves

X-ray source X-rays are used to locate the kidney stone precisely

Destroying kidney stones
Carefully focused shock waves crumble the kidney stone without damaging surrounding tissue. About 3,000 shock waves are delivered during a one-hour treatment session.

Ultrasound scanning

An ultrasound machine consists of a scanner head, which is both the source and receiver, a computer, and a display unit. Piezoelectric (see p.227) transducers in the scanner head vibrate under an applied voltage, generating ultrasound. A computer controls the voltage applied, regulating and focusing the ultrasonic pulses. The pulses travel at different speeds through different tissues and organs, and are partially reflected as they cross boundaries between them. Returning echoes are received by the transducers, causing them to vibrate and produce a voltage. A computer analyzes the strength and delay of the echo signal and constructs an image on a monitor screen.

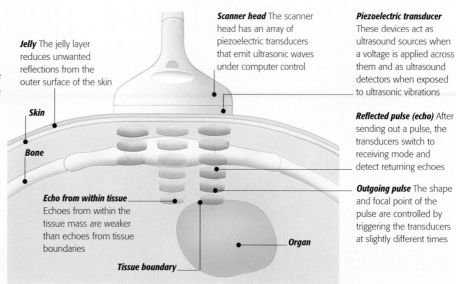

Jelly The jelly layer reduces unwanted reflections from the outer surface of the skin

Skin

Bone

Scanner head The scanner head has an array of piezoelectric transducers that emit ultrasonic waves under computer control

Piezoelectric transducer These devices act as ultrasound sources when a voltage is applied across them and as ultrasound detectors when exposed to ultrasonic vibrations

Reflected pulse (echo) After sending out a pulse, the transducers switch to receiving mode and detect returning echoes

Echo from within tissue Echoes from within the tissue mass are weaker than echoes from tissue boundaries

Outgoing pulse The shape and focal point of the pulse are controlled by triggering the transducers at slightly different times

Organ

Tissue boundary

Prenatal scanning

Ultrasound scanning is relatively simple, inexpensive, and safe. Since no radioactivity or electromagnetic radiation is involved, the technique is suitable for scanning a fetus and is a routine diagnostic procedure during pregnancy. Prenatal ultrasound is used to measure fetal growth, to check for multiple pregnancies, and to screen for abnormalities such as spina bifida and Down's syndrome.

Layer of jelly Since ultrasound is reflected at air-tissue junctions, a layer of jelly is applied between the scanner head and the abdomen to reduce echoes from the skin surface

Abdomen

Fetus

Uterus

Cable to computer

Scanner head The head contains hundreds of transducers, each less than one millimeter wide

Ultrasound beam Under computer control, the transducers emit ultrasonic pulses that combine to form a tightly focused spot of sound

Scanning plane The focused spot of ultrasound is electronically scanned across a plane to give a 2-D "slice" view through the uterus and fetus

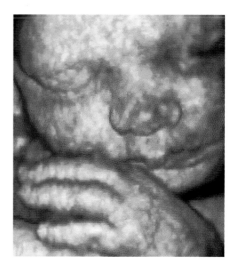

3-D fetal ultrasound
The most common type of prenatal ultrasound imaging produces 2-D images. However, these 2-D "slices" may be combined, using sophisticated computers, to make an accurate 3-D image, which can help to diagnose certain prenatal abnormalities more easily.

Monitoring the progress of pregnancy
An ultrasound scan is usually carried out about seven weeks into a suspected pregnancy to confirm it. At 20 weeks, a second scan is carried out to check for abnormalities and multiple fetuses. A final scan is performed at 34 weeks to measure fetal size and assess growth.

SEE ALSO: NAVIGATION *p.132* | X-RAYS & CT SCANNING *p.196* | MRI & PET SCANNING *p.198*

X-rays and CT scanning

Knowledge of internal human anatomy advanced greatly after William Röntgen discovered in 1895 that X-rays, a type of electromagnetic radiation, were partially absorbed by matter and could be used to create images of the body. X-ray images are used to diagnose fractures and diseases and also to scan materials. X-rays may be used to treat cancer and to investigate the structure of crystals.

The first X-ray machines revealed shadows on the lungs, fractures in bones, and inflammation of arthritic joints. Later, radiopaque substances, which are deliberately introduced into the body to absorb X-rays, were developed to make radiographs (X-ray photographs) of soft tissues possible. Barium meals are ingested to reveal gastric ulcers, and dyes are injected to image the heart and kidneys. Today, advanced machines called computed-tomography (CT) scanners use multiple X-ray beams to build up detailed cross-sectional images of the body. CT scans clearly show the structure of all tissue and bone, greatly enhancing diagnosis and treatment.

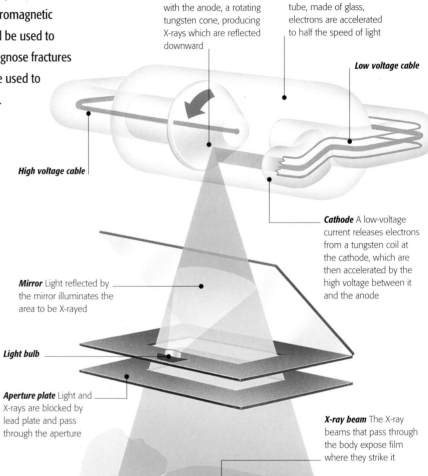

Anode Electrons collide with the anode, a rotating tungsten cone, producing X-rays which are reflected downward

X-ray tube In this vacuum tube, made of glass, electrons are accelerated to half the speed of light

Low voltage cable

High voltage cable

Cathode A low-voltage current releases electrons from a tungsten coil at the cathode, which are then accelerated by the high voltage between it and the anode

Mirror Light reflected by the mirror illuminates the area to be X-rayed

Light bulb

Aperture plate Light and X-rays are blocked by lead plate and pass through the aperture

X-ray beam The X-ray beams that pass through the body expose film where they strike it

Film tray Double-sided film sensitive to X-rays captures the image of the scan. Fluorescent screens placed either side of the film intensify the image

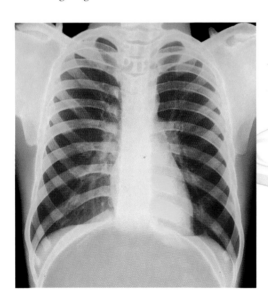

Chest X-ray
This color-enhanced radiograph of an 11-year-old boy's chest shows that his lungs (dark areas) and heart (yellow) are healthy. The X-ray also shows a few vertebrae of the spine, ribs (pink bands), diaphragm (green), and the clavicles, or collar bones.

Radiography

An X-ray tube projects a beam of X-rays through a body and onto a film where it makes a radiograph. The beam, which passes through a variable-sized aperture onto the body, is absorbed according to the density of the internal tissues. Bones absorb more of the X-ray beam than muscles, lungs, or other tissues and show up as light areas on the radiograph.

Computed tomography

A computed-tomography (CT) scanner projects a series of thin X-ray beams from different angles as its drum rotates through a 360-degree circle around a patient's body. A detectors opposite the tube send signals to a computer, which synthesizes the many views into one cross-sectional digital picture, showing a slice of the body. A 3-D accumulation of different slices allows entire organs to be viewed.

X-ray tube A tube emits a series of X-rays as the scanner rotates around the body

X-Ray beam The X-ray beams pass through the body and are sensed by a detector

Detector The X-ray beams strike light-sensitive crystals in the detector and are converted to electronic signals that are transmitted to a computer

Motorized bed The bed can move forward or backward so that any part of the body can be aligned with the scanner

Scanning drum During each scan, the scanning drum is rotated through 360 degrees. It can also be tilted between scans to change the angle of view

CT image
This 3-D CT image, which views part of the human digestive system from below, can be helpful to surgeons planning difficult operations close to vital organs. Clearly visible are the ribs and spine (yellow), stomach and duodenum (purple), pancreas (green), and kidneys (light blue).

MRI and PET scanning

Modern medical-imaging techniques allow doctors to look inside the human body with far more detail than that offered by X-rays. Techniques such as magnetic resonance imaging (MRI) and positron-emission tomography (PET) scanning image soft tissue inside the body and even chemical changes in tissues carrying out certain functions. Together, these techniques are revolutionizing many aspects of diagnosis, treatment, and research.

While X-ray imaging relies on the fact that different body tissues absorb X-rays aimed at them to different degrees, MRI and PET scanning turn the body into a radiation emitter. PET scanning detects radiation from a "tracer" substance introduced into the body. The tracer is taken up by areas of the body that are most active, so PET scanning can be used to map body functions. In MRI, the entire body is magnetized and then scanned with radio waves to build up highly detailed images of tissue structure and composition. These techniques provide more information at less risk to the body than X-ray imaging.

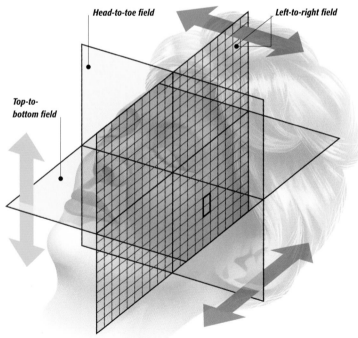

False-color MRI section through the head
MRI scans produce images based on the distribution of hydrogen atoms in the body. Hydrogen atoms are found in water and many organic molecules. Bodily fluids have the highest density of hydrogen atoms, then soft tissues, cartilage, and membrane. Bone is invisible to an MRI scan.

How an MRI scanner works

The human body is packed with hydrogen atoms, each of which acts like a tiny magnet. Normally, these atoms are randomly aligned, so the body has no net magnetic field. An MRI scanner forces all the hydrogen atoms to align and then probes the body with radio waves to "see" the density of hydrogen in different areas.

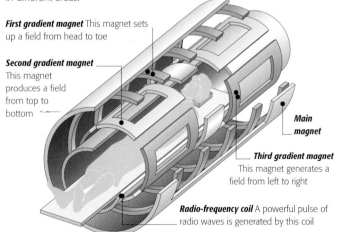

First gradient magnet This magnet sets up a field from head to toe

Second gradient magnet This magnet produces a field from top to bottom

Main magnet

Third gradient magnet This magnet generates a field from left to right

Radio-frequency coil A powerful pulse of radio waves is generated by this coil

Head-to-toe field

Left-to-right field

Top-to-bottom field

1 MRI scanners contain several magnetic coils, each of which has a separate function. A main electromagnet is on all the time, and three gradient magnets are turned on as needed. In the first stage of the imaging procedure, the patient enters the MRI scanner to become immersed in the extremely powerful field of the main magnet.

2 When any one of the three gradient magnets is activated, it alters the field of the main magnet along one axis, enabling a magnetic slice through the body to be selected. One magnet selects the slice, and the other two magnets read data up and down and across the slice. The selected slice is divided into a grid of small boxes or voxels by the scanner's computer.

How PET scanning works

Positron-emission tomography (PET) scanning maps the function of the body's organs and tissues by detecting the amount of metabolic or chemical activity in a particular part of the body. A small amount of a tracer substance, typically radioactive glucose, is introduced into the body. Glucose tracer molecules are "labeled" with radioactive fluorine atoms. The tracer becomes concentrated in those regions of the body that are most active and need most glucose. Radiation from the decay of tracer molecules is then mapped by a ring-shaped array of detectors.

Glucose molecule "labeled" with radioactive fluorine

Positron The radioactive tracer emits a positron which collides with an electron

Gamma ray The collision produces two gamma rays traveling in opposite directions

Gamma ray

Mapping the brain

A PET scanner produces a map of activity in brain tissues by detecting radiation emitted when positrons (produced by radioactive tracer molecules) collide with electrons in brain tissue.

Brain The part of the body to be imaged, in this case the brain, is surrounded by a ring of gamma-ray detectors

Gamma-ray detector When two gamma rays traveling in opposite directions are detected simultaneously at two detectors, a computer maps the point from which they originated

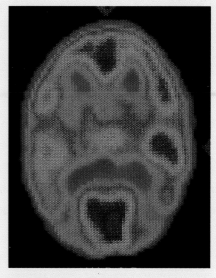

PET scan of brain activity

This image has been color coded to indicate regional variations in brain activity while listening to music. Red and yellow indicate high activity, blue and green lower activity. Researchers can use PET scans to identify the brain regions used in different types of thought processes, while surgeons can use it to identify regions of brain damage.

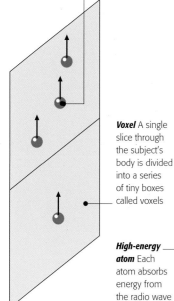

Hydrogen atom Each atom lines up with the magnetic field, but oscillates at the Larmor frequency

Incoming radio waves A precisely tuned radio pulse alters the orientation of the hydrogen atoms

Flipped atom The hydrogen atoms flip back into line with the magnetic field, giving off radio waves

Outgoing radio waves The frequency of waves produced as the hydrogen atoms flip back depends on the strength of the local magnetic field produced by the gradient magnet

Voxel A single slice through the subject's body is divided into a series of tiny boxes called voxels

High-energy atom Each atom absorbs energy from the radio wave

Magnetic field The gradient magnets switch on, giving each voxel a unique magnetic field

Strong signal In regions with high densities of hydrogen, the radio waves combine to produce a strong signal

Frequency difference Atoms in neighboring voxels emit waves of similar (but not identical) frequencies

Weaker signal from hydrogen-poor area

3 *The powerful main magnetic field forces hydrogen atoms to line up in one direction and oscillate around their magnetic poles at a frequency called the Larmor frequency.*

4 *When a pulse of radio waves at the Larmor frequency irradiates the body, the hydrogen atoms absorb its energy and are temporarily knocked into a different alignment.*

5 *The gradient magnets are switched on, and the hydrogen atoms flip back into alignment with the main field, emitting radio waves. The intensity and frequency of the waves produced depend on the strength of the local field. Different radio frequencies come from different locations in the slice selected, while the signal strength reveals the density of hydrogen atoms.*

SEE ALSO: COMPUTER OUTPUT DEVICES *p.68* | X-RAYS & CT SCANNING *p.196* | MATTER *p.258*

Designing modern drugs

Drugs are natural or manufactured substances that affect physiological processes in the body. In the past, their effects were often discovered by chance. Today, tailor-made drugs are created in the laboratory using 3-D computer graphics to recreate the molecular structures of the drugs themselves and of the natural chemicals in our bodies with which these drugs are designed to interact.

Body cells make substances called proteins that affect the body's function. Many proteins are produced by the interaction of a natural chemical messenger outside the cell, such as a hormone molecule, with a cell receptor (itself a type of protein) on the cell's surface. Cell receptors, like all protein molecules, consist of amino-acid chains with a characteristic shape. The messenger molecule, or ligand, fits the shape of a hollow region, or "pocket," in the receptor. The ligand binds with the receptor, like a key in a lock, sending a chemical signal into the cell so that protein production occurs. Computers can create an accurate image of the shape of a receptor pocket, on which the shapes of various organic molecules are overlaid and examined until one is found that fits exactly. This molecule can then be tested as a drug. Some such drugs block protein production, others enhance it. Many other types of drugs can also be designed using 3-D graphics software.

Computer modeling of drugs

The most modern drug-design techniques produce 3-D images of drug molecules and the receptor proteins in the body with which the drug will interact. The molecular structure of the receptor protein is revealed using an scanning technique called "X-ray crystallography," and a 3-D model is created using computer graphics programs. A model of the potential drug is then designed and manipulated on screen to fit the receptor molecule.

How drugs work

Most drugs work by binding to receptors—specialized proteins on the surface membranes of body cells or the cells of disease-causing organisms in the body—to enhance or inhibit a chemical process. Normally, natural chemical triggers, such as hormones, bind with receptors to produce changes within cells and, in this way, affect bodily processes. Drugs known as agonists mimic natural chemical triggers to enhance their effects within cells. Other drugs, known as antagonists, block the actions of chemical triggers so that no change occurs.

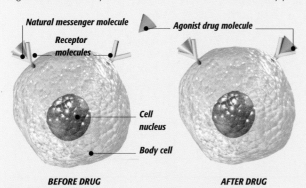

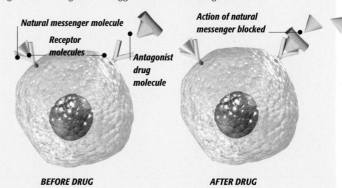

Agonist drug

Molecules of a natural chemical messenger bind with receptors on the surface of a cell to produce chemical changes in the cell. An agonist drug mimics the natural chemical messenger, thereby enhancing its effect.

Antagonist drug

An antagonist drug binds with the receptor, but produces no internal change in the cell. As the antagonist occupies the receptor sites, the action of the natural chemical messenger is blocked.

Modeling a modern drug

Modern drugs are designed with the aid of computer programs that model the molecular structures of both potential drugs and the receptor chemicals into whose hollow regions, or "pockets," the drug molecules must fit. The molecules must be able to bond chemically with the pockets. The receptor chemicals may belong either to human body cells or to disease-causing organisms that may invade the body, such as bacteria or viruses. A computer operator rotates 3-D images of the molecules on screen in order to find the closest fit between the receptor pocket and the potential drug molecule. When the best match has been found, the potential drug can be synthesized and clinically tested.

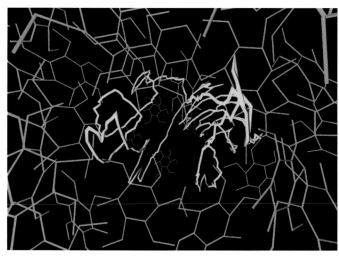

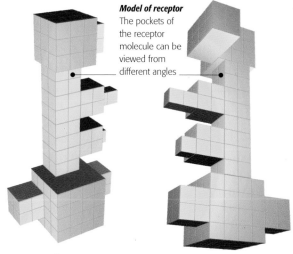

Model of receptor
The pockets of the receptor molecule can be viewed from different angles

1 A computer generates a model of the 3-D molecular structure of the targeted receptor protein (here shown in green) using the protein's known amino-acid sequence. The hollow region, or "pocket," which could accommodate a potential drug molecule, is shown in yellow.

2 The regions modeling the molecular structure of the protein are then removed, leaving the model of the pocket, which can be rotated on screen. The model provides a starting point for the design of a drug molecule that will inhibit or enhance a chemical change in a cell.

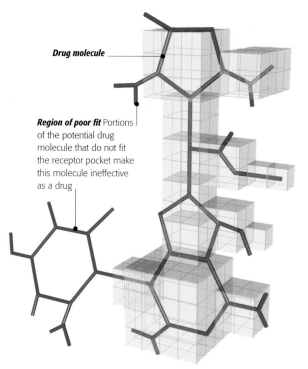

Drug molecule

Region of poor fit Portions of the potential drug molecule that do not fit the receptor pocket make this molecule ineffective as a drug

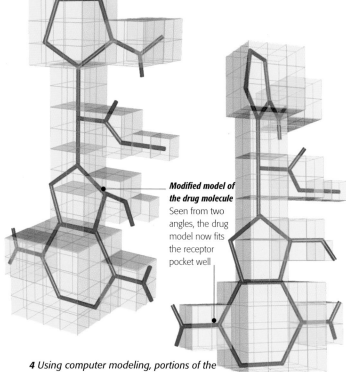

Modified model of the drug molecule
Seen from two angles, the drug model now fits the receptor pocket well

3 A model of a potential drug molecule is inserted into the model of the receptor pocket to see how well it fits. Models of many chemicals with slightly different molecular structures will be tested on screen until the best match is found.

4 Using computer modeling, portions of the molecule model that project from the pocket can be identified, and the drug molecule modified to make a better fit. Once the most appropriate molecular structure has been found, the drug can be synthesized in the laboratory, then tested for effectiveness and safety.

SEE ALSO: THE VIRTUAL WORLD *p.74* | DRUG DELIVERY SYSTEMS *p.202* | HUMAN GENOME PROJECT *p.214*

Drug delivery systems

Advances in drug delivery today make it possible to take medication without having to use pills or injections. The development of slow release skin patches and drug implants, internal pumps that automatically deliver precise doses of medicine at regular intervals, along with painless methods of transferring drugs directly into the bloodstream is creating a revolution in medical treatment.

Although the skin appears to be an impenetrable barrier around the human body, it is in fact permeable. The first research into slow release or transdermal (through the skin) patches took place in the 1950s, but interest did not really take off until the 1980s. One of the earliest uses was in treating motion sickness, but patches are now used for a growing number of applications, including hormone replacement therapy, treating angina and glaucoma, and assisting with dieting and smoking cessation.

Implantable drugs are a more recent development. Slow-release capsules inserted just under the skin are used to provide long-term contraceptive protection, while implantable pumps are used to treat illnesses such as diabetes and cancer.

Targeting drugs

While some modern drug delivery methods simply put medicines into the bloodstream painlessly and effectively, others make it possible to target the areas where treatment is required more accurately than ever before. Slow-release patches and capsules are used for drugs that need to be distributed around the body, but injection guns and internal pumps with catheters can target specific areas, such as the teeth or spine.

Powder injection High-pressure injection guns are used to administer anesthetics into the gums

Patch A slow-release patch attaches to a clean, dry, and hairless part of the body, such as the hip, torso, or upper arm

Slow-release capsule These capsules are often inserted into the arm

Slow-release patches

A typical slow-release patch has several layers. The patch works by the process of diffusion, by which a concentrated substance naturally disperses into areas where it is less concentrated. The rate of release can be controlled by an intervening membrane, or by suspending the drug in another material called a matrix, which lowers its initial concentration.

Backing The backing protects the patch from damage and prevents the drug from seeping out

Drug reservoir The drug is sandwiched between the other layers, in an easily absorbed liquid or gel form

Membrane The membrane limits the flow of drug from the reservoir, controlling the rate of release

Adhesive A layer of glue attaches the patch to the skin

Blood vessel

Drug

Dermis

Epidermis

Slow-release capsules

Capsules inserted underneath the skin can release drugs directly into the bloodstream. Each capsule has a thin permeable membrane surrounding a gel with the drug suspended in it. The drug slowly diffuses into the body, sometimes over several years.

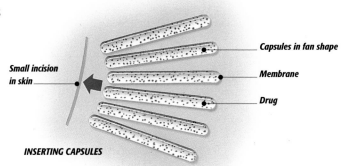

Small incision in skin

Capsules in fan shape

Membrane

Drug

INSERTING CAPSULES

Injection guns

Instead of allowing drugs to permeate gradually through the skin, new injection methods force them into the bloodstream at such high speeds that they pass straight through the outer layers of the skin. These pain-free injections do not use needles—they rely on gas pressure to inject the drug, which is prepared in a finely powdered form. Injection guns have many applications, including dentistry. Since the guns do not penetrate the skin, they do not have to be discarded after a single use for reasons of hygiene.

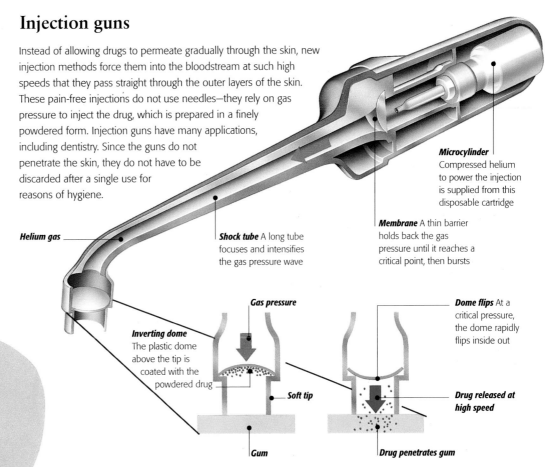

Microcylinder Compressed helium to power the injection is supplied from this disposable cartridge

Membrane A thin barrier holds back the gas pressure until it reaches a critical point, then bursts

Helium gas

Shock tube A long tube focuses and intensifies the gas pressure wave

Gas pressure

Dome flips At a critical pressure, the dome rapidly flips inside out

Inverting dome The plastic dome above the tip is coated with the powdered drug

Soft tip

Drug released at high speed

Gum

Drug penetrates gum

Pump Pumps placed inside the body, have to be made of materials that the body will not reject, such as titanium

Externally controlled pumps

Implantable pumps deliver a specified amount of a liquid drug on a regular basis through a tube called a catheter, which leads directly to the site where the drug is needed. They are used for treating cancer, diabetes, and chronic pain. Inside the pump is a delivery device, drug reservoir, power source, and electronics that control the rate of delivery. The pump is monitored through an external controller and regularly refilled.

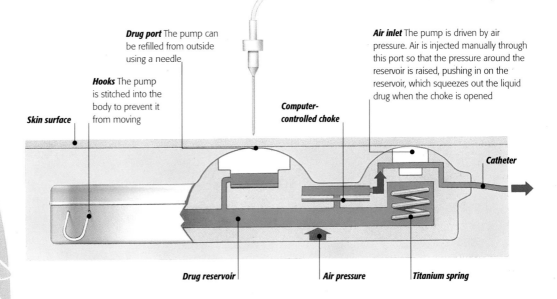

Drug port The pump can be refilled from outside using a needle

Hooks The pump is stitched into the body to prevent it from moving

Skin surface

Computer-controlled choke

Air inlet The pump is driven by air pressure. Air is injected manually through this port so that the pressure around the reservoir is raised, pushing in on the reservoir, which squeezes out the liquid drug when the choke is opened

Catheter

Drug reservoir

Air pressure

Titanium spring

SEE ALSO: DESIGNING MODERN DRUGS *p.200* | GENETIC ENGINEERING *p.212*

Endoscopic surgery

Endoscopic surgery is a revolutionary procedure in which surgeons operate without making large incisions. It relies on the use of an endoscope—a viewing tube that is inserted through a tiny incision or a natural body opening such as the mouth—through which small surgical instruments may be passed. Endoscopic surgery produces less scarring than open surgery, is quicker, and reduces healing time.

Endoscopes are given different names according to the body part being viewed. For example, a laparoscope is used for operations on the abdomen, a bronchoscope for the windpipe and lungs, and an arthroscope for joints. The endoscope is commonly used as a diagnostic tool, but an examination with an endoscope will often be the precursor to endoscopic surgery. Imaging technology enables surgeons to identify precisely the areas requiring surgery before operating. Instruments and laser attachments passed down the endoscope can cut tissue, drain fluid, destroy diseased cells, and seal wounds.

Monitor image The monitor provides a magnified view of the surgical site, enabling the entire surgical team to observe the progress of an operation

Video monitor
The image of the surgery site is carried by optical fibers to the eyepiece of the endoscope handset and also to a video monitor.

Eyepiece

Brake for up/down control

Up/down control

Brake for left/right control

Left/right control

Principal tool control The surgeon uses this to manipulate the endoscopic tools when operating on a patient

Suction pump control Suction removes fluid from the endoscope tip

Air and water pump control Air is pumped into the site to distend an organ, allowing better access for viewing, while water is used to clean the endoscope tip

Surgical endoscope handset

The surgeon uses control wheels on the handset to maneuver the endoscope inside the body. The handset also has controls for delivering air and water to the endoscope tip and for removing waste by suction. The handset is linked to the video monitor.

Endoscope tube

Tool raiser control

Secondary control The assisting surgeon manipulates other tools with this control

Optical fiber link to video monitor

Light, water, and suction cable

Endoscope tube

Endoscopic instruments

Specialized surgical tools controlled by a cable are passed down the endoscope to enable the surgeon to operate inside the body. Forceps are used for holding and removing body parts undergoing surgery. An electromagnetic wire loop can sever or seal tissue using a high-frequency electric current. Biopsy forceps remove small tissue samples, and cytology brushes scrape cells from tissue surfaces. These samples can then be tested for infection or disease. Other instruments include laser attachments, which are used in many operations to cut or seal tissue and to stop bleeding. Today, endoscopic surgery has replaced traditional open surgery for many operations.

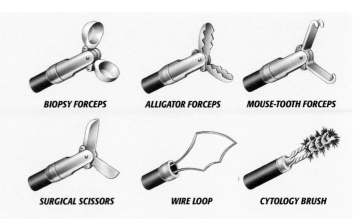

BIOPSY FORCEPS   **ALLIGATOR FORCEPS**   **MOUSE-TOOTH FORCEPS**

SURGICAL SCISSORS   **WIRE LOOP**   **CYTOLOGY BRUSH**

Endoscope tube

Inside the tube are channels that house control wires and bundles of optical fibers. These light-transmitting glass fibers, as thin as human hair, both illuminate the inside of the body and transmit images. Other channels carry air, water, and a range of surgical instruments to the operation site.

Right control wire This control wire allows the endoscope to be angled to the right

Image guide This bundle of optical fibers carries an image of the surgical site to the eyepiece

Biopsy channel Surgical tools and their control wires run through this channel

Up control wire

Left control wire

Tool-raiser wire Some tools are operated by the tool-raiser wire

Light guide Light is provided inside the body through this bundle of optical fibers

Focusing wire This wire is used to adjust a lens and focus the image

Light guide

Metal braid The metal covering is flexible enough to twist its way around any organ

Intertwined helical metal strips A second metallic layer protects the inside of the endoscope tube

Air pipe Air is pumped into the site and sucked from it through this tube

Water pipe Water is carried by this pipe to clean the endoscope tip and to remove waste material

Down control wire

Endoscope tube

Endoscopic removal of a polyp

A colonoscope is an endoscope used to examine the colon—the upper part of the large intestine—for abnormalities such as tumors and small lumps called polyps, which often cause bleeding and can become cancerous. To curve its way through the full length of the large intestine, the endoscope has to be extremely flexible. If a polyp is found, it is snared by a wire loop, cut off using a high-frequency electric current passed through the snare, then removed by tiny forceps attached to the end of the colonoscope.

SEE ALSO: CABLE TECHNOLOGY p.46 | COMPUTER OUTPUT DEVICES p.68

Ceramics

The term ceramic was once used solely to refer to pottery and to articles made by firing (heating) materials extracted from the Earth. Archeologists have found evidence for this type of ceramic manufacture dating back to about 24,000 BCE. Today, however, ceramic has a much broader definition, covering everything from glass to silicon wafers used in chips. Ceramics include a huge variety of materials with a broad range of useful properties.

Space Shuttle
When NASA's Space Shuttle returns to Earth, thousands of heat-resistant tiles protect its exterior from overheating due to friction with the upper atmosphere. Current shuttles use layers of carbon, ceramic, and silica composite tiles, but future spacecraft may use a new slimline ceramic tile made from hafnium and zirconium, able to withstand temperatures up to 4,300°F (2,400°C).

What are ceramics?

The best-known ceramics are pottery, glass, brick, porcelain, and cement. The word ceramic comes from the Greek *keramos*, "burnt stuff," but this is far too narrow a definition to cover the huge range of modern ceramic materials. At one end of the scale, ceramics include simple materials such as graphite and diamond, made up from different crystalline arrangements of the element carbon. But at the other end, complex crystals of yttrium, barium, copper, and oxygen make up advanced ceramics used in so-called high-temperature superconductors (materials with almost no electrical resistance). Most ceramics fall somewhere between these extremes. Many are metal oxides, crystalline compounds of a metal element and oxygen. Others are silicides, borides, carbides, and nitrides, which are respectively compounds of metals and silicon, boron, carbon, and nitrogen. Some of the most advanced are combinations of ceramics with other materials, known as ceramic matrix composites (CMCs). A definition to cover them all has to be deliberately vague—a ceramic is a nonmetallic substance, often made from several elements (some of which may be metals) chemically combined to produce certain desirable properties.

The classic properties of ceramics include durability, strength and brittleness, high electrical and thermal resistance, and ability to withstand the damaging effects of acids, oxygen, and other chemicals because of their inertness (lack of chemical reactivity). But not all ceramics behave in this way. For example, diamond is a good conductor of heat, while ceramics called ferrites are good conductors of electricity. A few ceramics, such as yttrium barium copper oxide, are used as superconductors and have almost no electrical resistance at low temperatures. CMCs, made by embedding strengthening fibers into a ceramic, are not at all brittle.

The properties of a ceramic depend not just on the materials from which it is made, but also on the way they are bonded together. Diamond is strong because all of its carbon atoms are bonded tightly to their neighbors. Graphite (pencil "lead") shears easily because it is made from layers held together by much weaker bonds. China clay (kaolin) behaves in a similar way to

Superconductor
Superconductors—materials with almost no electric resistance at very low temperatures—have been known since 1911. Since the 1980s, advances in ceramics have led to new, high-temperature superconductors that operate at manageable temperatures of about -220°F (-140°C). These materials are expected to find many applications in super-fast computer circuits and for magnets in high-speed maglev trains.

graphite. Its aluminum, silicon, oxygen, and hydrogen atoms are tightly bonded into flat sheets, but the weak bonds between the sheets are easily broken when water surrounds them. It is this fragility that makes wet clay so easy to mold. When china clay is fired, heat drives off the water, and the atoms inside the clay rearrange themselves into a much stronger crystal structure.

How ceramics are made

Simple ceramics such as bricks and certain types of glass are still made using processes that would be recognized by people who lived thousands of years ago. Just as in ancient times, modern pottery is made by digging clay from the ground, mixing it with water to make it malleable, shaping it on a wheel or in a mold, then firing it in a kiln. Some of today's processes are more sophisticated than the techniques of past times. Machines have long been used in processes such as extrusion (forcing a material into shape by squeezing it like toothpaste through a tool), jiggering (laying material automatically into a rotating mold), or hot pressing (forcing powdered ceramic into a mold, then simultaneously heating and compressing it to fuse the material into shape).

The latest industrial ceramics sometimes demand more advanced production processes. For example, extremely tough ceramics made of silicon nitride are made by a method called reaction bonding. This involves forming silicon powder into the desired shape, then heating it with nitrogen gas.

The world of modern ceramics

Very few areas of modern life are not touched by ceramics. Our homes are made from brick walls held together by calcium-silicate cement, and glass windows are also made from silica. Interiors are plastered with ceramic gypsum, bathrooms are decorated with tiles made of clay and talc, and kitchens, which may also have ceramic floor tiles, are stacked with pottery and glass. Clay pipes link our homes to the sewage system. Electric circuits use capacitors and resistors made of ceramics, which are poor electrical conductors, and ceramic insulators are used in high-power electric grids. Ceramic magnets are used in the motors of appliances such as vacuum cleaners and food blenders. Telephone calls and cable television signals are piped to the home through glass fibers, while other kinds of glass fiber are used to insulate our homes.

But ceramics have not just proved useful in everyday situations. The properties of advanced ceramics have made them important for some much more extraordinary applications. For example, the toughened silicon carbide used in hip replacements is designed to be porous so that it stimulates natural bone growth and tissue formation around the artificial joint. Ceramic engine

Electrical insulators
Familiar insulators such as rubber and plastic can still conduct electricity when very high voltages are placed across them. This makes them unsuitable for use in electrical generators and transformers. Ceramics made from alumina and porcelain are widely used as insulators for these and other applications—they have been used as the insulation in automobile spark plugs since the early 20th century.

components are used in "lean-burn" engines that combust fuel more cleanly. Catalytic convertors, which convert pollutants into less harmful gases, are made from light but strong aluminosilicate ceramics that can withstand the high temperatures generated in automobile exhausts. The latest generation of lightweight, deep-sea submersibles are being built not from steel, as their predecessors, but from ceramics originally made for defense purposes. One of the most innovative uses of ceramics is a new kind of paint made from a piezoelectric ceramic. Like other piezoelectric materials, this paint produces a tiny electric current when it undergoes stresses and strains, and its Japanese inventors believe it could be used to detect metal fatigue or even earthquakes.

SEE ALSO: MAGLEV & MONORAILS *p.120* | BIONIC TECHNOLOGY *p.208* | SPACE SHUTTLE *p.240*

Bionic technology

Replacement body parts have been made for centuries, but until recently they were minimally functional devices, such as wooden legs and glass eyes. In recent decades, advances in electronics and materials science, along with an increased understanding of how our bodies work, have allowed the development of bionic devices that can fulfill many of the functions of the parts they replace or augment.

Bionics is the science of replacing or improving human tissue with artificial, often electronic, implants. Bionic devices can be external to the body, or partly or wholly implanted within it. They range from artificial arms and legs that respond to nerve signals through the wearer's skin, to cochlear implants that replace a nonfunctioning cochlea (a part of the middle ear) and stimulate the auditory nerve with electric signals.

Cochlear implants in particular are increasingly widely used, with thousands being fitted every year. However, implantation in adults who have been deaf for some time is not very successful because their auditory nerves have degenerated due to lack of stimulation. Implants in children are much more successful, and can prevent degeneration of the auditory nerve.

Implanted bionic devices should not irritate or be rejected by body tissues. Recent advances in the development of biocompatible materials have begun to address this problem.

Retinal implants

Bionic implants can restore limited sight to the blind by bridging the gap between a diseased eye and the optic nerve. In many cases, blindness is caused by disease in the eye's light receptors—the rod and cone cells on the back wall or retina. Limited vision can now be restored by a bionic system that consists of an electronic chip implanted in the retina, coupled with an image processor incorporated into a pair of glasses. The implanted chip generates electric signals to stimulate the optic nerve in response to signals from the glasses.

Image processing A miniaturized computer is housed in the glasses that simplifies the CCD image into a grid of 100 black or white elements

Object being viewed Light from an object is normally collected and focused by the eye

Camera Light is collected by lenses and focused onto an electronic light detector called a CCD

Bionic limbs
Artificial limbs have come a long way since wooden legs and arms. Modern designs incorporate sophisticated electronics that enable them to mimic the responses of a natural limb. Arms may have sensors which attach to the stump of the limb and detect nerve signals that would have passed to the hand. Legs have complex motion detectors and pneumatic actuators that allow them to respond as the user shifts his or her weight while walking.

Photocells The pulsating laser beam traces back and forth over the grid of light-detecting cell generating a current in the cell when the beam is on

Power supply The incident laser light provides power to the chip, so it does not need a bulky battery

Implant The implant converts the pulses of laser light into electric charges, which stimulate the ganglia or nerve cells behind the retina

Retina The laser beam passes through the eye until it hits the implant on the retina, at the back of the eye

Ganglia Nerve cells detect the electric charge on the plate, and convert it into signals to the main optic nerve

Electrodes Conducting electrodes on the back of each cell carry current to a plate, causing an electric charge to build up in some areas but not in others

Cone cells When functioning normally, three different types of cone cells detect red, green, and blue light

Rod cells When functioning normally, rod cells detect contrast but not color

Optic nerve Signals from the ganglia pass to the main optic nerve through the back of the eye to the brain, which interprets them as an image

Laser beam A beam fired from the back of the glasses rapidly traces back and forth across the implant, switching on and off to represent light and dark areas of the simplified CCD image

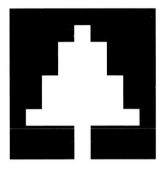

Synthesized image
The wearer sees a very simplified version of the original image. As technology improves, so should the resolution of the synthesized image. In an alternative retinal implant system, information is sent from the external camera to the implant by a radio signal—but in this system, the implant needs a separate power supply.

Glasses Part of the retinal technology is incorporated into a pair of glasses

Biocompatibility

Materials used for implants have to be biocompatible, so that they will not cause reactions in the surrounding tissue, and also very hardwearing, to survive for long periods in the hostile environment of the human body. Many biocompatible substances are made of metals such as titanium, composites, or ceramics. Composites are particularly useful since they can can be made in many forms ranging from fibers to molded shapes.

Bioglass
When sodium, calcium, and phosphorus are added to silica glass, the result is bioglass, a composite material that bonds with natural bone through a chemical reaction once it is placed inside the body. The bioglass pin shown here is just one application—bioglass is also used for middle-ear implants, inserting wedges into broken bones, and dentistry.

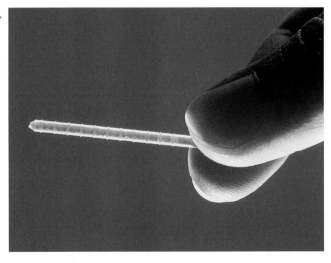

SEE ALSO: HOW COLOR WORKS *p.82* | CAMERAS *p.92* | VIDEO TECHNOLOGY *p.104* | CERAMICS *p.206* | LASERS *p.220*

Robotics

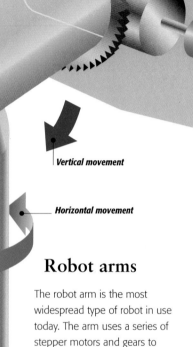

Stepper motor

Shoulder

Vertical movement

Horizontal movement

Robots are widespread in the manufacturing industry today but, as yet, they have not fulfilled the visions of the science fiction writers who conceived them. Most robots are designed to mimic the actions of certain parts of the human body—particularly the arm and hand. Robots that can duplicate the full body movement of a human being are called androids. For now, they remain purely experimental.

The word "robot" comes from the Czech *robota*, meaning "willing servant." A modern robot is effectively a mechanical extension of a computer, and as such it has all the advantages of computerized technology—precision, speed, and the ability to carry out repetitive tasks perfectly time after time. Apart from being labor-saving, robots can work in radioactive or toxic environments, and can perform hazardous tasks such as bomb disposal.

Some robots move using pneumatic (compressed air) or hydraulic systems, but most use electric stepper motors that rotate in tiny increments to allow highly precise positioning. The different ranges of movement available are called degrees of freedom. A robot joint that rotates around a single axis has one degree of freedom; more complex joints may have two or three degrees of freedom.

Robot arms

The robot arm is the most widespread type of robot in use today. The arm uses a series of stepper motors and gears to achieve many degrees of freedom in its joints. The end of the arm can be fitted with a variety of tools, called end effectors, which range from claws to drills. End effectors can be fitted with touch sensors or other detectors to provide feedback to the controlling computer.

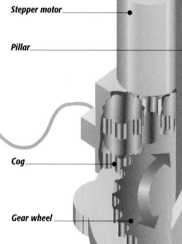

Stepper motor

Pillar

Cog

Gear wheel

Computer control A computer script (program) controls complex movements of the robot by issuing a series of simple commands, and also controls the operation of the various end effectors

Base The pillar of the arm rotates on a base that is securely anchored to the floor. A stepper motor turns a cog that drives a gear wheel attached to the pillar to rotate it

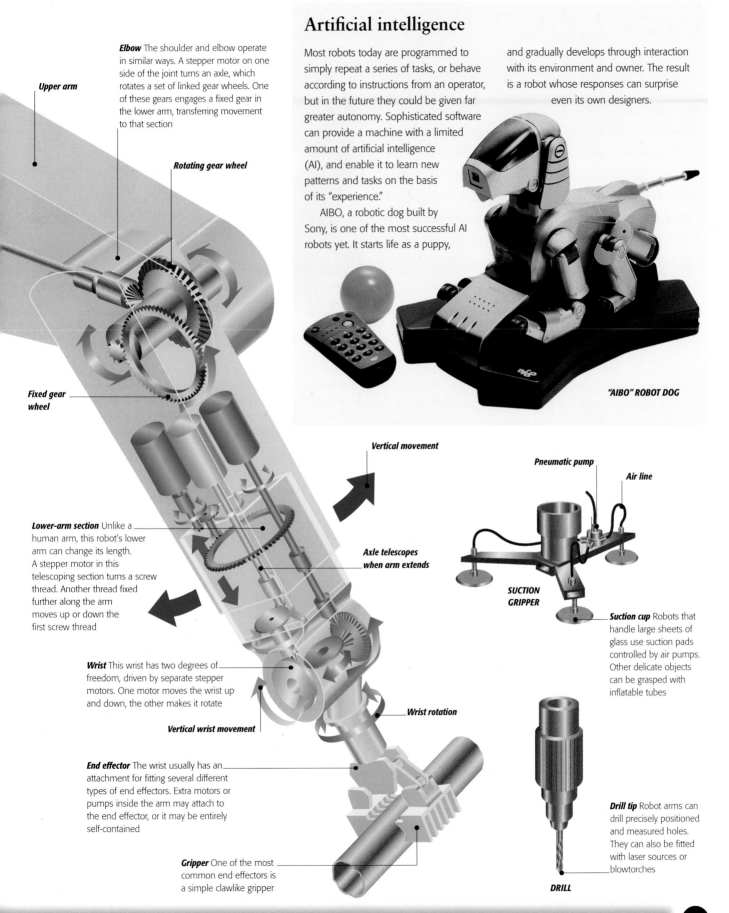

Upper arm

Elbow The shoulder and elbow operate in similar ways. A stepper motor on one side of the joint turns an axle, which rotates a set of linked gear wheels. One of these gears engages a fixed gear in the lower arm, transferring movement to that section

Rotating gear wheel

Fixed gear wheel

Lower-arm section Unlike a human arm, this robot's lower arm can change its length. A stepper motor in this telescoping section turns a screw thread. Another thread fixed further along the arm moves up or down the first screw thread

Wrist This wrist has two degrees of freedom, driven by separate stepper motors. One motor moves the wrist up and down, the other makes it rotate

Vertical wrist movement

End effector The wrist usually has an attachment for fitting several different types of end effectors. Extra motors or pumps inside the arm may attach to the end effector, or it may be entirely self-contained

Gripper One of the most common end effectors is a simple clawlike gripper

Vertical movement

Axle telescopes when arm extends

Wrist rotation

Artificial intelligence

Most robots today are programmed to simply repeat a series of tasks, or behave according to instructions from an operator, but in the future they could be given far greater autonomy. Sophisticated software can provide a machine with a limited amount of artificial intelligence (AI), and enable it to learn new patterns and tasks on the basis of its "experience."

AIBO, a robotic dog built by Sony, is one of the most successful AI robots yet. It starts life as a puppy, and gradually develops through interaction with its environment and owner. The result is a robot whose responses can surprise even its own designers.

"AIBO" ROBOT DOG

Pneumatic pump

Air line

SUCTION GRIPPER

Suction cup Robots that handle large sheets of glass use suction pads controlled by air pumps. Other delicate objects can be grasped with inflatable tubes

Drill tip Robot arms can drill precisely positioned and measured holes. They can also be fitted with laser sources or blowtorches

DRILL

Genetic engineering

Genetic engineering is the modification of an organism's genetic material for practical use. Genes are units of information that instruct living cells to make specific proteins, chemicals that affect all life processes. By modifying an organism's genes, scientists can produce various effects, such as inhibiting its protein production or causing it to manufacture a modified or entirely new protein. Genetically engineered products have many uses in industry and in medicine.

Genetic engineering can be used to make enzymes for biological washing powders, to manufacture insulin for people with diabetes, and to generate antibodies that can fight cancer. Genetically engineered products can often work better and more safely than natural ones—for example, genetically engineered blood-clotting factor carries no risk of bloodborne infections. They can also be produced in large quantities; insulin is now produced in bacteria or in yeast, which multiply rapidly, allowing the production of the protein in abundance.

Subtilisin fermenters

Subtilisin is a protein-breaking enzyme produced naturally by certain bacteria. Genetically modified (GM) subtilisin that can work in hot, alkaline conditions is used in detergents to break down protein-based stains, such as blood. GM subtilisin is produced by GM bacteria grown in fermenters.

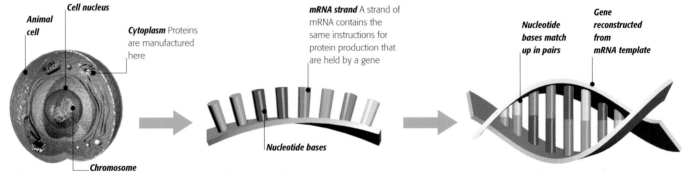

The cell

Genes are sections of DNA, a double-stranded molecule that is coiled up in structures called chromosomes and found in the cell nucleus. Genes encode instructions for protein production.

Messenger RNA

Messenger ribonucleic acid (mRNA) is naturally manufactured in the cell nucleus. It is a complementary copy of one strand of the section of DNA that comprises a gene.

Reverse transcription

The mRNA is used as a template by geneticists to synthesize a complementary strand of DNA (cDNA) that binds to it. From this complementary strand, the original gene can be reconstructed.

Extracting a gene

Genetic engineering involves two processes. First, the gene is extracted and isolated and, if necessary, modified. Then, it is inserted into an animal or plant cell (creating a "transgenic" species) or a microbial cell. The protein encoded by the inserted gene is produced by the animal or plant cell, or (if the gene was inserted into a microbe) by culturing microbial cells in a fermenter.

Each gene is simply a set of chemical instructions, encoded as a sequence of four chemical bases (abbreviated as A, C, G, and T) that make up a long, double-stranded molecule called DNA. A complete molecule of

DNA, called a chromosome, contains many distinct genes. The first step in genetic engineering is to identify the gene of interest in the midst of a chromosome. Scientists often do this by isolating the gene's "messenger RNA" (mRNA), a molecule that carries instructions for protein manufacture from the gene in the cell nucleus to protein-making units in the cytoplasm. mRNA is a complementary copy of one strand of the section of DNA comprising a gene and can be used as a template for reconstructing the gene. If desired, the gene can be modified to produce an improved protein.

Genetic engineering in animal cells

One method of incorporating a foreign or modified gene into an animal is to inject copies of the gene into a newly fertilized egg. The gene then becomes integrated into the chromosomes in the nucleus of the egg, which is then implanted into a foster mother. As the cell starts dividing, the gene is copied along with every other gene in the chromosomes. In this way, the gene is present in every body cell of the "transgenic" animal.

Fertilizing an egg cell
To create a transgenic animal, egg cells are first removed from a donor. These are then fertilized by sperm cells that have been collected from males of the same species.

Fertilized egg
The nuclei of the egg and sperm fuse. Copies of the foreign gene are injected into the chromosomes.

Cell division
As the cell divides, the new gene is incorporated into every cell of the animal.

Genetic engineering in microorganisms and plant cells

Some bacteria contain extra chromosomal loops of DNA called plasmids, which replicate independently of the chromosome, both within the cell and also when a bacterium divides. By inserting foreign genes into plasmids using special "restriction" enzymes that cut the DNA at specific points, bacteria can be used as protein factories on an industrial scale. Plasmid DNA can also be inserted into plant cells in order to produce genetically modified crops.

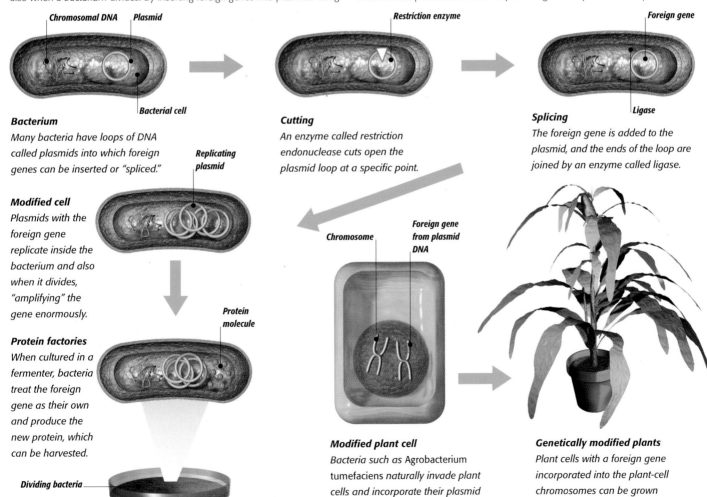

Bacterium
Many bacteria have loops of DNA called plasmids into which foreign genes can be inserted or "spliced."

Cutting
An enzyme called restriction endonuclease cuts open the plasmid loop at a specific point.

Splicing
The foreign gene is added to the plasmid, and the ends of the loop are joined by an enzyme called ligase.

Modified cell
Plasmids with the foreign gene replicate inside the bacterium and also when it divides, "amplifying" the gene enormously.

Protein factories
When cultured in a fermenter, bacteria treat the foreign gene as their own and produce the new protein, which can be harvested.

Modified plant cell
Bacteria such as Agrobacterium tumefaciens *naturally invade plant cells and incorporate their plasmid DNA into plant-cell chromosomes.*

Genetically modified plants
Plant cells with a foreign gene incorporated into the plant-cell chromosomes can be grown into new plants.

SEE ALSO: AGRICULTURAL TECHNOLOGY *p.188* | **GM** FOODS *p.190* | HUMAN GENOME PROJECT *p.214*

Human Genome Project

Most human body cells contain about 90,000 pairs of genes (units of hereditary data) arranged on two sets of 23 chromosomes (structures of the molecule DNA). This genetic information forms the genome. The Human Genome Project is an international program launched in 1990 to map the sequence of nucleotide bases (the building blocks of DNA) in the genome and determine the function of every gene.

The mapping of the human genome is a scientific achievement comparable with putting people on the Moon. The genomes of some organisms, including a worm and a fruit fly, have already been sequenced (mapped), but these genomes are small compared to the human genome, which contains about 3 billion nucleotide bases. By sequencing the human genome, scientists will better understand the genes involved in body processes and diseases. The genes in each of an individual's cells are identical, but the genome varies slightly between individuals (which, in part, makes them individuals), so the Human Genome Project maps genes from a representative group of people. Although we now have the DNA sequence of the human genome, it will take many more years to identify the individual genes and determine their function.

The function of genes

A gene is a section of double-stranded DNA containing data that enables an organism to function. Most genes contain instructions for making a protein. These instructions are encoded within the sequence of nucleotide bases on one of the gene's strands. An elaborate chemical machinery exists in body cells to make proteins when they are needed, using the instructions encoded by the genes. Proteins play many roles in the body, so genes ultimately affect every aspect of an individual's development and function. For the same reason, genetic defects can be a cause of ill health and disease.

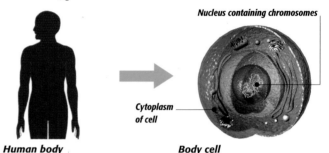

Nucleus containing chromosomes

Cytoplasm of cell

Human body
The adult human body contains about 5,000 billion cells, which combine to make up various organs and tissues.

Body cell
Most cells contain 46 chromosomes (22 pairs of chromosomes plus two sex chromosomes). Sex cells have only one set of 23 chromosomes.

DNA sequencing

DNA is sequenced to work out the order of its four nucleotide bases (T, A, C, and G), which spell out the genetic information. By examining the sequence of a specific length of DNA, scientists can see whether it contains a gene. Only about 10 percent of the human genome contains genes (the remaining 90 percent is known as junk DNA). There are several methods for sequencing. The one shown here is the Sanger Method, devised by Frederick Sanger, a biochemist who used it to sequence the genome of a virus during the 1970s. The Human Genome Project used new techniques, computers, and robotic machinery that greatly increased the speed of DNA sequencing.

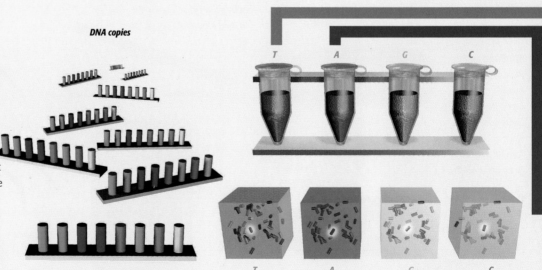

DNA copies

1 Amplification *A small fragment of DNA is cut out of a chromosome and "cloned," or amplified, to produce millions of copies.*

2 Controlled replication *The copies are divided into four solutions and are copied again. In each solution, some of the molecules of a particular base are modified. These molecules act as a special chemical fixer and halt the replication of a growing strand of DNA if they are added it.*

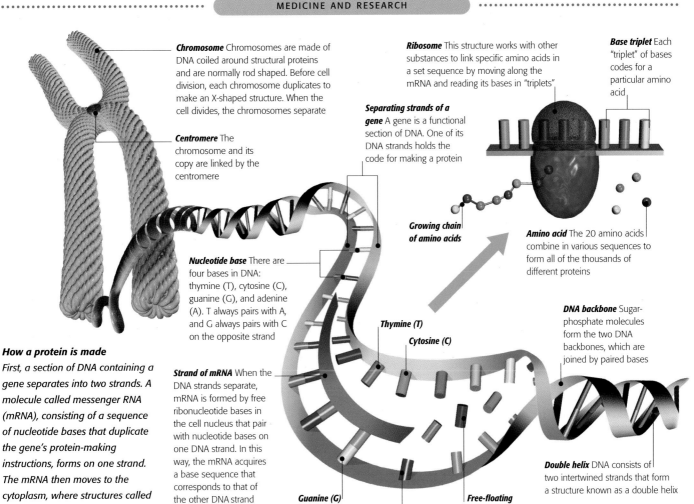

Chromosome Chromosomes are made of DNA coiled around structural proteins and are normally rod shaped. Before cell division, each chromosome duplicates to make an X-shaped structure. When the cell divides, the chromosomes separate

Centromere The chromosome and its copy are linked by the centromere

Ribosome This structure works with other substances to link specific amino acids in a set sequence by moving along the mRNA and reading its bases in "triplets"

Base triplet Each "triplet" of bases codes for a particular amino acid

Separating strands of a gene A gene is a functional section of DNA. One of its DNA strands holds the code for making a protein

Growing chain of amino acids

Amino acid The 20 amino acids combine in various sequences to form all of the thousands of different proteins

Nucleotide base There are four bases in DNA: thymine (T), cytosine (C), guanine (G), and adenine (A). T always pairs with A, and G always pairs with C on the opposite strand

Thymine (T)

Cytosine (C)

DNA backbone Sugar-phosphate molecules form the two DNA backbones, which are joined by paired bases

How a protein is made

First, a section of DNA containing a gene separates into two strands. A molecule called messenger RNA (mRNA), consisting of a sequence of nucleotide bases that duplicate the gene's protein-making instructions, forms on one strand. The mRNA then moves to the cytoplasm, where structures called ribosomes process the instructions.

Strand of mRNA When the DNA strands separate, mRNA is formed by free ribonucleotide bases in the cell nucleus that pair with nucleotide bases on one DNA strand. In this way, the mRNA acquires a base sequence that corresponds to that of the other DNA strand

Guanine (G)

Adenine (A)

Free-floating nucleotide base

Double helix DNA consists of two intertwined strands that form a structure known as a double helix

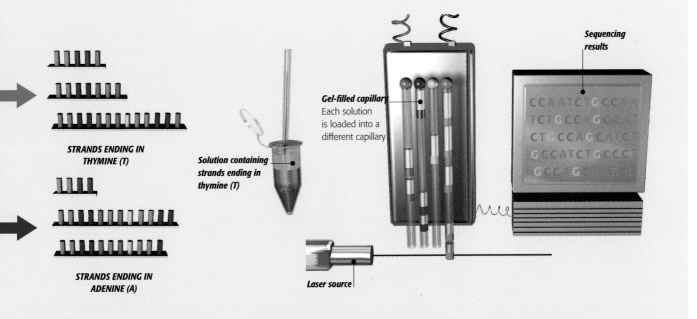

Sequencing results

Gel-filled capillary Each solution is loaded into a different capillary

STRANDS ENDING IN THYMINE (T)

Solution containing strands ending in thymine (T)

STRANDS ENDING IN ADENINE (A)

Laser source

3 Reaction products Many partial copies of the DNA with differing lengths are produced. Depending on the solution, they all end in the same base (T, A, G, or C).

4 Gel electrophoresis The partial copies in the four solutions are separated according to their size by a technique called gel electrophoresis, in which a voltage drives them along gel-filled capillaries (fine tubes). The shorter the strand, the farther it travels along the capillary. A laser then reads the order of the strands. A computer generates the DNA sequence using this data.

SEE ALSO: GENETICALLY MODIFIED FOODS *p.190* | GENETIC ENGINEERING *p.212* | LASERS *p.220* | MATTER *p.258*

Dating technology

Dating technology describes several techniques for establishing the ages of various ancient natural objects and artefacts. Different methods are used, depending on the nature of the objects, which range from 4 billion-year-old rocks at one end of the spectrum to 19th-century wooden timbers at the other. These methods are indispensable research tools for geologists, anthropologists, and archaeologists.

Many dating methods are based on the phenomenon of radioactivity. Some chemical elements have unstable forms, called radioactive isotopes, whose atoms spontaneously "decay" over time. Decay usually involves an atom of the isotope changing into an atom of a different element and simultaneously releasing an alpha or beta particle (radiation). The rate at which a given isotope decays is constant and is expressed as its half-life, which is the time taken for half of its atoms to decay. The half-life provides an in-built clock within any object that was formed with a radioactive isotope incorporated into its structure. Some such objects are dated by measuring the ratio of the isotope to its decay products (the method used in radiometric rock dating),

others by establishing how much of the isotope remains in relation to the stable form of the same element (the method used in radiocarbon dating). A further method relies on examining the direct effects of radioactive decay in mineral grains. When atoms of the isotope uranium-238 (U-238) decay, the particles emitted produce trails of damage, called fission tracks, in the mineral's structure. The number of tracks increases over time, so the age of a mineral can be calculated by measuring the density of its fission tracks and calibrating against the measured amount of U-238 in the mineral. Luminescence dating is also radioactivity-related, being based on measuring the accumulation of "trapped" electrons in objects that have been subjected to natural radiation.

Of methods unrelated to radioactivity, archaeomagnetic dating is based on reversals that occur from time to time in the Earth's magnetic field. These produce alternating bands of magnetic orientation within minerals of sedimentary rock that are akin to a date "bar code" within the rock. Dendrochronology, or tree-ring dating, is possible because of annual variations in tree growth caused by changes in climatic factors such as rainfall. These variations are visible as ring patterns within the tree trunks, which can be used somewhat like a date "fingerprint".

Radiocarbon dating

Radiocarbon dating is used to date objects of organic origin (those derived from animals or plants) up to 70,000 years old. The method relies on the fact that all living organisms absorb the radioactive isotope carbon-14 (C-14) into their cells along with other carbon atoms. (C-14 is generated by cosmic-ray bombardment of carbon compounds in the upper atmosphere and forms a tiny but constant proportion of the carbon in air.) Once an organism dies, no more carbon of any sort is taken in, and the existing C-14 gradually disappears by radioactive decay (C-14 has a half-life of 5,730 years). Dating an object relies on measuring how much C-14 remains as a proportion of total carbon.

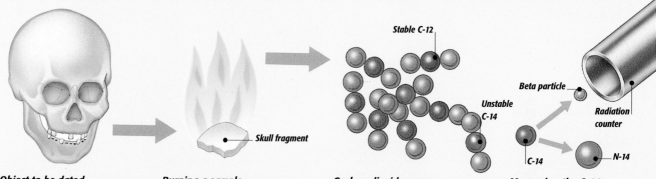

Object to be dated
The object to be dated may consist of bone (as in the example shown above), wood, shell, organic fibres used in cloth, or other organic remains.

Burning a sample
A small fragment of the object is burned. During the combustion process, all of the carbon in the sample combines with oxygen to form carbon dioxide gas.

Carbon dioxide gas
The carbon dioxide includes some molecules that contain stable carbon-12 (C-12) and some that contain the unstable radioactive isotope carbon-14 (C-14).

Measuring the C-14
The C-14 atoms decay to form nitrogen-14 (N-14), releasing beta particles (electrons), which can be detected using a radiation counter.

Luminescence dating

Luminescence dating is used to date objects that contain minerals such as quartz and have been buried for hundreds or thousands of years. Over time, the minerals accumulate a measurable property known as latent luminescence, which is caused by exposure to natural radiation. This latent luminescence is released when the object is exposed to heat or light, so by measuring the amount of latent luminescence the object has, scientists calculate when the object was last exposed to heat or light. With artefacts such as pottery or flints used in cooking, the method can indicate how long ago the people who made and used the objects lived.

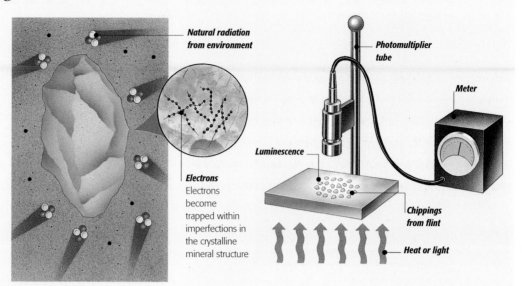

Natural radiation from environment

Electrons
Electrons become trapped within imperfections in the crystalline mineral structure

Photomultiplier tube

Meter

Luminescence

Chippings from flint

Heat or light

Ancient flint
A buried flint absorbs radiation from its surroundings. Energy from the radiation frees electrons, which are trapped by imperfections within mineral crystals in the flint.

Measuring luminescence
Heating or illuminating a sample from the flint releases some of the trapped electrons. As the electrons escape, the crystals emit light, which can be measured with a photomultiplier tube. The amount of light released is the latent luminescence.

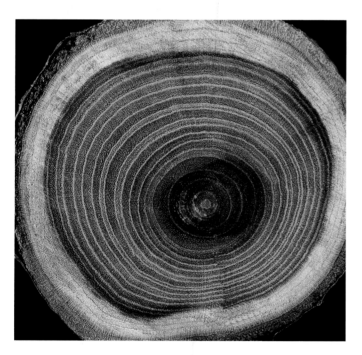

Tree rings
By studying wood from recently felled trees, old buildings, and archaeological sites, scientists have compiled tree-ring data stretching back thousands of years for certain tree species, such as the bristlecone pine in the USA and European oaks. Many ancient wooden objects can now be dated with great precision by matching their ring patterns to these records.

Radiometric rock dating

Certain rocks, such as granite and basalt, are formed when minerals in magma (molten rock) crystallize. These rocks can be dated if a suitable radioactive isotope was incorporated into their structure during formation. In order to date rocks that may be billions of years old, only isotopes with extremely long half-lives (such as uranium-238) are suitable. Dating relies on measuring the ratio between the radioactive isotope and its stable decay product. The lower the ratio, the older is the rock.

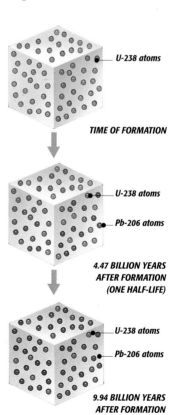

U-238 atoms

TIME OF FORMATION

U-238 atoms

Pb-206 atoms

4.47 BILLION YEARS AFTER FORMATION (ONE HALF-LIFE)

U-238 atoms

Pb-206 atoms

9.94 BILLION YEARS AFTER FORMATION

Uranium/lead dating
Uranium-238 (U-238) has a half-life of 4.47 billion years, gradually being transformed to stable lead-206 (Pb-206). Dating relies on measuring the ratio between the uranium and lead.

SEE ALSO: ATOMIC CLOCKS *p.224* | MATTER *p.258* | ELECTRICITY & MAGNETISM *p.260*

Geological exploration

Many natural resources lie buried beneath the ground, locked in the Earth's rocky crust. Locating metal ores, minerals, coal, oil, and gas was once a matter of providence, but a multibillion-dollar industry is now dedicated to it. The Earth's hidden treasures are finite, but as ready reserves become depleted, ever more sophisticated techniques are employed to find new sources.

In order to locate reserves of minerals, ores, and fossil fuels, geologists need to discover what types of rocks lie beneath the surface, their age relationships, and their 3-D structure. Although such resources are found in entirely different geological settings, many of the techniques used to find them are the same. Remote-sensing satellites and radar are used to survey the landscape and its geological make-up, while gravitational and magnetic surveys determine the properties of the rocks underground. Taken together, these methods identify target areas for more detailed mapping of subsurface rock formations. Seismic surveying produces detailed 3-D maps of rock formations using shock waves. The shock waves travel through rocks and are partially reflected at the boundaries between different types of rock. The length of time that the reflections take to return to the surface gives the depth of the various rock layers. Exploratory drilling is then used to confirm the presence of a reserve.

Surveying techniques

Large-scale data gathering, such as satellite and airborne surveys, is undertaken to look at regional geology and indicate the possibility of ore, mineral, or oil deposits. Other, more directed exploration tools are then used to determine the exact position, nature, extent, and size of the reserves.

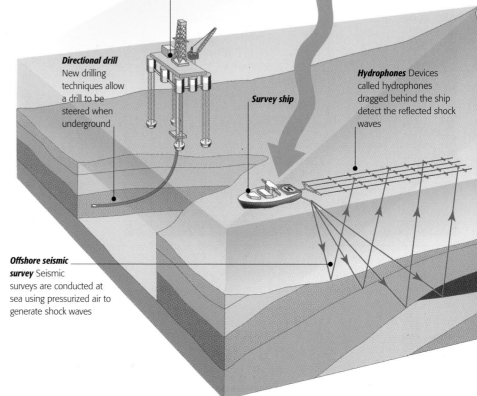

GPS satellite Geologists use the Global Positioning System to correlate measurement data to precise location data

Offshore test drilling When an oilfield is located offshore, an oil rig is towed out to sea to conduct exploratory drilling

Directional drill New drilling techniques allow a drill to be steered when underground

Survey ship

Hydrophones Devices called hydrophones dragged behind the ship detect the reflected shock waves

Offshore seismic survey Seismic surveys are conducted at sea using pressurized air to generate shock waves

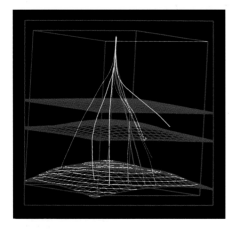

3-D seismic surveying
Seismic surveys are the backbone of the petroleum exploration industry. They are used to locate oil and produce 3-D maps of the oilfield. The picture above shows alternative drilling pathways plotted onto such a map. These routes are used to guide the directional drilling process.

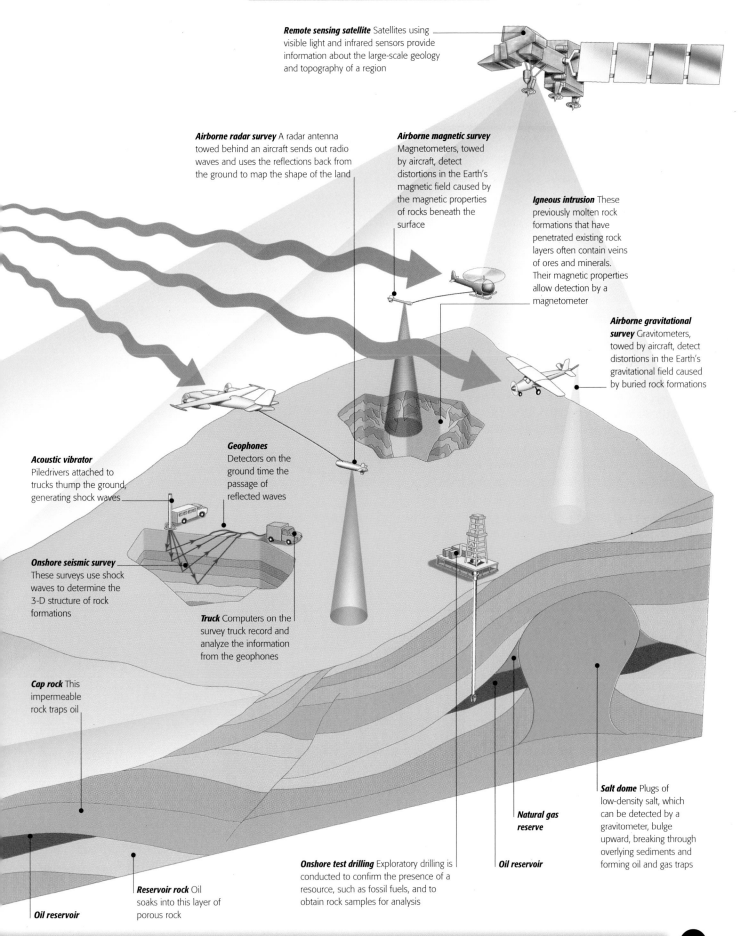

Remote sensing satellite Satellites using visible light and infrared sensors provide information about the large-scale geology and topography of a region

Airborne radar survey A radar antenna towed behind an aircraft sends out radio waves and uses the reflections back from the ground to map the shape of the land

Airborne magnetic survey Magnetometers, towed by aircraft, detect distortions in the Earth's magnetic field caused by the magnetic properties of rocks beneath the surface

Igneous intrusion These previously molten rock formations that have penetrated existing rock layers often contain veins of ores and minerals. Their magnetic properties allow detection by a magnetometer

Airborne gravitational survey Gravitometers, towed by aircraft, detect distortions in the Earth's gravitational field caused by buried rock formations

Acoustic vibrator Piledrivers attached to trucks thump the ground, generating shock waves

Geophones Detectors on the ground time the passage of reflected waves

Onshore seismic survey These surveys use shock waves to determine the 3-D structure of rock formations

Truck Computers on the survey truck record and analyze the information from the geophones

Cap rock This impermeable rock traps oil

Natural gas reserve

Salt dome Plugs of low-density salt, which can be detected by a gravitometer, bulge upward, breaking through overlying sediments and forming oil and gas traps

Oil reservoir

Onshore test drilling Exploratory drilling is conducted to confirm the presence of a resource, such as fossil fuels, and to obtain rock samples for analysis

Reservoir rock Oil soaks into this layer of porous rock

Oil reservoir

SEE ALSO: SATELLITE COMMUNICATIONS *p.50* | GLOBAL POSITIONING SYSTEM *p.136* | OIL PRODUCTION *p.180*

Lasers

A laser is a device that amplifies light to produce an intense, narrow beam of very pure color in which the light waves are "in step." Lasers have a wide range of applications, including cutting or welding metal, cutting tissue during surgery, transmitting signals down optical fibers, measuring distances, and making holograms. Lasers are also used in bar-code scanners, printers, photocopiers, and CD players.

Light from the Sun and from most artificial sources contains a mixture of wavelengths, (or, in terms of particles, photons of different energies). This light also spreads out in many directions from the source. Laser light, in contrast, is coherent—it has a very small range of wavelengths, and all the waves are in phase (in step). Moreover, a laser beam spreads out far less than ordinary light.

Lasers contain a substance called a lasing medium, which may be solid, liquid, or gas. An energy source is used to excite the atoms in the lasing medium. These atoms then emit light, which is trapped and amplified by mirrors in an optical cavity before being released as a beam. Depending on the type of laser, the beam may be continuous or pulsed; it may be visible, infrared, or ultraviolet; and it may range in power from a few milliwatts to trillions of watts.

Power supply A high voltage is supplied to the trigger electrode and the xenon flash tube

Xenon flash tube This tube emits a bright flash of white light that excites chromium atoms in the ruby crystal

Totally reflective mirror All the laser light is reflected back along the ruby crystal by this mirror

Trigger electrode This electrode supplies a high voltage that primes the laser

Power switch

Laser surgery

Lasers are used increasingly in surgery. Unlike scalpels, lasers cut by generating heat in body tissue. This heat cauterizes (seals) blood vessels as it cuts, minimizing bleeding. Laser light can be directed with great precision, so it is ideal for use in delicate eye surgery to "weld" detached retinas, seal burst blood vessels, or reshape the cornea to correct vision defects. Laser scalpels generally use carbon dioxide gas as a lasing medium, while lasers used in eye surgery are often based on argon gas.

Laser eye surgery
During eye surgery, the patient's head is secured to prevent movement and the laser is aimed using a device called a retinal camera. Laser treatments for conditions such as detached retinas and retinopathy do not usually require anesthesia or an overnight stay in hospital.

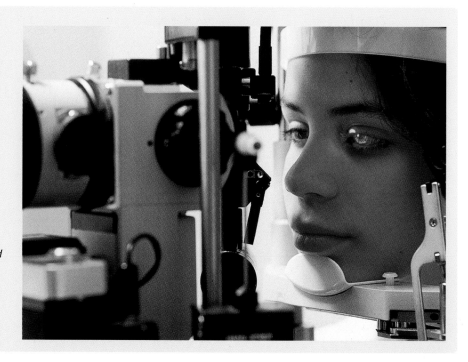

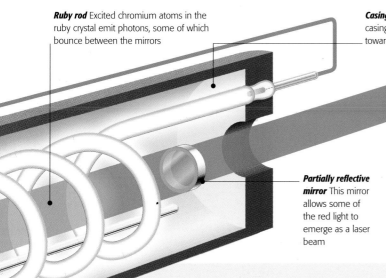

Ruby rod Excited chromium atoms in the ruby crystal emit photons, some of which bounce between the mirrors

Casing The inside of the aluminum casing is polished to reflect light toward the ruby crystal

Laser beam Beam of red light emitted

Partially reflective mirror This mirror allows some of the red light to emerge as a laser beam

Ruby laser

The first laser built was a ruby laser. Ruby is a crystal of aluminum oxide, containing some chromium atoms. A flash tube coiled around a ruby rod supplies light energy, which excites the chromium atoms. Light emitted by these atoms bounces between mirrors at the ends of the rod, stimulating more chromium atoms to emit light.

Amplifying light

Laser is an acronym for "Light Amplification by Stimulated Emission of Radiation." In a ruby laser, energy from a flash tube raises chromium atoms from their ground state. These excited atoms spontaneously emit photons of red light and return to their ground state. If one of these photons hits an excited atom, an identical photon is released by a process called stimulated emission. Photons that travel along the ruby rod, bouncing between two mirrors, cause a cascade of stimulated emission, producing a laser beam.

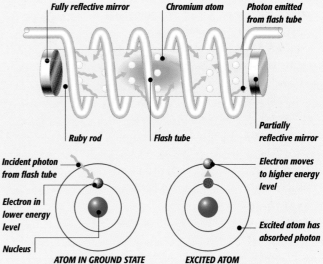

Fully reflective mirror *Chromium atom* *Photon emitted from flash tube*

Ruby rod *Flash tube* *Partially reflective mirror*

Incident photon from flash tube *Electron moves to higher energy level*

Electron in lower energy level

Nucleus

Excited atom has absorbed photon

ATOM IN GROUND STATE **EXCITED ATOM**

1 Ordinarily, most chromium atoms are in the ground state. For lasing to occur, most of the atoms have to become excited (have their electrons raised to a higher energy level). This is called "population inversion."

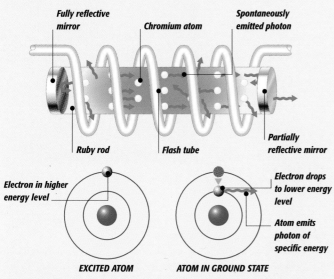

Fully reflective mirror *Chromium atom* *Spontaneously emitted photon*

Ruby rod *Flash tube* *Partially reflective mirror*

Electron in higher energy level

Electron drops to lower energy level

Atom emits photon of specific energy

EXCITED ATOM **ATOM IN GROUND STATE**

2 An excited atom emits light when its electron spontaneously drops from a higher energy level to a lower one. The photon emitted has the same energy as the difference in energy levels. This spontaneous emission occurs in random directions and at random times.

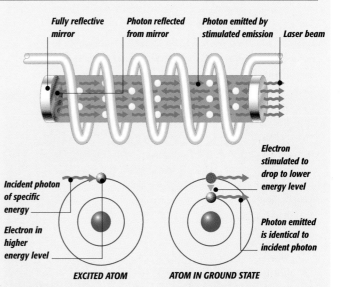

Fully reflective mirror *Photon reflected from mirror* *Photon emitted by stimulated emission* *Laser beam*

Incident photon of specific energy

Electron in higher energy level

Electron stimulated to drop to lower energy level

Photon emitted is identical to incident photon

EXCITED ATOM **ATOM IN GROUND STATE**

3 Stimulated emission occurs when an excited atom is hit by a photon of specific energy (one emitted by another excited atom). This causes the atom to drop to its ground state and emit a photon with the same energy (wavelength), phase, and direction of travel as the incident photon.

SEE ALSO: LIGHTING *p.28* | SOUND REPRODUCTION *p.80* | HOLOGRAMS *p.222* | OPTICAL TELESCOPES *p.228*

Holography

Holograms are familiar as 3-D photographs and as the multicolored security emblems on credit cards and CD packages. The theory of holography was developed in 1948 by Dennis Gabor, who coined the word "hologram" from the Greek words *holos* ("whole") and *gramma* ("message"), but holography became a practical possibility only after the development of the laser in the 1960s.

A hologram is a photograph of an interference pattern created by the interaction of light beams. Unlike a standard photograph, which is a direct copy of its subject, the image on a holographic plate looks nothing like its subject. Instead, it is made of light and dark fringes, which contain information about the light that traveled from the object being imaged to the plate, including its phase (whether the light waves were in step or not). The phase information is what gives holograms their depth, and allows the viewer to "see around" parts of the image. A laser is used to make holograms because it provides the required coherent light (light whose waves are all in phase). When a hologram is viewed, the fringes diffract (bend) light waves striking them, so that they appear to have come from the original object.

Holography is widely used in scientific research and testing. Holographic labels are used as security devices because they are difficult to forge. Other applications include aircraft head-up displays, which project images of instruments into the pilot's field of vision, and bar-code scanners in stores. And if optical computers, based on optical pulses instead of electric signals, become a reality, they will rely on holographic technology.

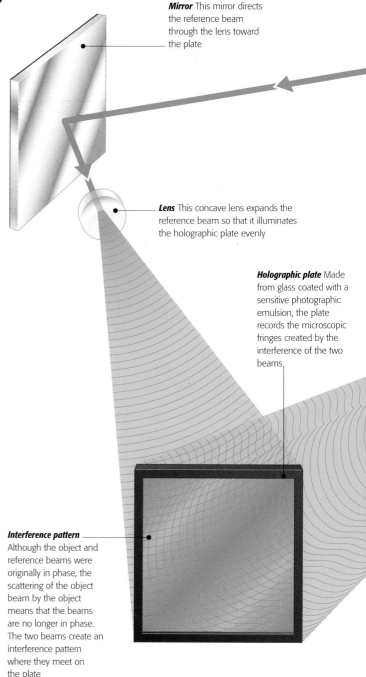

Mirror This mirror directs the reference beam through the lens toward the plate

Lens This concave lens expands the reference beam so that it illuminates the holographic plate evenly

Holographic plate Made from glass coated with a sensitive photographic emulsion, the plate records the microscopic fringes created by the interference of the two beams

Interference pattern Although the object and reference beams were originally in phase, the scattering of the object beam by the object means that the beams are no longer in phase. The two beams create an interference pattern where they meet on the plate

Holography in art and entertainment

Apart from its use in critical scientific research and sensitive commercial applications such as security, holography has been developed into an art form, and holograms are found in museums and galleries worldwide. Holograms can be mass-produced cheaply enough to be used for gifts and novelties. The entertainment industry is awaiting the development of full-motion holography, but the technology to make holographic TV programs, motion pictures, and computer games has not reached maturity yet.

Viewing holograms

The interference pattern on a holographic plate diffracts (bends) light shone on it. This creates a 3-D image by reconstructing light rays as originally scattered by the object. Reflection holograms are lit from the front of the plate. Transmission holograms are usually lit from behind the plate, but some, such as those on credit cards, are embossed on reflective film so that they can be lit from the front.

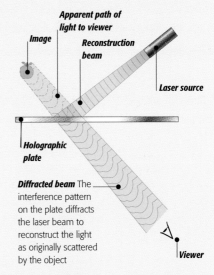

Transmission hologram
Most transmission holograms are illuminated by shining laser light through the plate and are single-color. White-light transmission holograms are lit by ordinary light and produce multicolored images.

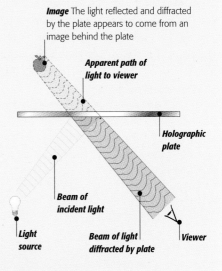

Reflection hologram
Reflection holograms are viewed using white light to illuminate the plate from the same side as the viewer. They are usually a single color and can contain great detail.

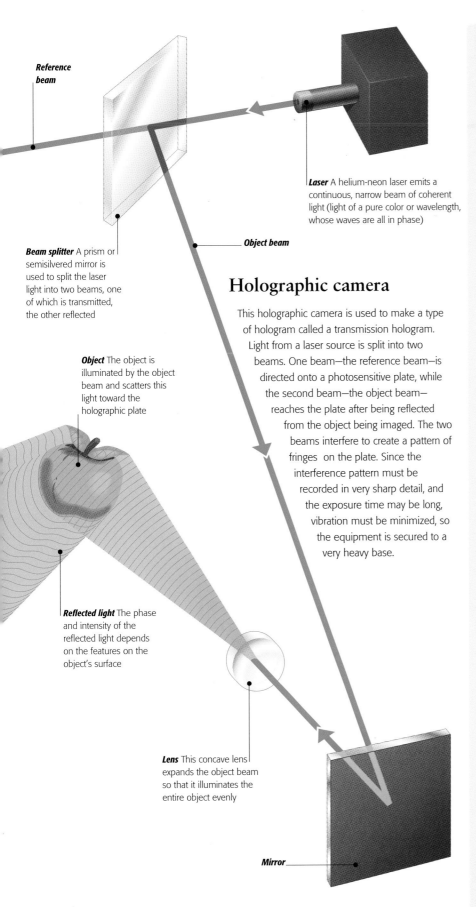

Reference beam

Beam splitter A prism or semisilvered mirror is used to split the laser light into two beams, one of which is transmitted, the other reflected

Object The object is illuminated by the object beam and scatters this light toward the holographic plate

Reflected light The phase and intensity of the reflected light depends on the features on the object's surface

Lens This concave lens expands the object beam so that it illuminates the entire object evenly

Mirror

Laser A helium-neon laser emits a continuous, narrow beam of coherent light (light of a pure color or wavelength, whose waves are all in phase)

Object beam

Holographic camera

This holographic camera is used to make a type of hologram called a transmission hologram. Light from a laser source is split into two beams. One beam—the reference beam—is directed onto a photosensitive plate, while the second beam—the object beam—reaches the plate after being reflected from the object being imaged. The two beams interfere to create a pattern of fringes on the plate. Since the interference pattern must be recorded in very sharp detail, and the exposure time may be long, vibration must be minimized, so the equipment is secured to a very heavy base.

Atomic clocks

Clocks measure time by counting off regularly repeated events, such as the Sun's motion across the sky, a pendulum's swing, or a quartz crystal's oscillations. The more frequent the event, the greater the potential accuracy of the clock. The most accurate timekeeping devices are atomic clocks, which are regulated by microwaves oscillating billions of times each second.

Most atomic clocks use atoms of cesium-133, an isotope of cesium. These atoms have two "spin states," according to whether the outermost electron is spinning in the same (parallel) or opposite (antiparallel) direction as the atom as a whole. These spin states have slightly different energies, and if microwave radiation of precisely the right frequency hits the atoms, they will flip from one state to the other. Cesium clocks are designed to "lock" an oscillator to this frequency, called the transition frequency, of 9,192,631,770 Hz.

In a cesium clock, atoms are separated according to their spin state, then exposed to radiation from a microwave oscillator. A detector counts how many atoms have changed spin state. The greatest number of atoms change spin states when the microwaves are precisely tuned to the transition frequency. A computer uses signals from the detector to maintain the correct oscillator frequency, and pulses from the oscillator can be used to calibrate other clocks. Atomic clocks have revolutionized timekeeping, to the extent that one second is now defined in terms of the spin transition frequency of cesium-133 atoms.

Atomic clocks provide the timekeeping standard not just for scientific research, but also for the world's communication and transportation networks. The Global Positioning System, a satellite-based navigation system, relies on the accuracy of atomic clocks to fix a user's position to within a few inches.

Cesium clock

In a cesium atomic clock, a sample of cesium-133 is heated in an oven. Cesium atoms boil off the sample and emerge from the oven at high speed. These are focused by magnets into separate beams according to their spin state. The atoms pass through a microwave cavity, where most of them are flipped into the opposite spin state. The atoms are then deflected by a second set of magnets so that only the flipped atoms pass into a detector. A computer uses the detector signals to "lock" the microwave frequency. The entire clock is contained in an evacuated chamber.

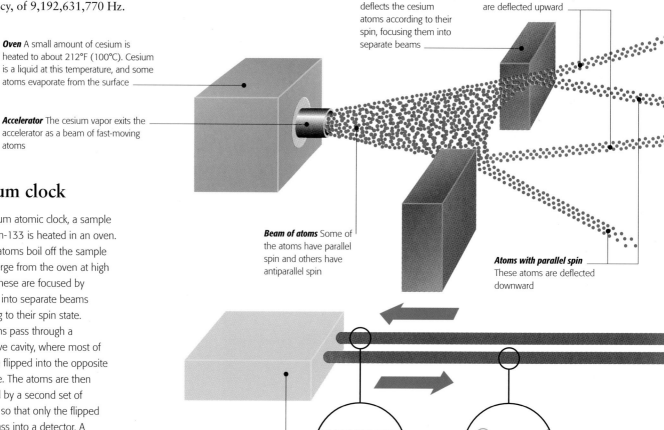

Oven A small amount of cesium is heated to about 212°F (100°C). Cesium is a liquid at this temperature, and some atoms evaporate from the surface

Accelerator The cesium vapor exits the accelerator as a beam of fast-moving atoms

Magnet A magnetic field deflects the cesium atoms according to their spin, focusing them into separate beams

Atoms with antiparallel spin These atoms are deflected upward

Beam of atoms Some of the atoms have parallel spin and others have antiparallel spin

Atoms with parallel spin These atoms are deflected downward

Frequency divider This device receives a high-frequency electric signal from the oscillator and steps it down to lower-frequency pulses that drive the digital time display

High-frequency signal

Low-frequency signal

NIST-7 atomic fountain
Currently, the most accurate timekeeping device is a type of atomic clock called the "atomic fountain," which tunes into the spin transition frequency of cesium-133 atoms, as do other atomic clocks. Its greater accuracy is obtained by using lasers to trap and cool a "packet" of cesium atoms to just above absolute zero (the lowest possible temperature). The latest model, the NIST F-1, is accurate to one second in 20 million years.

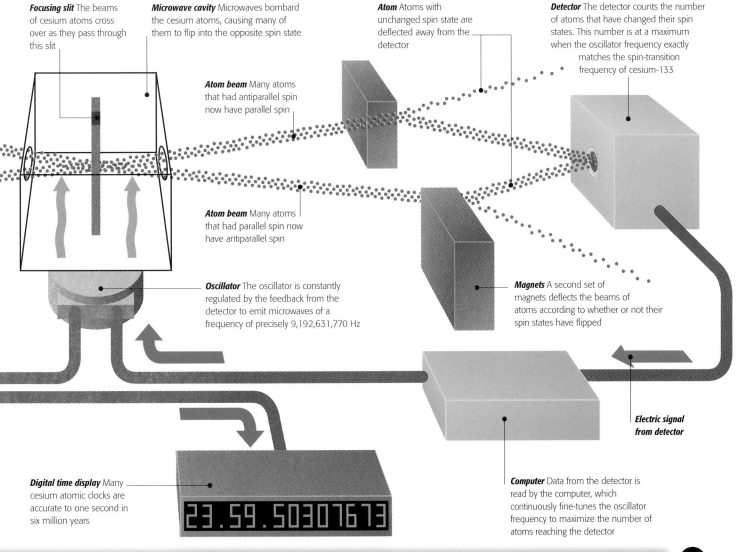

Focusing slit The beams of cesium atoms cross over as they pass through this slit

Microwave cavity Microwaves bombard the cesium atoms, causing many of them to flip into the opposite spin state

Atom Atoms with unchanged spin state are deflected away from the detector

Detector The detector counts the number of atoms that have changed their spin states. This number is at a maximum when the oscillator frequency exactly matches the spin-transition frequency of cesium-133

Atom beam Many atoms that had antiparallel spin now have parallel spin

Atom beam Many atoms that had parallel spin now have antiparallel spin

Oscillator The oscillator is constantly regulated by the feedback from the detector to emit microwaves of a frequency of precisely 9,192,631,770 Hz

Magnets A second set of magnets deflects the beams of atoms according to whether or not their spin states have flipped

Electric signal from detector

Digital time display Many cesium atomic clocks are accurate to one second in six million years

23.59.50307673

Computer Data from the detector is read by the computer, which continuously fine-tunes the oscillator frequency to maximize the number of atoms reaching the detector

SEE ALSO: CLOCKS & WATCHES *p.40* | NAVIGATION *p.132* | GLOBAL POSITIONING SYSTEM *p.136* | MATTER *p.258*

Electron microscopes

Optical microscopes have been used since the early 1600s to view objects too small to be seen with the naked eye, and over 400 years of refinement have brought them to the limits of their physical capabilities. The drive to see further into the microscopic world led to the development of the electron microscope, a technology capable of revealing details that are invisible to an optical microscope.

What limits the power of a microscope is not simply its capacity to magnify—it is the degree of detail that it can reveal under magnification (its resolution). Optical microscopes, which image objects using visible light, have resolutions up to 2,000 times greater than that of the human eye. Electron microscopes use beams of accelerated electrons to image objects. Like light, a beam of matter such as electrons can be considered to have a wavelength. An electron beam has a much shorter wavelength than visible light, which gives it a much greater resolution—about two million times that of the human eye.

There are three types of electron microscopes: the transmission electron microscope (TEM), the scanning electron microscope (SEM), and the scanning tunneling microscope (STM). TEMs and SEMs use an electron gun to generate a high-energy electron beam and electromagnets to focus it. STMs, the most sensitive type, use a tiny electron current to resolve objects on an atomic scale.

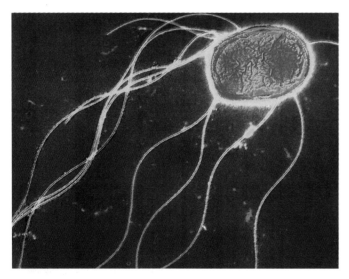

Salmonella bacterium
TEMs take cross-sectional images that reveal details too small to be seen with visible light. Histologists (plant and animal tissue specialists) use TEMs to examine cells such as this bacterium that causes food-poisoning.

Transmission electron microscope

The TEM fires an electron beam through a specimen placed in the center of the microscope. Electrons are absorbed or scattered by denser parts of the specimen, while less dense parts allow the beam through. This casts a shadow of the specimen onto an imaging plate directly underneath it.

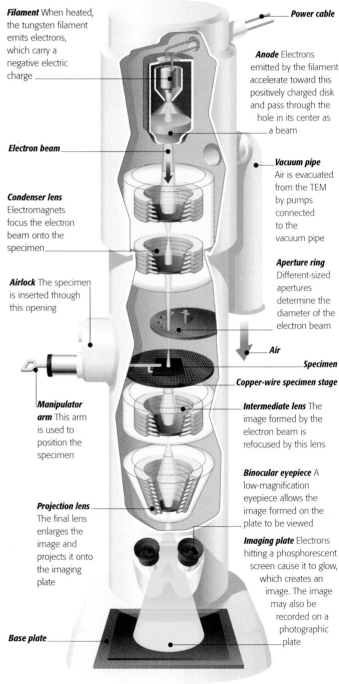

Filament When heated, the tungsten filament emits electrons, which carry a negative electric charge

Electron beam

Condenser lens Electromagnets focus the electron beam onto the specimen

Airlock The specimen is inserted through this opening

Manipulator arm This arm is used to position the specimen

Projection lens The final lens enlarges the image and projects it onto the imaging plate

Base plate

Power cable

Anode Electrons emitted by the filament accelerate toward this positively charged disk and pass through the hole in its center as a beam

Vacuum pipe Air is evacuated from the TEM by pumps connected to the vacuum pipe

Aperture ring Different-sized apertures determine the diameter of the electron beam

Air

Specimen

Copper-wire specimen stage

Intermediate lens The image formed by the electron beam is refocused by this lens

Binocular eyepiece A low-magnification eyepiece allows the image formed on the plate to be viewed

Imaging plate Electrons hitting a phosphorescent screen cause it to glow, which creates an image. The image may also be recorded on a photographic plate

Scanning electron microscope

In an SEM, an electron beam is focused onto the specimen and scanned across it, rather like in a television cathode-ray tube. When struck by the beam, atoms on the surface of the specimen release "secondary electrons." An Everhart-Thornley detector collects these electrons and converts them into electric pulses, which are processed by a computer to give a 3-D image of the specimen's surface.

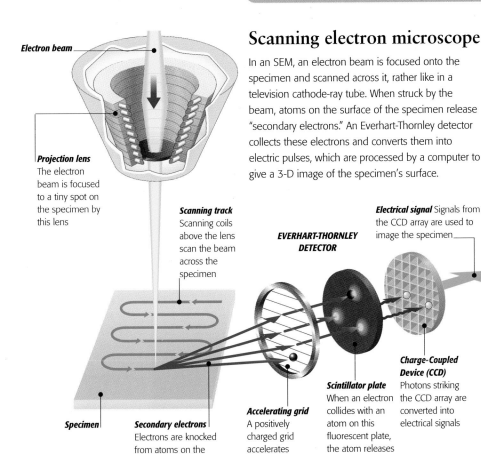

Electron beam

Projection lens
The electron beam is focused to a tiny spot on the specimen by this lens

Scanning track
Scanning coils above the lens scan the beam across the specimen

EVERHART-THORNLEY DETECTOR

Electrical signal Signals from the CCD array are used to image the specimen

Specimen

Secondary electrons
Electrons are knocked from atoms on the surface of the specimen by the electron beam

Accelerating grid
A positively charged grid accelerates electrons into the detector

Scintillator plate
When an electron collides with an atom on this fluorescent plate, the atom releases a photon

Charge-Coupled Device (CCD)
Photons striking the CCD array are converted into electrical signals

Red and white blood cells
SEMs are widely used by cell biologists to examine the surfaces of tissues and cells because they produce highly detailed and vivid 3-D images. In the false-color micrograph shown here, platelike red blood cells and a spherical white blood cell can be clearly seen.

Scanning tunneling microscope

A STM works by passing a sharp, electrically charged stylus so close to the surface of a specimen that electrons jump the gap between them. The amount of current produced is determined by the size of the gap. A computer maintains the current, and hence the gap, at a constant level as the stylus rides over atoms on the surface of the specimen. The vertical movement of the stylus is used to produce a 3-D map of the surface of the specimen.

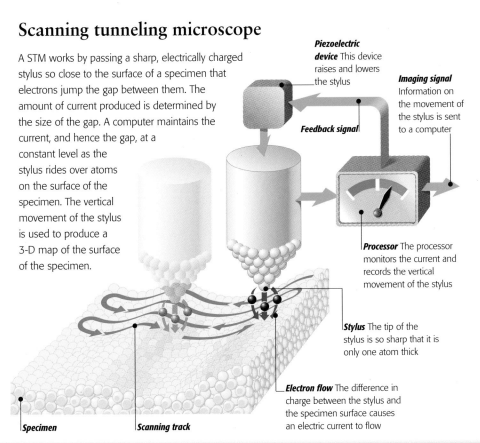

Piezoelectric device This device raises and lowers the stylus

Imaging signal
Information on the movement of the stylus is sent to a computer

Feedback signal

Processor The processor monitors the current and records the vertical movement of the stylus

Stylus The tip of the stylus is so sharp that it is only one atom thick

Electron flow The difference in charge between the stylus and the specimen surface causes an electric current to flow

Specimen

Scanning track

Gold atoms
STMs have the power to resolve on an atomic scale, enabling individual atoms to be seen. This false-color STM image illustrates gold atoms, shown in red and yellow, resting on top of evenly spaced graphite atoms, shown in green.

SEE ALSO: TELEVISION *p.100* | PARTICLE ACCELERATORS *p.232* | MATTER *p.258*

Optical telescopes

Astronomers' yearning to magnify the light from the night sky into increasingly detailed images has inspired engineers to invent more and more powerful optical telescopes. Early astronomers, such as the Italian Galileo, used lenses to focus the light of the stars. Refracting (lens-based) telescopes have now been superseded by reflecting telescopes, which use parabolic mirrors to collect and focus light.

The main advantage mirrors have over lenses is that they are comparatively less heavy and can be supported from behind, allowing larger telescopes to be built that collect more light. They are also able to collect some infrared and ultraviolet rays as well as visible light. Further increases in the size of optical telescopes has been made possible by the development of multisectional mirrors. These mirrors have active control systems that enable their shape to be adjusted to overcome distortions caused by their own weight and temperature differences in the atmosphere.

NGC 1232
This image of the NGC 1232 spiral galaxy was captured by the Very Large Telescope in Chile. It is a combination of three exposures, showing ultraviolet, blue, and red light. The reddish color of the central region comes from older stars, while the spiral arms are populated by younger stars, shown in blue.

Adaptive optics

Air temperature in the atmosphere continually changes, causing local changes in atmospheric density that distort light received (on Earth) from celestial objects. The result is blurring of images from telescopes. Adaptive optic systems iron out many of these distortions. In one system, a laser beam is focused to form a "beacon" high in the atmosphere. Light from this beacon is collected by the telescope and routed to a sensor that measures any distortions. Mirrors are then used to correct these distortions, and in doing so correct the distortions of light from the stars.

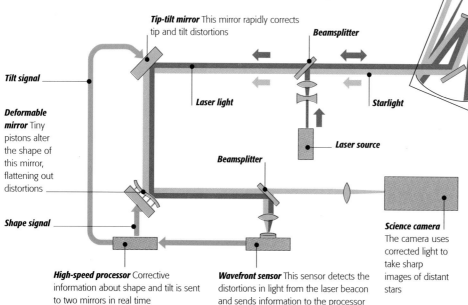

Tip-tilt mirror This mirror rapidly corrects tip and tilt distortions

Beamsplitter

Tilt signal

Deformable mirror Tiny pistons alter the shape of this mirror, flattening out distortions

Laser light

Starlight

Shape signal

Beamsplitter

Laser source

Science camera The camera uses corrected light to take sharp images of distant stars

High-speed processor Corrective information about shape and tilt is sent to two mirrors in real time

Wavefront sensor This sensor detects the distortions in light from the laser beacon and sends information to the processor

Laser beacon Laser beams are focused from 6 miles (10 km) to 60 miles (100 km) up in the atmosphere

Light from star Incoming light from a star is distorted by perturbations in the atmosphere

How adaptive optics work
A high-speed processor rapidly calculates how to correct the wavefront distortions and drives two flexible mirrors that make the correction in real time. The corrected light is then routed to a camera, which captures a very sharp image of the star or other object being viewed.

KEY

▬ LASER LIGHT	◁▷ CONVEX LENS
▬ STAR LIGHT	⋈ CONCAVE LENS

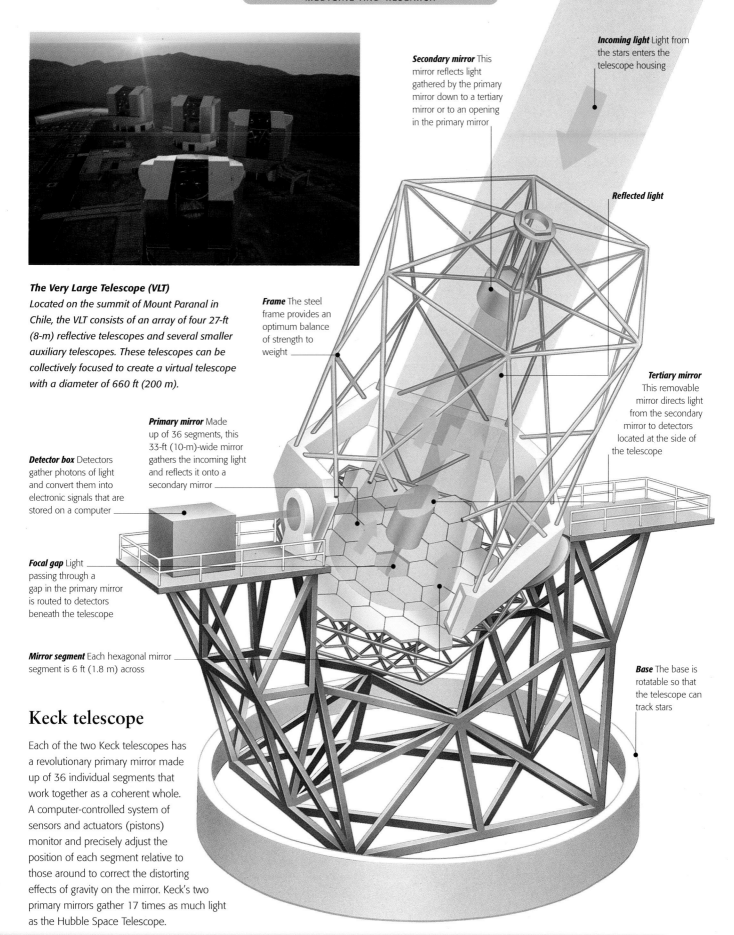

Incoming light Light from the stars enters the telescope housing

Secondary mirror This mirror reflects light gathered by the primary mirror down to a tertiary mirror or to an opening in the primary mirror

Reflected light

The Very Large Telescope (VLT)
Located on the summit of Mount Paranal in Chile, the VLT consists of an array of four 27-ft (8-m) reflective telescopes and several smaller auxiliary telescopes. These telescopes can be collectively focused to create a virtual telescope with a diameter of 660 ft (200 m).

Frame The steel frame provides an optimum balance of strength to weight

Tertiary mirror This removable mirror directs light from the secondary mirror to detectors located at the side of the telescope

Primary mirror Made up of 36 segments, this 33-ft (10-m)-wide mirror gathers the incoming light and reflects it onto a secondary mirror

Detector box Detectors gather photons of light and convert them into electronic signals that are stored on a computer

Focal gap Light passing through a gap in the primary mirror is routed to detectors beneath the telescope

Mirror segment Each hexagonal mirror segment is 6 ft (1.8 m) across

Base The base is rotatable so that the telescope can track stars

Keck telescope

Each of the two Keck telescopes has a revolutionary primary mirror made up of 36 individual segments that work together as a coherent whole. A computer-controlled system of sensors and actuators (pistons) monitor and precisely adjust the position of each segment relative to those around to correct the distorting effects of gravity on the mirror. Keck's two primary mirrors gather 17 times as much light as the Hubble Space Telescope.

SEE ALSO: LASERS *p.220* | RADIO TELESCOPES *p.230* | SPACE TELESCOPES *p.254* | LIGHT *p.262*

Radio telescopes

Astronomers investigating the nature of the Universe often use telescopes to detect radio waves emitted from objects in space. These generate images of objects, such as pulsars and nebulae, that optical telescopes may be unable to detect at all. Radio telescopes are a powerful tool for probing the deepest reaches of space, being capable of resolving celestial bodies as far away as 15 billion light years. They are also helping to unravel the secrets of the big bang.

The hiss of radio signals from the Milky Way that Karl Jansky detected in 1932 began the era of radio astronomy. His fellow American, Grote Reber, invented the radio telescope and revealed elements of the radio sky for the first time. In order to locate the source of a radio signal, the diameter of the dish must be longer than the wavelength of the radio wave it is designed to detect. Radio telescopes are much larger than optical telescopes because radio waves have wavelengths that may be more than 30 ft (10 m) long, while light has wavelengths of only hundreds of nanometers. The first giant dishes were constructed at Jodrell Bank, England, in 1957. The Very Large Array (VLA), built by the U.S. National Radio Observatory in 1980, uses a technique called interferometry to produce much sharper images from an array of dishes. Radio telescopes can transmit radio waves into space and detect the returning echoes. Telescopes such as the Arecibo in Puerto Rico, which spans 1,000 ft (305 m) across, making it the largest dish in the world, can be used to map and measure the rotation of planets and moons in our Solar System.

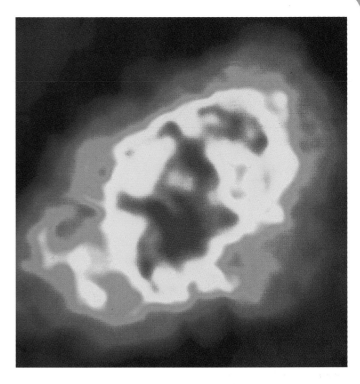

Crab Nebula
This radio image of the Crab Nebula in the constellation of Taurus shows the same supernova explosion that Chinese astronomers recorded in 1054 CE. The nebula was formed by enormous quantities of gas expelled during the supernova event and by interstellar gas sucked in and ionized by the subsequent shock wave. The Crab Pulsar, a dense neutron star at the core of the nebula, gives off intense bursts of radiation at regular intervals.

Interferometry

The quest for more detailed astronomical images led radio astronomers to apply the technique of interferometry to their instruments. Two dishes focused on the same radio source in the sky have greater resolving power than a single dish: the greater the distance between the dishes, or baseline, the sharper the image. Linking dishes on different continents by computer can generate images as much as a thousand times as sharp as the image from an optical telescope. The Very Large Baseline Array (VLBA), which links ten dishes from Hawaii to the Virgin Islands, produces very high-quality images. A baseline twice the diameter of the Earth has been created by linking Earth- and space-telescopes.

The Very Large Array (VLA)
The 27 dishes of the VLA near Socorro, New Mexico, can be positioned on the tracks of a Y-shaped railway. Each track is about 12 miles (20 km) long and each dish is 80 ft (25 m) in diameter.

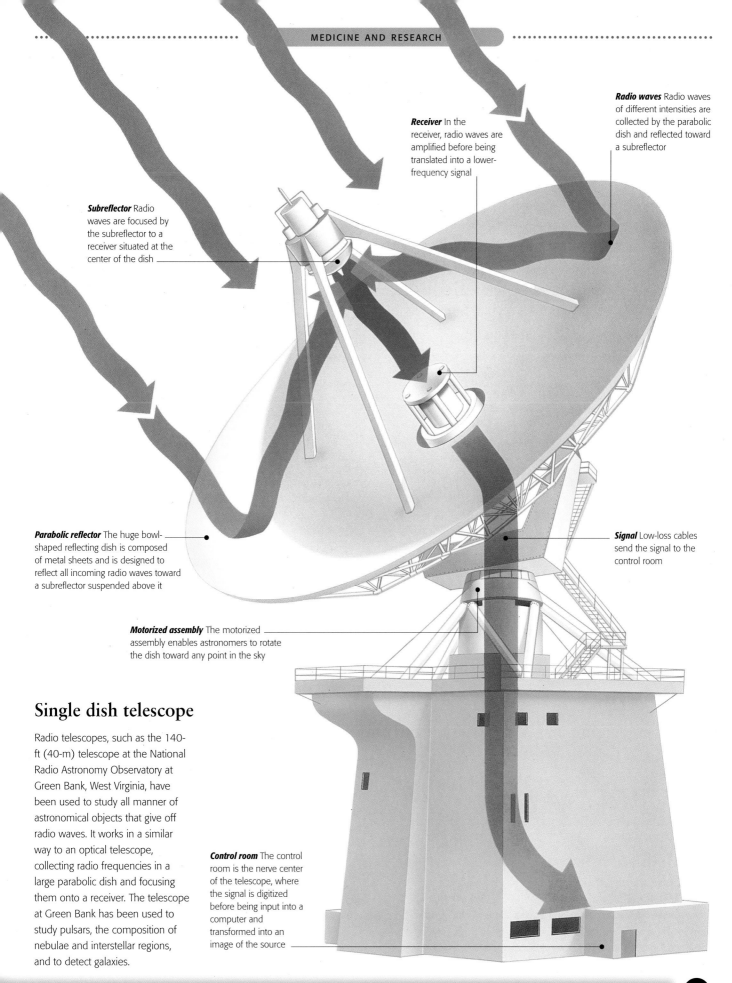

Receiver In the receiver, radio waves are amplified before being translated into a lower-frequency signal

Radio waves Radio waves of different intensities are collected by the parabolic dish and reflected toward a subreflector

Subreflector Radio waves are focused by the subreflector to a receiver situated at the center of the dish

Parabolic reflector The huge bowl-shaped reflecting dish is composed of metal sheets and is designed to reflect all incoming radio waves toward a subreflector suspended above it

Signal Low-loss cables send the signal to the control room

Motorized assembly The motorized assembly enables astronomers to rotate the dish toward any point in the sky

Single dish telescope

Radio telescopes, such as the 140-ft (40-m) telescope at the National Radio Astronomy Observatory at Green Bank, West Virginia, have been used to study all manner of astronomical objects that give off radio waves. It works in a similar way to an optical telescope, collecting radio frequencies in a large parabolic dish and focusing them onto a receiver. The telescope at Green Bank has been used to study pulsars, the composition of nebulae and interstellar regions, and to detect galaxies.

Control room The control room is the nerve center of the telescope, where the signal is digitized before being input into a computer and transformed into an image of the source

SEE ALSO: RADIO *p.48* | OPTICAL TELESCOPES *p.228* | SPACE TELESCOPES *p.254* | LIGHT *p.262*

Particle accelerators

Particle accelerators, some of the largest scientific instruments built, are used to probe the smallest pieces of matter that exist. By speeding up subatomic particles and then colliding them together, physicists are able to explore what matter is made of and the forces that hold it together, and to produce unstable, exotic varieties of matter that have not existed since the beginning of the Universe.

Particle accelerators speed up charged particles, such as electrons, positrons, and protons, along straight or circular vacuum tubes and make them collide with each other or with stationary targets. The energy released by the collisions is turned into different forms of matter, some previously unseen but predicted by theory, revealing information about the fundamental forces of nature and the history of the Universe. Particle accelerators are not just tools for research; hospitals use small accelerators to create radioisotopes for imaging, and nearly every house has a simple accelerator—the cathode-ray tube inside a TV—which accelerates electons and collides them with a phospor screen.

The Large Electron-Positron (LEP) collider at CERN, near Geneva, Switzerland, is the largest existing accelerator. It is being modified to make a new machine—the Large Hadron Collider (LHC)—which will accelerate and collide protons and ions.

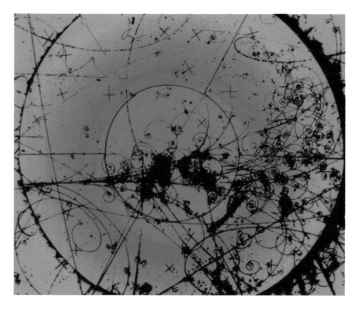

Color-enhanced picture from bubble chamber
Particle collisions cannot be observed directly; so special detectors are needed to record them. One type of detector, called a bubble chamber, contains liquid hydrogen that is held close to its boiling point. Particles passing through the liquid, produce ions around which bubbles of boiling hydrogen form. DIfferent particles produce characteristic tracks of bubbles that can be photographed.

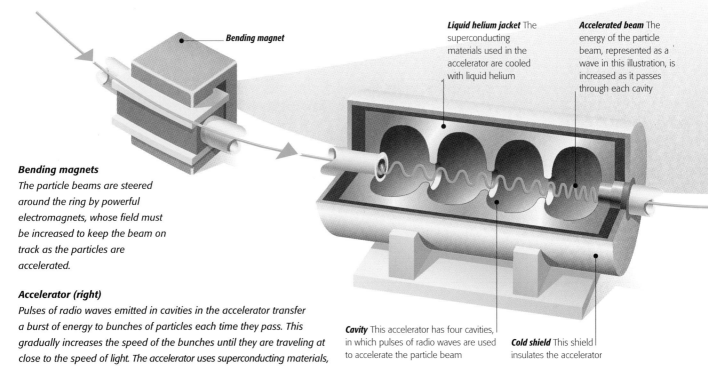

Bending magnet

Liquid helium jacket The superconducting materials used in the accelerator are cooled with liquid helium

Accelerated beam The energy of the particle beam, represented as a wave in this illustration, is increased as it passes through each cavity

Bending magnets
The particle beams are steered around the ring by powerful electromagnets, whose field must be increased to keep the beam on track as the particles are accelerated.

Accelerator (right)
Pulses of radio waves emitted in cavities in the accelerator transfer a burst of energy to bunches of particles each time they pass. This gradually increases the speed of the bunches until they are traveling at close to the speed of light. The accelerator uses superconducting materials, which must be kept extremely cold.

Cavity This accelerator has four cavities, in which pulses of radio waves are used to accelerate the particle beam

Cold shield This shield insulates the accelerator

KEY

 FOCUSING MAGNETS

BENDING MAGNETS

 ACCELERATOR CAVITY

COLLISION DETECTOR

 ACCELERATOR PIPE

BYPASS SECTION

 PATH OF ELECTRONS

PATH OF POSITRONS

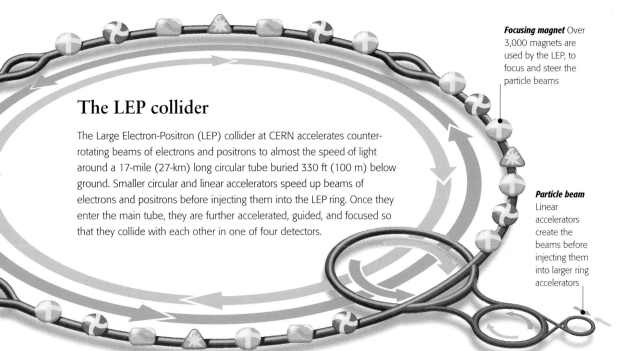

The LEP collider

The Large Electron-Positron (LEP) collider at CERN accelerates counter-rotating beams of electrons and positrons to almost the speed of light around a 17-mile (27-km) long circular tube buried 330 ft (100 m) below ground. Smaller circular and linear accelerators speed up beams of electrons and positrons before injecting them into the LEP ring. Once they enter the main tube, they are further accelerated, guided, and focused so that they collide with each other in one of four detectors.

Focusing magnet Over 3,000 magnets are used by the LEP, to focus and steer the particle beams

Particle beam Linear accelerators create the beams before injecting them into larger ring accelerators

DELPHI detector This detector, which is wrapped around the vacuum tube, has several layers that each detect different types of particles produced in collisions

Collision An electron and a positron annihilate each other in a head-on collision, producing energy that turns back into other particles, just as matter must have formed from energy at the birth of the universe

Positron beam Only one collision in 40,000 occurs head-on as required, so the particle beams are circulated and collided millions of times over several hours

Quadrupole focusing magnet

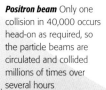

Focusing magnets
Quadrupole and sextupole magnets (with four and six poles respectively) focus the beams down to the thickness of a human hair, in much the same way as a lens focuses light.

Collision detector (left)
Collisions are recorded by complex detectors run by international teams of hundreds of scientists. The DELPHI detector is as large as a four-story house.

Electron beam This beam is made up of 100 billion electrons, which travel in bunches around the ring over 10,000 times a second

Vacuum tube Particles travel through a tube with a diameter of 6 in (15 cm), which is evacuated of air molecules to avoid unwanted collisions

SEE ALSO: TELEVISION *p.100* | MRI & PET SCANNING *p.198* | ELECTRON MICROSCOPES *p.226* | MATTER *p.258*

Space

Since the launch of the first satellite in 1957, expeditions to outer space have affected all our lives. The Space Age has seen a transformation in our understanding of Earth and the Universe, through both orbiting satellites and probes to other planets, and has revealed the extent of our fragility and isolation in the vast expanses of space. As well as contributing to pure science, space exploration has also produced many technological spin-offs with more mundane uses. This chapter shows how the technology of space travel, which often pushes ingenuity and engineering skill to the limit, has enabled humanity to conquer this new frontier.

History of space exploration

The Space Age began on October 4, 1957, with the launch of the first satellite, *Sputnik 1*, by the Soviet Union. Since then, space technology has developed rapidly, and had a major influence on our understanding of the Universe, as well as our everyday lives.

The Space Race

The drive to explore space originated in the Cold War of the 1950s. The United States and the Soviet Union both had nuclear weapons, and were racing to build missiles that could deliver them to distant targets. Many of those at the forefront of this research were also keen spaceflight enthusiasts. They argued that space programs would be an effective way of demonstrating missile technology, and could also be a tremendous propaganda weapon. Satellites in orbit could have practical uses as well.

Both the U.S. and U.S.S.R. announced they would orbit satellites in 1957, but wrangles between military and civilian engineers delayed the U.S. program, and the U.S.S.R. took the lead. In 1958, the U.S. satellite *Explorer 1* was launched, and NASA (the National Aeronautics and Space Administration) was founded. The Space Race was underway, but NASA had much catching up to do.

By the early 1960s, satellite launches were becoming common. The first communications satellites were launched, though they could not yet reach the high geostationary orbits that most use today. Other satellites carried cameras into orbit on spying and weather forecasting missions, or studied near-Earth space. The first probes were also launched to the Moon and nearby planets.

The obvious next step was to put a human into space. The Soviet Union won this race as well, launching Yuri Gagarin aboard *Vostok 1* for a brief, single-orbit flight on April 12, 1961. NASA launched *Mercury 1* a month later, but because their rockets were smaller and less powerful than Soviet ones, the capsule carrying Alan Shepard could not even be put into a stable orbit.

Men on the Moon

Shortly after the first Mercury flight, President John F. Kennedy announced that the U.S. would aim to put astronauts on the Moon by the end of the 1960s. The Apollo program he launched was one of humanity's greatest accomplishments, sending astronauts 250,000 miles (400,000 km) to the Moon and returning them safely. It involved construction of the largest rocket ever

Sputnik 1
The first artificial satellite was a small steel sphere weighing 185 lb (85 kg), and fitted with four radio antennae to transmit a simple signal back to Earth. More sophisticated satellites soon followed.

Lunar missions
An astronaut stands in front of the lunar module during the Apollo 15 mission. Twelve astronauts walked on the Moon between 1969 to 1972, but a full-scale return to the Moon is still probably some decades away.

Space Station Mir

Mir was the most ambitious Soviet space station, launched in 1986, and assembled over several years. The station long outlasted its intended lifetime, but ran into a series of problems in the late 1990s, and was eventually abandoned. In 2000, cosmonauts returned to the station and revived it as part of a privately financed space program.

built, the 360-ft (110-m) Saturn V, and the design of a complex three-part, three-man space vehicle. Later missions even carried a jeep—the Lunar Roving Vehicle.

The Apollo missions came to fruition when Neil Armstrong and Buzz Aldrin stepped onto the Moon on July 20, 1969. Their mission, *Apollo 11*, was followed by six more, before waning public interest and budget cuts brought the project to an end.

Commercializing space

By the end of the Apollo missions, space was a very different place. Commercial satellite launches were becoming common, as communications satellites in particular began to pay their way. Europe and Japan were also developing space programs, and as the Cold War faded, profit, not politics, became the main motivation for space exploration. If not directly commercial, then most space missions would at least need to produce practical scientific data and possible technological spin-offs.

After difficulties with their lunar program, the Soviets had switched their efforts to establishing space stations in orbit, where cosmonauts could stay for long periods, and carry out useful research. From the early 1970s onward, they launched a series of stations, and established many space endurance records, capped by Dr. Valeri Polyakov's 438-day stay aboard *Mir* in 1994–95.

NASA, meanwhile, concentrated on developing the Space Shuttle—a reusable launch vehicle. But the Shuttle, finally launched in 1981, did not cut launch costs as much as hoped. As a result, Europe's Ariane rockets took the largest share of the satellite launch market. NASA's Shuttle program had another setback in 1986 when *Challenger* exploded shortly after its launch.

The Space Station and beyond

The collapse of the Soviet Union and NASA's inability to finance the construction of its own space station has brought about an era of international cooperation. NASA is now working alongside Russia and other space powers to build the International Space Station (ISS), which will provide a permanent crewed laboratory in space. The high projected cost of transporting materials and crew to the ISS, has prompted NASA to start developing a new generation of economical reusable launch vehicles that should reduce payload costs and may finally open up space travel to the masses.

X-33 VentureStar

The next generation of launch vehicles will be reusable "single stage to orbit" designs, capable of carrying cargos into space at about ten percent of current costs. Several governments and private corporations are working on aerospace planes that combine features of aircraft and rockets. NASA's X-33 is a reusable vehicle that takes off like a rocket and lands like a plane.

SEE ALSO: SPACE ROCKETS p.238 | SPACE SHUTTLE p.240 | ASTRONAUTS' EQUIPMENT p.242 | SPACE STATIONS p.244

Space rockets

Conquering space was probably humanity's greatest technological achievement of the 20th century. It required the development of immensely powerful rocket-propelled vehicles controlled by highly precise guidance systems. Since the Soviet Union launched Sputnik—the first satellite into space—in 1957, thousands of successful manned and unmanned space missions have been launched by rockets.

Unlike most engines, which combust fuel with oxygen from the atmosphere, rocket engines must work in the vacuum of space and so need to carry their own oxygen. Rockets are reaction engines—the hot, expanding gases produced by combustion of liquid or solid fuel are forced out of the engine nozzle by an action force, and produce an equal and opposite push (the reaction force) on the rocket. Rocket vehicles must reach speeds of around 18,000 mph (29,000 km/h) to reach low-altitude orbit, and around 25,000 mph (40,000 km/h) to escape the Earth's gravitational field. The enormous thrust required is provided by the combustion of tons of fuel in multistage vehicles.

Blast off
Since 1957, Atlas rockets have launched about 100 unmanned space missions, including voyages to the Moon, the Pioneer missions that flew past Jupiter and Venus, and the Voyager space probe that landed on Mars. The latest version, Atlas V, will be used as a launch vehicle for government and commercial satellites.

Atlas IIAS rocket

Atlas IIAS rockets carried out 14 successful space missions between December 1993 and April 1999. Over 150 ft (45 m) tall, they are composed of three stages: the *Atlas*, *Centaur*, and payload stages. The *Atlas* stage, which launches the rocket into space, is powered by liquid-fuel engines and four solid-fuel boosters. This stage is then jettisoned and the payload moved into orbit by the *Centaur* stage.

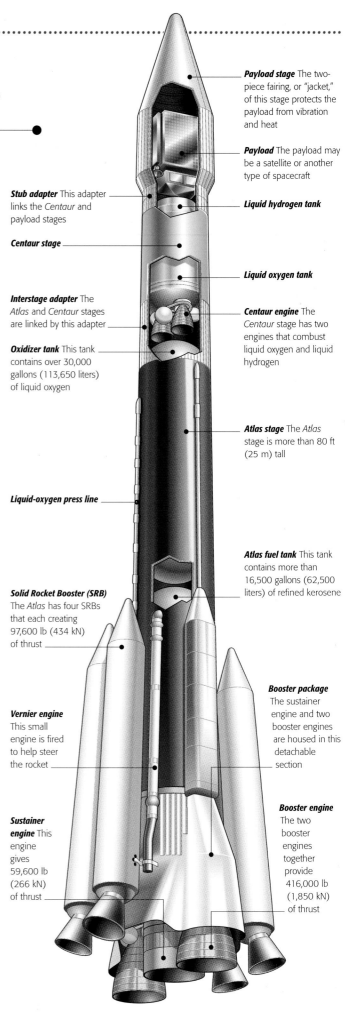

Payload stage The two-piece fairing, or "jacket," of this stage protects the payload from vibration and heat

Payload The payload may be a satellite or another type of spacecraft

Stub adapter This adapter links the *Centaur* and payload stages

Liquid hydrogen tank

Centaur stage

Liquid oxygen tank

Interstage adapter The *Atlas* and *Centaur* stages are linked by this adapter

Centaur engine The *Centaur* stage has two engines that combust liquid oxygen and liquid hydrogen

Oxidizer tank This tank contains over 30,000 gallons (113,650 liters) of liquid oxygen

Atlas stage The *Atlas* stage is more than 80 ft (25 m) tall

Liquid-oxygen press line

Atlas fuel tank This tank contains more than 16,500 gallons (62,500 liters) of refined kerosene

Solid Rocket Booster (SRB) The *Atlas* has four SRBs that each creating 97,600 lb (434 kN) of thrust

Booster package The sustainer engine and two booster engines are housed in this detachable section

Vernier engine This small engine is fired to help steer the rocket

Booster engine The two booster engines together provide 416,000 lb (1,850 kN) of thrust

Sustainer engine This engine gives 59,600 lb (266 kN) of thrust

Launching a satellite

The *Atlas* rocket that launches satellites into space is made up of separate stages. Once the fuel in one stage is burned up, the stage is jettisoned, reducing the weight of the vehicle and the amount of fuel that needs to be carried by subsequent stages. The *Atlas* stage propels the vehicle out of the Earth's atmosphere. The *Centaur* stage then maneuvers the payload stage into the correct orbit.

Centaur separates from Atlas Once clear of the Earth's atmosphere, the *Atlas* stage is jettisoned

Changing orbit *Centaur* uses its twin engines to propel itself into the correct orbit

Satellite launched After establishing the correct orbit, *Centaur* releases the satellite from a vertical position

Mission complete Once the satellite is in orbit, Centaur propels itself toward the Earth, where it burns up as it reenters the atmosphere

Payload fairing jettisoned After about three-and-a-half minutes, spring-operated latches open to jettison the protective payload fairing

Booster engines jettisoned Less than one minute later, the booster engines cut out and the booster package is jettisoned by releasing ten pneumatic (air-operated) latches. The sustainer engine continues to fire, providing forward thrust for the *Atlas* stage

SRBs jettisoned Both pairs of solid rocket boosters use up their fuel during the first two minutes of flight before being jettisoned

Liftoff Three liquid-propellant engines and two solid rocket boosters (SRBs) are used to launch the rocket. The second pair of SRBs are engaged about one minute later.

Rocket engines

Many rocket vehicles use a combination of liquid-propellant engines and solid rocket boosters (SRBs). SRBs use solid fuel and an oxidizer (which supplies oxygen necessary for combustion), and are used mainly at launch since they cannot be turned off after the fuel is ignited. Liquid-propellant engines typically use hydrogen and oxygen, stored at high pressure. The hydrogen and oxygen mix in a combustion chamber, burning at high temperature to produce expanding exhaust gases that exit through a nozzle at speeds of around 10,000 mph (16,000 km/h). The propellant flow is controlled, enabling liquid-propellant engines to be turned on and off.

Explosive power
Turbopumps draw liquid hydrogen and liquid oxygen through a network of pipes to a combustion chamber. Here the liquids react explosively, generating huge amounts of thrust as they leave the engine nozzle.

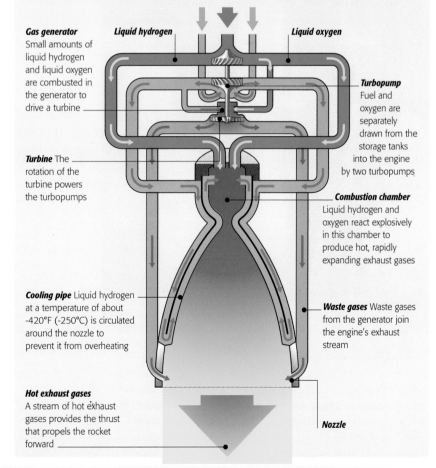

Gas generator Small amounts of liquid hydrogen and liquid oxygen are combusted in the generator to drive a turbine

Liquid hydrogen

Liquid oxygen

Turbopump Fuel and oxygen are separately drawn from the storage tanks into the engine by two turbopumps

Turbine The rotation of the turbine powers the turbopumps

Combustion chamber Liquid hydrogen and oxygen react explosively in this chamber to produce hot, rapidly expanding exhaust gases

Cooling pipe Liquid hydrogen at a temperature of about -420°F (-250°C) is circulated around the nozzle to prevent it from overheating

Waste gases Waste gases from the generator join the engine's exhaust stream

Hot exhaust gases A stream of hot exhaust gases provides the thrust that propels the rocket forward

Nozzle

Space Shuttle

Rudder The rudder provides lateral control and acts as a speed brake on landing

The development of NASA's Space Shuttle launched a new age of space exploration. Previously, spacecraft lasted for only one mission, but the Shuttle, which lifts off like a rocket and returns to Earth like a glider, can be reused up to 100 times. The Shuttle has launched and repaired numerous satellites, including the Hubble Space Telescope, as well as carrying the onboard Spacelab into orbit.

The main component of the Shuttle is a spacecraft, about two-thirds the size of a Boeing 747 airplane, called the Orbiter. In order to achieve liftoff and travel into space, the Orbiter is "mated" to a giant external tank (ET), which feeds liquid fuel to its three main engines. This is, in turn, mated to two Solid Rocket Boosters (SRBs), which burn solid fuel. The Orbiter also carries liquid propellant on board to fuel its own orbital maneuvering engines. Six Orbiters have been built in total: *Enterprise* (prototype), *Columbia*, *Challenger*, *Discovery*, *Atlantis*, and *Endeavor*. *Columbia*'s maiden voyage on April 12, 1981, confirmed that a Shuttle could return successfully from space. On January 28, 1986, *Challenger* exploded 73 seconds after takeoff, killing all seven crew members. The *Endeavor* replaced the *Challenger*; flights resumed in late 1988. Shuttles are now being used to assemble the new International Space Station.

The Orbiter

The Orbiter has its own engine system for maneuvering in space, which is fueled by liquid hydrogen. In order for the hydrogen to burn in space, it is combusted with liquid oxygen stored in separate oxidizer tanks. At 60 ft (18 m) long and 15 ft (4.5 m) wide, the Orbiter's payload bay is big enough to hold a bus. It contains a 50-ft (15-m) manipulator arm, used for launching and retrieving satellites. The reflective inside doors of the bay, known as heat radiators, act as heat shields to protect the cargo from solar radiation.

Main engine (SSME) The Orbiter's three main engines are fed by the ET

From takeoff to landing

On a typical voyage to release a satellite, the Shuttle is launched from its base at Kennedy Space Center (KSC), Florida, and enters orbit around the Earth. When its mission is complete, it returns to Earth, reentering the Earth's atmosphere at a speed of 17,500 mph (28,000 km/h). It then makes an unpowered landing, like a glider, either at KSC or at Edwards Airforce Base in California.

Launching the Space Shuttle
A vast amount of fuel is needed to accelerate the Shuttle to a speed of roughly 17,500 mph (28,000 km/h) to reach an orbit of 190–330 miles (300–530 km) above Earth. The external tank (ET) itself contains 528,000 gallons (2.4 million liters) of liquid fuel. When firing together at launch, the SRBs and the ET produce some 7.3 million pounds (33 million N) of thrust—over 150 times as much as a typical jet engine.

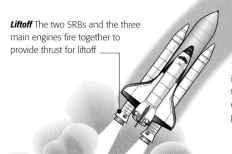

Liftoff The two SRBs and the three main engines fire together to provide thrust for liftoff

Two minutes After the fuel in the SRBs is exhausted, they detach at an altitude of 30 miles (50 km) and parachute back to Earth

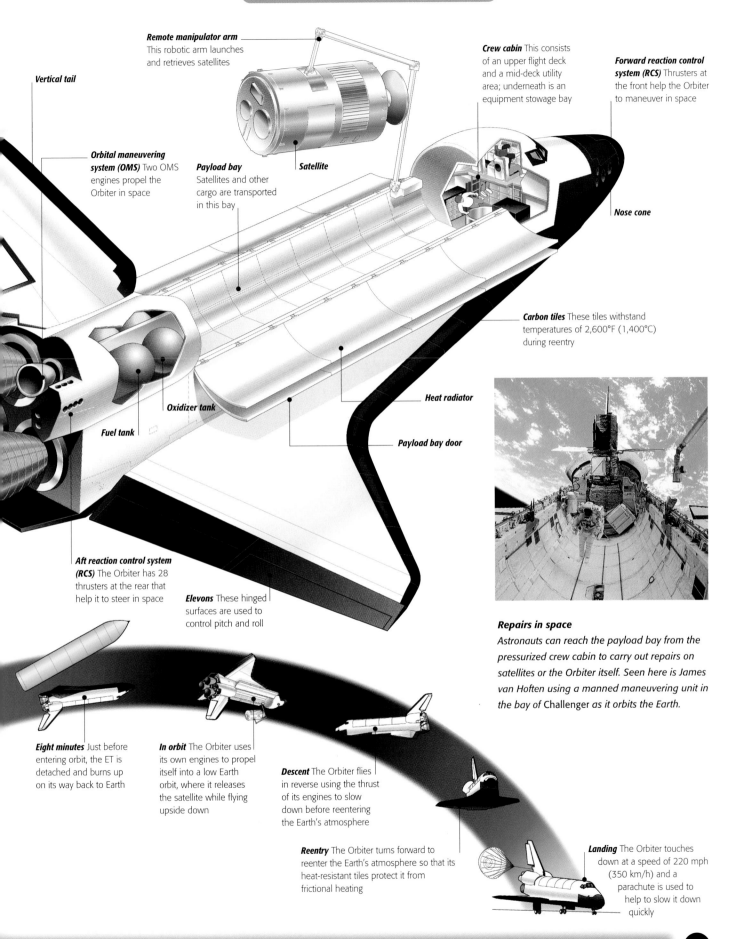

Remote manipulator arm
This robotic arm launches and retrieves satellites

Vertical tail

Orbital maneuvering system (OMS) Two OMS engines propel the Orbiter in space

Payload bay
Satellites and other cargo are transported in this bay

Satellite

Crew cabin This consists of an upper flight deck and a mid-deck utility area; underneath is an equipment stowage bay

Forward reaction control system (RCS) Thrusters at the front help the Orbiter to maneuver in space

Nose cone

Carbon tiles These tiles withstand temperatures of 2,600°F (1,400°C) during reentry

Heat radiator

Fuel tank

Oxidizer tank

Payload bay door

Aft reaction control system (RCS) The Orbiter has 28 thrusters at the rear that help it to steer in space

Elevons These hinged surfaces are used to control pitch and roll

Repairs in space
Astronauts can reach the payload bay from the pressurized crew cabin to carry out repairs on satellites or the Orbiter itself. Seen here is James van Hoften using a manned maneuvering unit in the bay of Challenger as it orbits the Earth.

Eight minutes Just before entering orbit, the ET is detached and burns up on its way back to Earth

In orbit The Orbiter uses its own engines to propel itself into a low Earth orbit, where it releases the satellite while flying upside down

Descent The Orbiter flies in reverse using the thrust of its engines to slow down before reentering the Earth's atmosphere

Reentry The Orbiter turns forward to reenter the Earth's atmosphere so that its heat-resistant tiles protect it from frictional heating

Landing The Orbiter touches down at a speed of 220 mph (350 km/h) and a parachute is used to help to slow it down quickly

SEE ALSO: CERAMICS *p.206* | SPACE ROCKETS *p.238* | SPACE STATIONS *p.244* | SPACE TELESCOPES *p.254*

Astronauts' equipment

The first space suits were simple pressure suits designed to protect astronauts from the stresses of liftoff and reentry and could not be used outside the spacecraft. Modern space suits and equipment, used by Shuttle astronauts, provide a complete life support system in space and, unlike earlier extravehicular suits, allow plenty of maneuverability for space-based construction and repair work.

The space suit, or extravehicular mobility unit (EMU), worn by Shuttle astronauts is much more than a protective garment—it incorporates complex systems and equipment to provide a pressurized environment, oxygen supply, effective screening against tiny particles of fast-moving space debris, and thermal control, in a situation where temperatures can vary from -250°F to +250°F (-155°C to +120°C) in minutes. The earliest types of space suits often provided life-support systems through a cable attached to the spacecraft, but modern EMUs incorporate these functions as part of the suit, allowing greater freedom of movement.

Although most extravehicular activities (EVAs) make use of a secure tether to the spacecraft, astronauts must also carry their own propulsion systems. These have developed from simple jet guns, through bulky backpacks, to the relatively compact units in use today.

The Shuttle extravehicular mobility unit (EMU)

The EMU is made up of 18 separate items, including a 14-layer suit, and provides a safe environment for an astronaut working outside the spacecraft. The EMU is tethered to the Shuttle or a space station to prevent the astronaut from drifting away. All parts of the suit are secured using locks that require three separate motions to open, so that no part of the astronaut can be accidentally exposed in space.

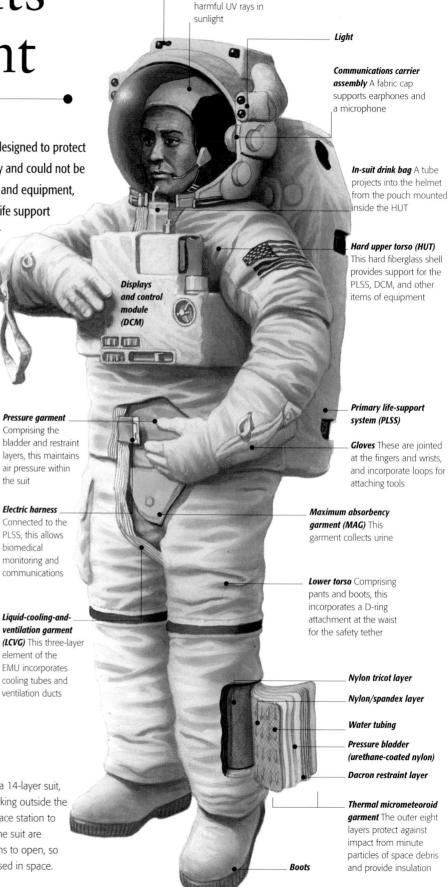

Video camera

Helmet and visor The visor gives protection against harmful UV rays in sunlight

Light

Communications carrier assembly A fabric cap supports earphones and a microphone

In-suit drink bag A tube projects into the helmet from the pouch mounted inside the HUT

Hard upper torso (HUT) This hard fiberglass shell provides support for the PLSS, DCM, and other items of equipment

Displays and control module (DCM)

Pressure garment Comprising the bladder and restraint layers, this maintains air pressure within the suit

Electric harness Connected to the PLSS, this allows biomedical monitoring and communications

Liquid-cooling-and-ventilation garment (LCVG) This three-layer element of the EMU incorporates cooling tubes and ventilation ducts

Primary life-support system (PLSS)

Gloves These are jointed at the fingers and wrists, and incorporate loops for attaching tools

Maximum absorbency garment (MAG) This garment collects urine

Lower torso Comprising pants and boots, this incorporates a D-ring attachment at the waist for the safety tether

Nylon tricot layer

Nylon/spandex layer

Water tubing

Pressure bladder (urethane-coated nylon)

Dacron restraint layer

Thermal micrometeoroid garment The outer eight layers protect against impact from minute particles of space debris and provide insulation

Boots

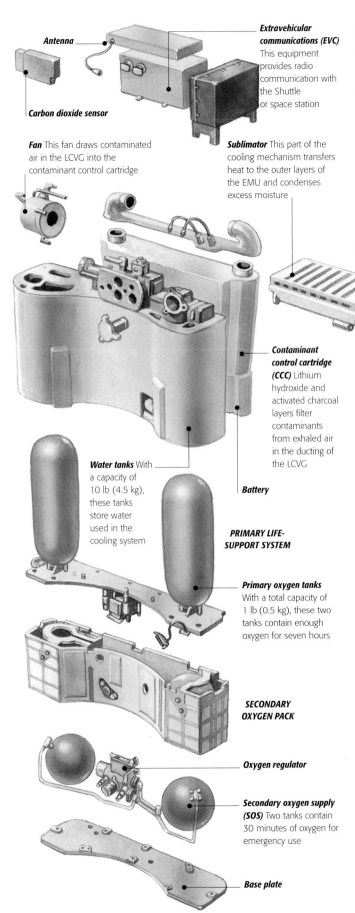

Antenna

Carbon dioxide sensor

Extravehicular communications (EVC) This equipment provides radio communication with the Shuttle or space station

Fan This fan draws contaminated air in the LCVG into the contaminant control cartridge

Sublimator This part of the cooling mechanism transfers heat to the outer layers of the EMU and condenses excess moisture

Contaminant control cartridge (CCC) Lithium hydroxide and activated charcoal layers filter contaminants from exhaled air in the ducting of the LCVG

Water tanks With a capacity of 10 lb (4.5 kg), these tanks store water used in the cooling system

Battery

PRIMARY LIFE-SUPPORT SYSTEM

Primary oxygen tanks With a total capacity of 1 lb (0.5 kg), these two tanks contain enough oxygen for seven hours

SECONDARY OXYGEN PACK

Oxygen regulator

Secondary oxygen supply (SOS) Two tanks contain 30 minutes of oxygen for emergency use

Base plate

Primary life-support system

The primary life-support system (PLSS) is an integral part of the EMU. It is a self-contained backpack carrying a supply of oxygen that is circulated through the suit, a water-pumping system to keep the EMU and astronaut cool, a filter system to prevent the buildup of carbon dioxide, various safety monitors, a battery, a ventilating fan, and communications equipment to keep the astronaut safely in touch with the spacecraft.

SAFER for spacewalking

The SAFER (simplified aid for EVA rescue) propulsion module is the successor to units used on the Skylab missions of the 1970s, and the MMU (manned maneuvering unit) used on early Space Shuttle flights in the 1980s. First used in 1994 on the Mir space station, SAFER is designed to bring an astronaut back to the spacecraft if he or she becomes accidentally separated from it—the system cannot provide propulsion over long distances. The module fits over the PLSS and contains nitrogen gas in pressurized tanks.

Walking in space
A propulsive force is created by firing jets of gas from 24 nozzles distributed around the SAFER unit. As the jets of gas escape in one direction, they create a reaction force that pushes the astronaut the other way. Movement is controlled by a joystick unit that selects which nozzles are activated.

SEE ALSO: Space Shuttle *p.240* | Space stations *p.244* | Space-age materials *p.246*

Space stations

Space stations provide a workplace for astronauts in orbit about 200 miles (320 km) above the Earth's surface. They are used for astronomy, Earth observation, and conducting experiments that require zero gravity or exposure to the vacuum of space. Astronauts can now spend many months in space, gaining experience of long-duration spaceflight that may be useful in future deep-space exploration.

The first space stations were developed by the Soviet Union in the late 1960s, when the Russians realized they were not going to win the race to the Moon. *Salyut* (*Salute*) *1*, launched in 1971, had a laboratory module, living quarters, solar panels to provide electricity, and a docking point for a *Soyuz* (*Union*) space capsule. In 1973, NASA launched its own small *Skylab* station, but development of a full-scale space station was continually delayed. After several more *Salyuts*, the Soviet Union launched *Mir* (*Peace*) in 1985. *Mir* had individual laboratory modules for astronomy, biology, and manufacturing materials. The station was an immense success but suffered a series of setbacks in the late 1990s. Russia and the U.S. are now partners in building the new *International Space Station* (*ISS*).

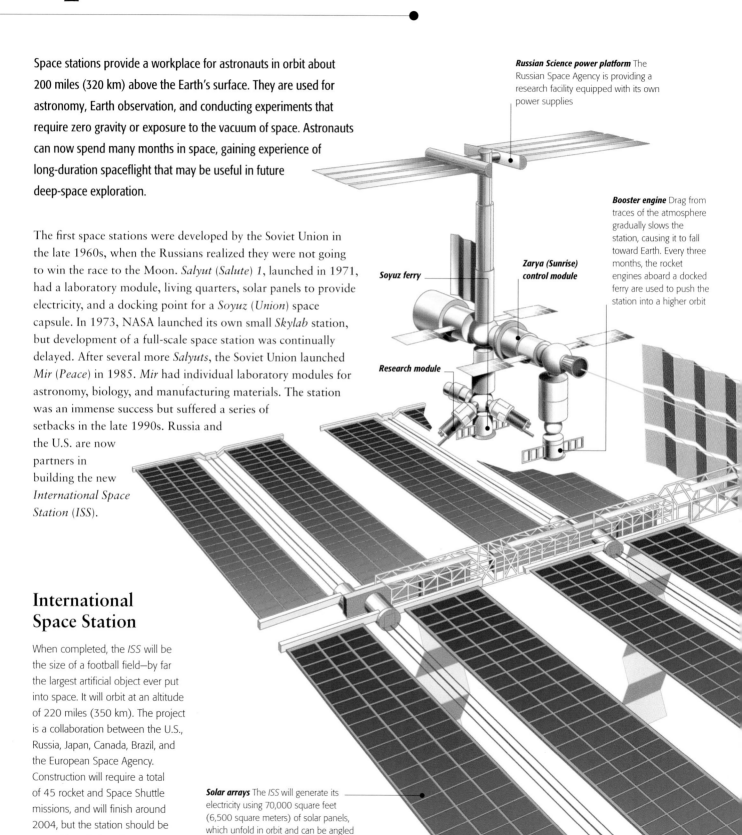

Russian Science power platform The Russian Space Agency is providing a research facility equipped with its own power supplies

Booster engine Drag from traces of the atmosphere gradually slows the station, causing it to fall toward Earth. Every three months, the rocket engines aboard a docked ferry are used to push the station into a higher orbit

Soyuz ferry

Zarya (Sunrise) control module

Research module

International Space Station

When completed, the *ISS* will be the size of a football field—by far the largest artificial object ever put into space. It will orbit at an altitude of 220 miles (350 km). The project is a collaboration between the U.S., Russia, Japan, Canada, Brazil, and the European Space Agency. Construction will require a total of 45 rocket and Space Shuttle missions, and will finish around 2004, but the station should be crewed long before then.

Solar arrays The *ISS* will generate its electricity using 70,000 square feet (6,500 square meters) of solar panels, which unfold in orbit and can be angled to catch the sunlight

Personal satellite assistant

Personal satellite assistants are intelligent robots that will fly aboard the completed *International Space Station*, acting as extra pairs of eyes and ears for the crew and helping to monitor the station's environment and to trace faults. Measuring about 6 in (15 cm) across, the assistant is a sphere that floats in zero gravity and propels itself around using six small fans. When accompanying an astronaut, the assistant may act as a mobile videophone connecting different parts of the station or linking the astronaut directly to Mission Control on Earth. In automatic mode, assistants will float around the station conducting routine maintenance checks and may transmit video and sound to Mission Control. Their small size allows them access to parts of the station difficult for astronauts to reach, and they can work in teams to zero in on a fault.

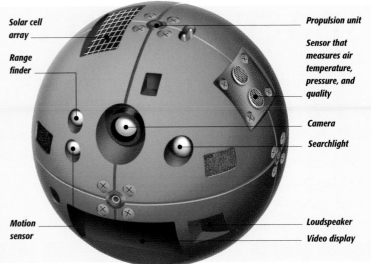

Solar cell array

Range finder

Motion sensor

Propulsion unit

Sensor that measures air temperature, pressure, and quality

Camera

Searchlight

Loudspeaker

Video display

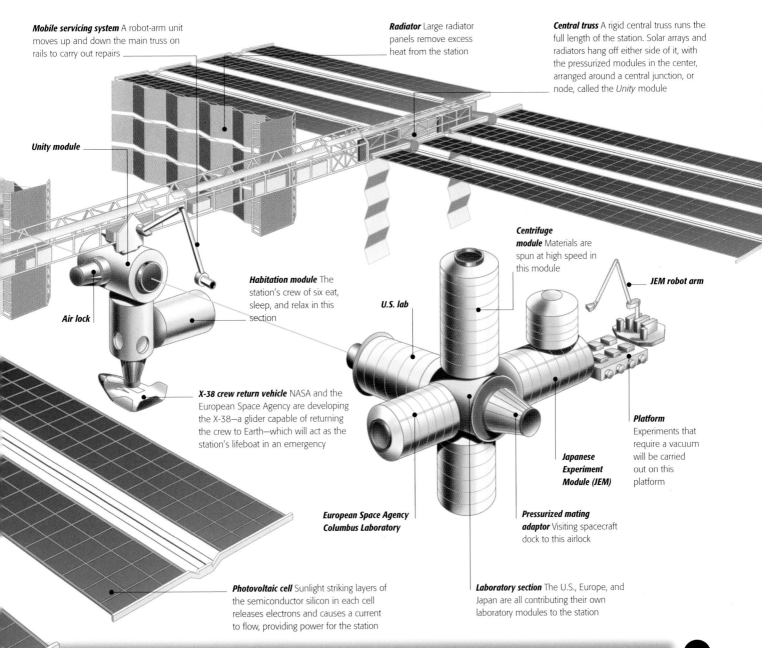

Mobile servicing system A robot-arm unit moves up and down the main truss on rails to carry out repairs

Radiator Large radiator panels remove excess heat from the station

Central truss A rigid central truss runs the full length of the station. Solar arrays and radiators hang off either side of it, with the pressurized modules in the center, arranged around a central junction, or node, called the *Unity* module

Unity module

Air lock

Habitation module The station's crew of six eat, sleep, and relax in this section

X-38 crew return vehicle NASA and the European Space Agency are developing the X-38—a glider capable of returning the crew to Earth—which will act as the station's lifeboat in an emergency

Centrifuge module Materials are spun at high speed in this module

JEM robot arm

U.S. lab

Platform Experiments that require a vacuum will be carried out on this platform

Japanese Experiment Module (JEM)

European Space Agency Columbus Laboratory

Pressurized mating adaptor Visiting spacecraft dock to this airlock

Photovoltaic cell Sunlight striking layers of the semiconductor silicon in each cell releases electrons and causes a current to flow, providing power for the station

Laboratory section The U.S., Europe, and Japan are all contributing their own laboratory modules to the station

SEE ALSO: SOLAR ENERGY *p.174* | ROBOTICS *p.210* | SPACE SHUTTLE *p.240* | SPACE-AGE MATERIALS *p.246*

Space-age materials

Space technology has had a huge effect on the development of materials science. In order to withstand the enormous stresses of launch, the harsh vacuum of space, bombardment with cosmic rays and micrometeorites, extreme temperature ranges and sudden temperature changes, low gravity, and the heat of reentry into the Earth's atmosphere, spacecraft engineers have adopted and developed many advanced materials from other applications, as well as invented materials specifically for spaceflight which later found uses elsewhere. Increasing laboratory work done in space has led to insights into the structure of materials in zero gravity and further applications in everyday life.

Materials for spaceflight

At the dawn of the space age, engineers had only the vaguest idea of the extreme conditions spacecraft materials might have to withstand, but they knew enough to form a list of basic requirements. The most important of these was lightness—rockets of the time (especially in this country) were relatively low-powered, and the weight of any spacecraft had to be kept to a minimum, without sacrificing strength and reliability, if it was to reach orbit safely. In order to achieve these criteria, the earliest spacecraft borrowed strong and light materials from the highest-performance vehicles of the time—fighter aircraft.

One of the most popular new materials was the light metal titanium, used in space capsules from the first NASA Mercury missions onward. During the 1950s and 1960s, titanium was being developed as a material for aircraft parts that needed strength, lightness, and resistance to high temperatures. Other advanced aircraft materials in development at the time included composites that provided even greater strength and durability. Although manufacturers were slow to adopt these costly and exotic materials, spacecraft engineers embraced them early.

Heat-resistant materials

Heat resistance is another vital requirement for spaceflight materials. Spacecraft need to resist the heat of reentry into Earth's atmosphere, while rocket engines must be able to with-

Satellite structures
Satellite casings are made from "honeycomb sandwich" panels—composite sheets on either side of a rigid honeycomb structure—bolted around a light metal chassis. Heat pipes containing low-pressure gases are used to conduct heat from hot to cold parts of the spacecraft.

stand the enormous temperatures of the fuel burning inside them, without melting or catching fire. The first spacecraft heat shields were made from beryllium and phenolic epoxy resin compounds poured into a stainless-steel honeycomb matrix. These formed "ablative" heat shields—cones mounted on the base of a space capsule, which were designed to burn and break away during reentry, carrying heat away from the spacecraft itself. The Space Shuttle and other reusable spacecraft have replaced ablative shields with heat-resistant ceramic tiles, which remain intact after reentry.

Inside rocket engines, resin cloths impregnated with glass fibers were used to wrap pipes and electric wiring exposed to

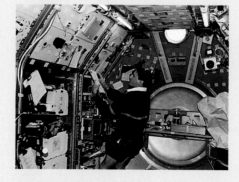

Spacelab
The European-built Spacelab module fits into the Space Shuttle's cargo bay and allows it to carry a fully equipped laboratory with hundreds of experiments into orbit. On this 1997 mission, scientists investigated the internal structure of drops of liquid in a microgravity environment.

Insulating materials

Thermal-blanket materials are now widely used in ski gloves, boots, and other winter clothing. A thermal blanket is a woven web of metal alloy and ceramic fibers, capable of blocking heat transfer by radiation or conduction.

hot gases around the engine exhaust and combustion chamber. Descendants of these materials are widely used today, providing highly efficient thermal insulation that keeps heat away from, or trap heat within, both domestic and industrial equipment.

The latest generation of space-age insulating products consists of aerogels—extremely lightweight dry silica gels, with just three times the density of air, that are incredibly strong. Aerogels were first developed in the 1930s, but NASA only adopted them recently and uses them as insulating materials for space vehicles including the 1997 Sojourner Mars Rover.

Materials made in space

As the duration of spaceflights increased, scientists were able to study the effects of weightlessness on materials. They developed ideas about an entirely new range of exotic products that could be made from materials processed in space.

Scientists had known for a long time that gravity distorts the mixture of materials in an alloy or a crystal. On Earth, when materials are melted for casting or to create an alloy in a crucible, the heavier elements sink to the bottom of the crucible, and solidifying materials are subjected to convection currents that shift them around. The result is that alloys are never perfectly blended, and metals and other crystals always form with minute fractures and imperfections running through them.

Although an orbiting spacecraft is not really beyond Earth's gravity, but instead is in a state of permanent free fall, the effects of gravity onboard are reduced to a point where they become negligible. This makes orbiting spacecraft and space stations ideal platforms for "microgravity" manufacture of materials. Alloys and crystals can form from molten materials without suffering distortions because of differences in local density during solidification. As long as they are solidified before returning to Earth, they will retain their near-perfect structure. In this way, completely new materials can be made in space, much lighter and stronger than anything alloyed together on the surface of the Earth.

Early experiments onboard NASA's Skylab space station showed that alloys made in space could have up to ten times the strength of comparable materials on Earth. Astronauts aboard Russian space stations and the Space Shuttle also showed that semiconductor crystals grow bigger in space and have greater purity, increasing their conductivity and opening up new possibilities for electronics. In 1996, scientists manufactured completely transparent and greatly improved aerogels in space for the first time. As well as using microgravity to grow large, perfect crystals of semiconductors and proteins, some experiments have used the effects of microgravity to prevent crystal growth completely, manufacturing extremely transparent glass for use in optical fibers.

The next great challenge is to move from these promising experiments to full-scale factories in orbit. The permanently crewed International Space Station is the first stepping stone to this ultimate goal.

Protecting Liberty

The Statue of Liberty in New York Harbor is just one of many monuments painted with a tough, weather-resistant coating called IC531. Originally developed to protect the launch pad for the Apollo lunar missions from rocket engine blasts and corrosion from the nearby sea, IC531 is a ceramic paint that provides a high level of protection with just a single coat.

SEE ALSO: COMPOSITE MATERIALS *p.144* | CERAMICS *p.206* | HISTORY OF SPACE EXPLORATION *p.236*

Inner planetary probes

Since the late 1950s, robot spacecraft have left Earth to explore the other worlds of our Solar System and the space that lies between them. The first missions aimed simply to fly past and photograph our nearest neighboring planets, Venus and Mars, but later missions were more ambitious, putting probes in orbit or landing them on planets, and using a wide range of instruments to collect information.

The first space probes were launched toward the Moon in 1958 and 1959. As rockets became more powerful, probes were soon able to escape completely from Earth's gravity and fly toward other planets. NASA was first to reach Venus with *Mariner 2* in 1962, Mars with *Mariner 4* in 1965, and Mercury with *Mariner 10* in 1974.

As our knowledge of the other planets improved, probes could be designed to answer specific questions. For example, the *Magellan* Venus probe (1990–92) carried high-resolution radar to peer through the planet's thick clouds, and the forthcoming Mars *Express* mission (2003) will search for traces of liquid water on the red planet.

Mariner 10

The first probe to fly past two planets, and the only probe to have visited Mercury, Mariner 10 was launched in November 1973. As the probe passed Venus, it allowed itself to be briefly captured by the planet's gravity, then fired its rockets to break free and travel toward Mercury, establishing an orbit that intercepted Mercury three times in 12 months.

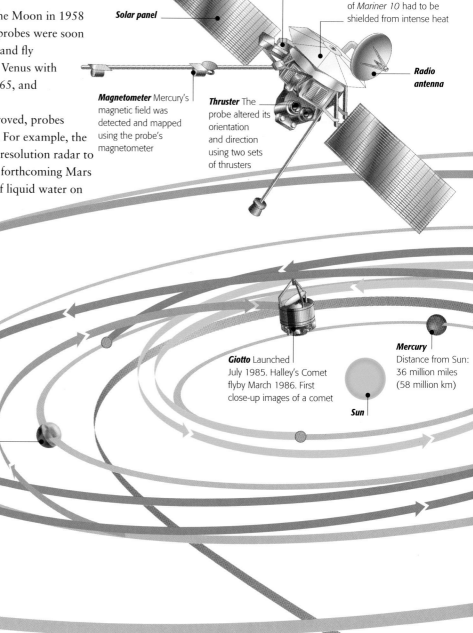

Star tracker The probe navigated by tracking the positions of bright stars

Sunshade The sunward side of *Mariner 10* had to be shielded from intense heat

Solar panel

Radio antenna

Magnetometer Mercury's magnetic field was detected and mapped using the probe's magnetometer

Thruster The probe altered its orientation and direction using two sets of thrusters

Mars
Distance from Sun: 142 million miles (228 million km)

Vikings 1 & 2
Both launched 1975
Arrived Mars 1976
Orbital survey and the first landers on Mars

Venus
Distance from Sun: 67 million miles (108 million km)

Giotto Launched July 1985. Halley's Comet flyby March 1986. First close-up images of a comet

Mercury
Distance from Sun: 36 million miles (58 million km)

Sun

KEY
- ⦿ **PLANETARY ORBITS**
- ⦿ **HALLEY'S PATH**
- ⦿ **VIKINGS 1 & 2**
- ⦿ **VENERA 13**
- ⦿ **GIOTTO**
- ⦿ **MARINER 10**
- ⦿ **MAGELLAN**
- ⦿ **STOPOVER**

Magellan map of Venus
The Magellan *probe carried technology normally
used in Earth-orbiting satellites to Venus, where
it peered through the thick cloudy atmosphere to
map the surface. By measuring the time taken
for radar pulses to echo back from the surface,
the elevation of different areas on the planet
could be mapped. False coloring shows high
areas as bright, and low areas as dark.*

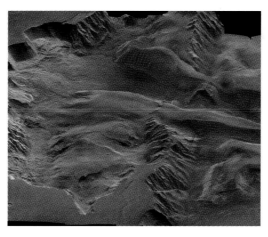

Viking relief map of Martian canyons
Radar and photographic data from the Viking *Mars
probes were combined to produce this image of the Valles
Marineris canyon. A computer wrapped photographs
taken from orbit around a radar map of the region, and
changed the perspective to create a realistic view.*

Giotto image of Halley's Comet
Europe's Giotto *probe sent back the first
pictures from the heart of a comet.
Probes use video cameras to capture
images, then transmit them back to
Earth using radio signals.*

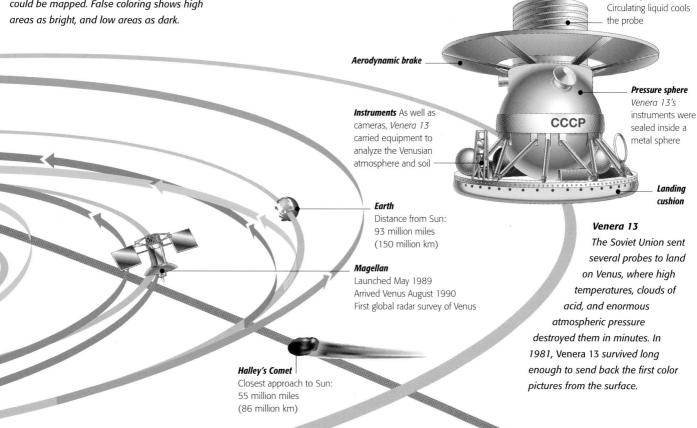

Coolant system
Circulating liquid cools
the probe

Aerodynamic brake

Pressure sphere
Venera 13's
instruments were
sealed inside a
metal sphere

Instruments As well as
cameras, *Venera 13*
carried equipment to
analyze the Venusian
atmosphere and soil

CCCP

**Landing
cushion**

Earth
Distance from Sun:
93 million miles
(150 million km)

Magellan
Launched May 1989
Arrived Venus August 1990
First global radar survey of Venus

Venera 13
The Soviet Union sent
several probes to land
on Venus, where high
temperatures, clouds of
acid, and enormous
atmospheric pressure
destroyed them in minutes. In
1981, Venera 13 survived long
enough to send back the first color
pictures from the surface.

Halley's Comet
Closest approach to Sun:
55 million miles
(86 million km)

Important probes

This map shows the orbits of the inner planets and Halley's Comet, along
with the paths that several important probes took to reach them from Earth.
The distance to a planet varies enormously depending on its alignment with
the Earth, so spacecraft launch windows are set well in advance—if they are
missed, the mission may be delayed for years.

SEE ALSO: SPACE ROCKETS *p.238* | SPACE SHUTTLE *p.240* | ORBITERS & LANDERS *p.252*

Outer planetary probes

Sending space probes beyond the neighborhood of Earth to the outer Solar System—realm of the giant planets Jupiter, Saturn, Uranus, and Neptune—is far more difficult than sending them to the inner planets. Nevertheless, several space probes have now made epic journeys beyond Pluto, and will ultimately leave the Solar System.

Because of the enormous distances to the outer planets, the first probes to visit them were built to travel at very high speeds with minimal equipment. *Pioneers 10* and *11*, launched in the early 1970s, took just a couple of years to reach Jupiter, but then had no way of slowing down, and made a single flyby of the system before leaving it behind. However, NASA scientists were able to use a gravitational slingshot to change *Pioneer 11*'s course and send it past Saturn in 1979. Both *Pioneers* were rehearsals for the *Voyager* probes, which took advantage of a rare planetary alignment to go on a "grand tour" of the outer Solar System. The latest probes, such as *Galileo*, *Cassini*, and the planned orbiter to Jupiter's moon Europa, are designed to carry out intensive studies of individual planetary systems. They must travel slowly enough to enter orbit around their destinations.

Radio wave detector

Ulysses
This joint project between NASA and the European Space Agency gave astronomers their first look at the poles of the Sun. Ulysses used Jupiter's gravity to slingshot it into an orbit at right angles to the plane of the Solar System.

Boom Sensitive instruments to measure magnetic fields and solar winds are mounted on this boom to keep them away from the body of the spacecraft

Antenna Commands to the spacecraft are received through this antenna, and information is sent back to Earth in the same way

Voyager 1 Launched September 1977. Jupiter flyby 1979, Saturn flyby 1981

Power source Heat from radioactive material is used to make electricity

Gravitational slingshot

All the probes shown on this map used gravitational slingshots to speed up and change direction on the way to their destinations. The slingshot involves swinging around a planet from behind. As the planet's gravity pulls the probe in, the probe's gravity also pulls very slightly on the planet, slowing it down in its orbit around the Sun. The momentum lost by the planet gets transferred to the probe as a tremendous boost in speed.

Uranus Distance from Sun: 1,783 million miles (2,871 million km)

Voyager 2 Launched August 1977 Jupiter flyby 1979, Saturn flyby 1981, Uranus flyby 1986, Neptune flyby 1989

Neptune Distance from Sun: 2,793 million miles (4,495 million km)

KEY
- PLANETARY ORBITS
- ULYSSES
- VOYAGER 1
- VOYAGER 2
- CASSINI
- GALILEO
- STOPOVER

Voyager image of Saturn's rings
Filters over the Voyager's *video camera* allowed it to photograph the rings of Saturn in enhanced colors. This image shows that the rings reflect light differently, producing a range of colors and indicating that the physical or chemical composition of the rings varies.

Voyager 2 image of Neptune
Neptune appears blue-green because a layer of methane gas in the atmosphere absorbs red light. This false-color image taken by Voyager 2 reveals a faint layer of haze invisible from Earth. It lies above the methane and does not reflect red light.

Cassini

NASA's Cassini *probe to Saturn, launched in October 1997, is designed to go into orbit around the planet in 2005. The probe is taking the slow route to Saturn, so that its rockets will be able to brake when it arrives.* Cassini *is a two-part mission—the spacecraft consists of the main orbiter and a European-built lander called* Huygens*. The lander will be released from the orbiter and parachuted into the atmosphere of Saturn's largest moon, the mysterious Titan.*

Radio antenna Three of these antennae pick up any radio emissions coming from Saturn

High-gain antenna This 12-ft (4-m) dish sends and receives radio signals

Huygens probe This shield-shaped probe will drop into Titan's atmosphere

Magnetometer A boom keeps this sensitive magnetic field detector away from the rest of the spacecraft

Radio antenna (1 of 3)

Television camera

Radioisotope Thermal Generator *Cassini* is powered by three radioactive power sources

Rocket engine This engine will slow down the space probe when it arrives at Saturn

Jupiter Distance from Sun: 483 million miles (778 million km)

Earth's orbit Distance from Sun: 93 million miles, (150 million km)

Galileo Launched October 1989. Arrived Jupiter 1995. Detailed survey of Jupiter and its moons

Saturn Distance from Sun: 886 million miles (1,426 million km)

SEE ALSO: SPACE ROCKETS *p.238* | INNER PLANETARY PROBES *p.248* | ORBITERS & LANDERS *p.252*

Orbiters and landers

Every space probe is specially designed from scratch for its particular mission. The destination of the probe and the information it will collect when it arrives there affect every aspect of the probe's design. Modern probes are either orbiters, which study a planet from high altitude and take measurements of the local environment, or landers, which descend to the surface, sending back pictures and carrying out chemical analysis of the rocks they find.

Probe engineers must consider a huge range of factors. The probe must be protected against the shocks and stresses of launch and landing, and against extremes of heat and cold in space. Power supplies must be kept away from delicate instruments, and sophisticated software must be developed so the probe can deal with problems immediately—probes are often so far away that radio signals take many minutes, and sometimes hours, to travel between the probe and mission control.

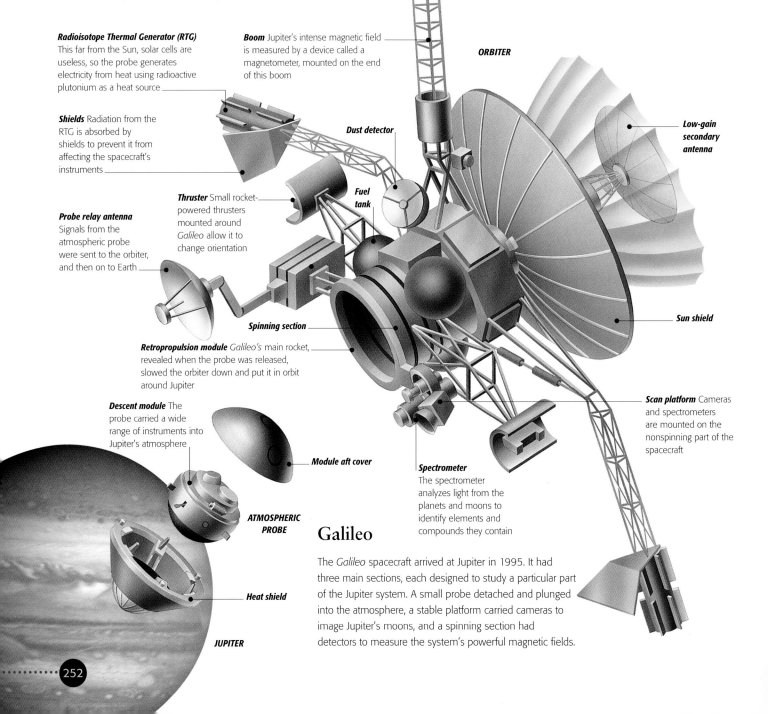

Radioisotope Thermal Generator (RTG) This far from the Sun, solar cells are useless, so the probe generates electricity from heat using radioactive plutonium as a heat source

Shields Radiation from the RTG is absorbed by shields to prevent it from affecting the spacecraft's instruments

Probe relay antenna Signals from the atmospheric probe were sent to the orbiter, and then on to Earth

Boom Jupiter's intense magnetic field is measured by a device called a magnetometer, mounted on the end of this boom

ORBITER

Dust detector

Thruster Small rocket-powered thrusters mounted around *Galileo* allow it to change orientation

Fuel tank

Spinning section

Retropropulsion module *Galileo's* main rocket, revealed when the probe was released, slowed the orbiter down and put it in orbit around Jupiter

Descent module The probe carried a wide range of instruments into Jupiter's atmosphere

Module aft cover

ATMOSPHERIC PROBE

Heat shield

JUPITER

Low-gain secondary antenna

Sun shield

Scan platform Cameras and spectrometers are mounted on the nonspinning part of the spacecraft

Spectrometer The spectrometer analyzes light from the planets and moons to identify elements and compounds they contain

Galileo

The *Galileo* spacecraft arrived at Jupiter in 1995. It had three main sections, each designed to study a particular part of the Jupiter system. A small probe detached and plunged into the atmosphere, a stable platform carried cameras to image Jupiter's moons, and a spinning section had detectors to measure the system's powerful magnetic fields.

Mars Pathfinder

The *Pathfinder* craft, which arrived at Mars in July 1997, was a surface probe. Folded into a pyramid for the journey, the probe unfolded after landing, revealing three solar panels around a central camera platform and weather station. Stereoscopic cameras allowed *Pathfinder* to capture 3-D images of the surface.

The probe also carried a tiny rover called *Sojourner*. This independent robot vehicle was fitted with cameras and a chemical "sniffer," with which it analyzed rocks near the landing site.

Pathfinder used batteries as its main power supply, supplemented by solar energy, so it had a limited lifetime. Nevertheless, it continued functioning for three months—a month longer than planned.

PATHFINDER

Windsock

Thermocouple *Pathfinder* measured air temperatures using electrical thermometers that produce a voltage that varies with temperature

Surface of Mars
From earlier missions, NASA knew that the Martian landscape was rocky and uneven, so they fitted Pathfinder *with airbags like those used in cars. The probe bounced around on the surface until it was upright, then deflated the airbags one by one until it was resting in a horizontal position.*

Imager for Mars Pathfinder (IMP)
This stereoscopic camera rotated to capture a 360° panorama of the landing site

Low-gain antenna The backup antenna could not handle as much information as the main dish

High-gain antenna This dish was *Pathfinder*'s main means of communication with Earth

Solar panel Mars is close enough to the Sun for space probes to generate electricity using solar cells

Rover antenna

Camera Another pair of stereo cameras was mounted at the rear of *Sojourner*

SOJOURNER ROVER

Securing clip In flight, *Pathfinder* was folded into a pyramid shape, then unfolded on landing

Electronic control systems The spacecraft's electronics were packed in plastic, which insulated them against the extremes of the Martian climate and protected them from shocks during launch and landing

Alpha proton X-ray spectrometer *Sojourner* nudged up against rocks and analyzed them using this sophisticated instrument

Suspension system *Sojourner* had six wheels and independent suspension to keep it upright on the rockiest surfaces

SEE ALSO: SATELLITE COMMUNICATIONS *p.50* | INNER PLANETARY PROBES *p.248* | OUTER PLANETARY PROBES *p.250*

Space telescopes

Earth-bound telescopes can answer only a fraction of our questions about the Universe because the Earth's atmosphere absorbs or distorts much of the radiation emitted by celestial bodies. To overcome this limitation, dozens of satellites equipped with telescopes have been put into orbit around the Earth. The first of these was the Orbiting Solar Observatory launched by NASA in 1962.

Telescopes collect electromagnetic radiation emitted by distant objects. Different kinds of space telescopes specialize in collecting radiation of different wavelengths. NASA's Compton Gamma-Ray Observatory (CGRO), launched in 1991, studies gamma-ray bursts emitted by pulsars and other high-energy sources such as the nucleus of the Milky Way. The Roentgen Satellite (ROSAT) was launched in 1990 to observe X-rays that come from very hot gases found in solar flares and other areas of intense cosmic activity. The International Ultraviolet Explorer (IUE), launched in 1978, obtained information about objects such as supernovas (very bright exploding stars) by analyzing short-wavelength ultraviolet rays. Lower down the electromagnetic spectrum, the Infrared Astronomical Satellite (IRAS), launched in 1983, can detect relatively cool objects, such as comets, and allows astronomers to observe stars being born through dense clouds of dust. This satellite has discovered more than 350,000 new

sources of infrared radiation. The Cosmic Background Explorer (COBE), launched in 1989, provided evidence for the big bang theory in the form of a microwave map, which showed that the Universe was cooling slightly as it expanded.

The two-billion dollar Hubble Space Telescope, launched in 1990, observes ultraviolet, infrared, and visible light. It can detect objects 100 times fainter than those detected by the best telescopes on Earth. A defective primary mirror meant that it did not live up to its potential until corrective optics were installed in 1993. Since then it has been sending back razor-sharp images. NASA's Next Generation Space Telescope (NGST), planned for launch in 2007, will study infrared sources to find out more about how the very first stars and galaxies were formed.

The Hubble Space Telescope

The 43-ft (13-m) Hubble Space Telescope is the largest reflecting telescope operating in space. It contains two cameras: one for photographing faint objects, the other for taking wide-angle shots. It also carries two spectroscopes that analyze the radiation given off by objects to determine their chemical composition. Fine-guidance sensors help the Hubble to lock onto and track stars and other cosmic phenomena.

Axial instruments unit This unit contains an array of light detectors, a faint image camera, a wide-field camera, and optics that correct the distortions caused by the spherical aberration of the primary mirror

The atmospheric filter
Celestial objects give off radiation at different wavelengths and frequencies, ranging from short-wavelength (high-frequency) gamma rays to long-wavelength (low-frequency) radio waves. Most types of electromagnetic radiation are absorbed by the Earth's atmosphere, while others, such as visible light and some radio waves, penetrate the atmosphere and reach the Earth's surface.

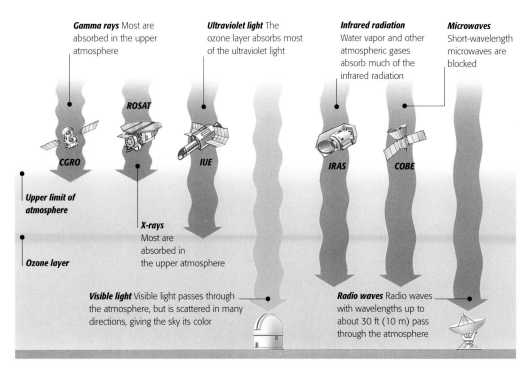

Gamma rays Most are absorbed in the upper atmosphere

Ultraviolet light The ozone layer absorbs most of the ultraviolet light

Infrared radiation Water vapor and other atmospheric gases absorb much of the infrared radiation

Microwaves Short-wavelength microwaves are blocked

ROSAT

CGRO

IUE

IRAS

COBE

Upper limit of atmosphere

Ozone layer

X-rays Most are absorbed in the upper atmosphere

Visible light Visible light passes through the atmosphere, but is scattered in many directions, giving the sky its color

Radio waves Radio waves with wavelengths up to about 30 ft (10 m) pass through the atmosphere

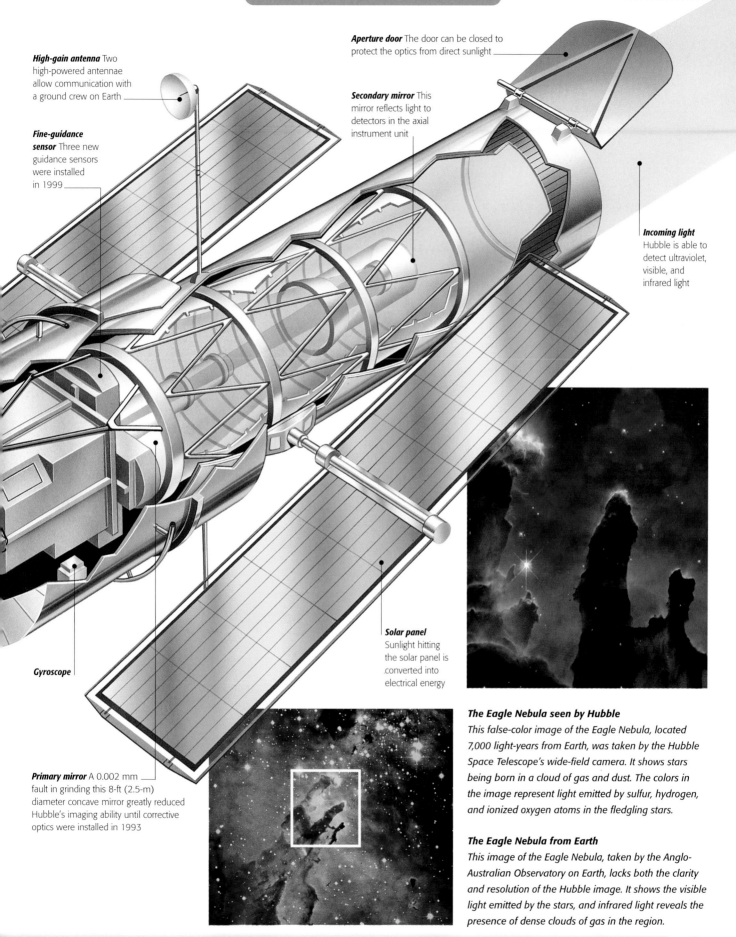

Aperture door The door can be closed to protect the optics from direct sunlight

High-gain antenna Two high-powered antennae allow communication with a ground crew on Earth

Secondary mirror This mirror reflects light to detectors in the axial instrument unit

Fine-guidance sensor Three new guidance sensors were installed in 1999

Incoming light Hubble is able to detect ultraviolet, visible, and infrared light

Gyroscope

Solar panel Sunlight hitting the solar panel is converted into electrical energy

Primary mirror A 0.002 mm fault in grinding this 8-ft (2.5-m) diameter concave mirror greatly reduced Hubble's imaging ability until corrective optics were installed in 1993

The Eagle Nebula seen by Hubble
This false-color image of the Eagle Nebula, located 7,000 light-years from Earth, was taken by the Hubble Space Telescope's wide-field camera. It shows stars being born in a cloud of gas and dust. The colors in the image represent light emitted by sulfur, hydrogen, and ionized oxygen atoms in the fledgling stars.

The Eagle Nebula from Earth
This image of the Eagle Nebula, taken by the Anglo-Australian Observatory on Earth, lacks both the clarity and resolution of the Hubble image. It shows the visible light emitted by the stars, and infrared light reveals the presence of dense clouds of gas in the region.

SEE ALSO: SATELLITE COMMUNICATIONS *p.50* | OPTICAL TELESCOPES *p.228* | RADIO TELESCOPES *p.230* | LIGHT *p.262*

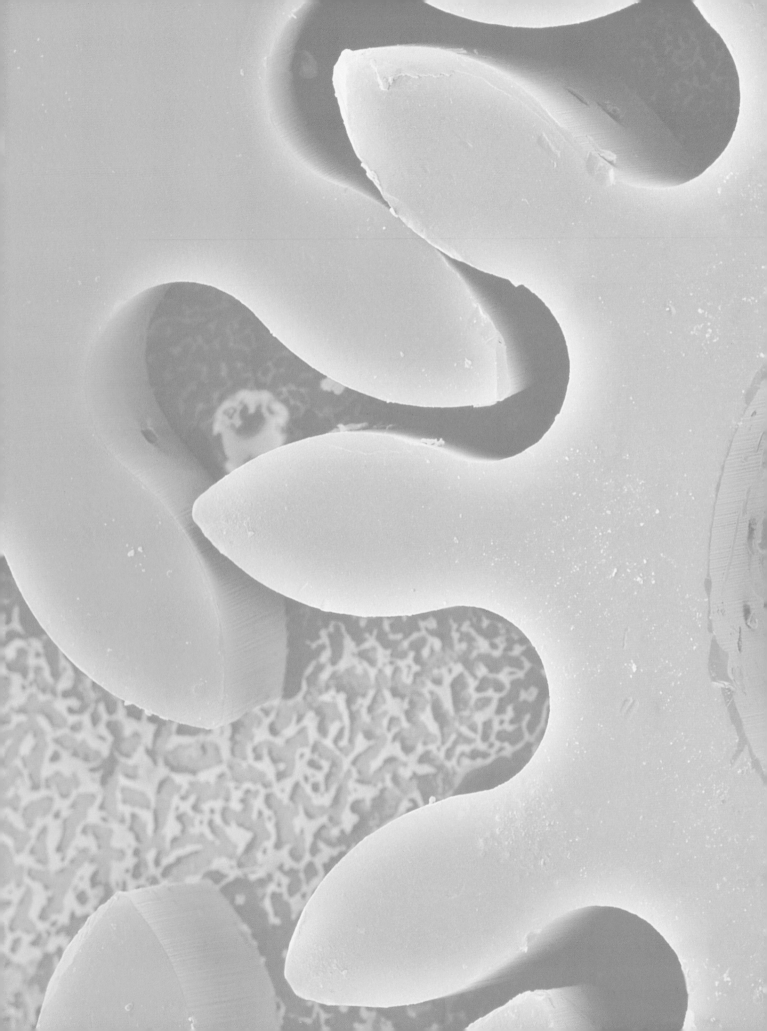

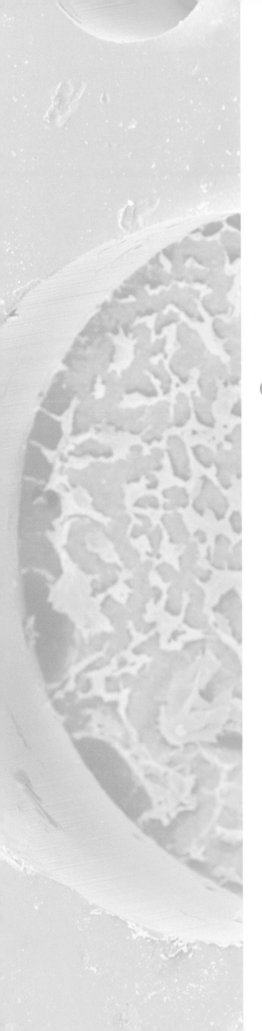

Reference

To understand many of the inventions and technologies in this book, it is necessary to have a grounding in some basic principles of both pure and applied science. Many of these principles are explained in the first section of this chapter, which also includes other reference materials you may find useful—a glossary with definitions of many terms used throughout the book, a list of useful Internet sites, and a thorough index.

Principles

To understand technology, it is necessary to have an insight into basic scientific principles. The following pages provide an introduction to the nature of matter and the properties of electricity, magnetism, and light, the principles of electronics, and the laws governing forces, motion, and simple machines.

Matter

Most matter—the physical material of the Universe—is made of atoms. There are 92 basic types of atoms that occur naturally in the Universe, corresponding to 92 different elements (pure, basic substances with distinctive properties). About 20 more elements have been created in nuclear reactors or particle colliders. An atom is the smallest piece of an element that can exist, typically measuring a few nanometers (billionths of a meter) across.

Much of the matter around us consists not of elements, but of compounds and mixtures. A compound consists of atoms of two or more elements bonded together in a fixed ratio, formed when the atoms react together under specific conditions. The properties of a compound are usually very different from those of the constituent elements. For example, common salt is a compound consisting of sodium, a soft metal, and chlorine, a greenish-yellow poisonous gas, in equal proportions.

The smallest unit of a compound is a molecule. Molecules may consist of just two or three atoms, or many thousands. For example, a water molecule contains one oxygen atom bonded to two hydrogen atoms. When solid, many compounds are made of macromolecules (giant structures consisting of thousands of atoms). A mixture is a combination of elements or compounds that retains many of the properties of its constituents. For example, a mixture of salt and water is both salty and wet.

Atomic and molecular structure

The word "atom" comes from a Greek word meaning "indivisible", but today we know that atoms are built from combinations of three smaller particles called protons, neutrons, and electrons. Atoms have

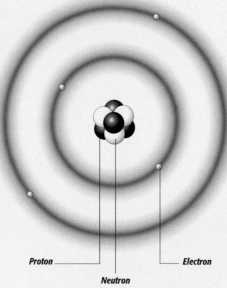

Proton ——
Neutron
—— *Electron*

Figure 1: ATOMIC STRUCTURE
This atom, of the element beryllium, has five neutrons and four protons in its nucleus, and four electrons distributed in two shells.

a heavy, compact central core called the nucleus. This contains the protons and neutrons, which are of similar mass. The much lighter electrons are found in regions called orbitals around the nucleus. Electron orbitals are arranged in "shells" at various distances from the nucleus (*see Figure 1*).

Protons have a positive electrical charge *(see p.260)*, electrons have an equal negative charge; neutrons have no charge—they are neutral. An atom has the same number of protons as electrons, so as a whole the atom is neutral. Atoms of a particular element have a specific number of protons; this is the unique atomic number of the element.

In some circumstances, the outer orbitals of an atom may gain or lose one or more electrons; this turns the neutral atom into a negatively or positively

charged particle called an ion. This process is fundamental to the chemical behavior of the atom. Movements of electrons between different atoms results in the creation of oppositely charged ions, which then attract one another. The resulting chemical bonds are called ionic bonds, and they form the basis of many crystalline substances. For example, in common salt, each sodium atom donates an electron to a chlorine atom, forming negatively charged chloride and positively charged sodium ions, which are held together in an ionic crystal of sodium chloride. Some compounds, such as water, are held together by covalent bonding, in which electrons are shared between atoms, rather than moving from one atom to another. In a water molecule, an oxygen atom shares a pair of electrons with each of two hydrogen atoms.

Solids, liquids, and gases

Matter can exist in three main states: solid, liquid, and gas (*see Figure 2*). A gaslike fourth state, plasma (*see p.277*), also exists but only at very high temperatures, such as in lightning flashes. Most substances can exist in more than one state, depending on the temperature and pressure, and can change state without becoming a different substance. For example, water exists as a liquid, as solid ice, and as gaseous steam.

In general, solids have a fixed shape and volume. Liquids have a fixed volume (they cannot easily be compressed), but no fixed shape (they flow). Gases have neither fixed shape nor fixed volume—they can be squeezed into small containers, or expand to fill large ones. These differences between states are due to differences in the forces holding the atoms, ions, or molecules together. In solids, the particles are held rigidly in place, usually in a regular, repeating structure called a crystal. Some solids have a highly regular structure without forming crystals. For example, plastics are made of thousands of identical molecules called monomers joined together in long chains (called polymers).

In liquids, the forces between particles are weaker than those in solids, allowing the liquid to flow and take the shape of its container. Particles in a liquid are more widely spaced than those in a solid, and can both vibrate and move limited distances, but the forces are strong enough to keep the

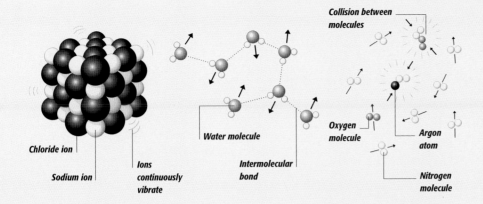

Figure 2: SOLID, LIQUID, AND GAS
The compound sodium chloride (left) exists as a solid crystal of sodium and chloride ions; liquid water (middle) consists of individual water molecules held together by intermolecular bonds; air (right) is a gaseous mixture of various atoms and molecules in continual rapid motion.

particles together. The strength of the forces determines how viscous (sticky) the liquid is.

Forces between particles in a gas are much weaker than those in a liquid, and the particles move around rapidly. The particles fill up whatever space is available and constantly bump into each other and into any other particles in their way, such as those of the container walls. These collisions provide the physical basis for the pressure exerted by the gas.

Substances change between solid, liquid, and gaseous states at specific temperatures—the melting and boiling points. These changes take place without a change of temperature, but they involve a transfer of heat energy, known as latent (hidden) heat. This is because even at the same temperature, the particles in a gas have on average more energy than those in a liquid, and those in a liquid have more energy than particles in a solid. Latent heat is absorbed when solids melt or liquids boil, and is emitted when gases condense or liquids freeze. Changes of state also result from pressure changes— lowering the pressure causes solids to melt and liquids to boil at a lower temperature than normal.

Electrons, orbitals, and energy

Electrons in an atom are found in orbitals arranged in shells at different distances from the nucleus. The attractive force between an electron and the protons in the nucleus varies according to which orbital and shell it is in. In order to move from an inner to an outer orbital, the electron has to absorb energy; similarly, to move to an inner orbital, the electron must emit energy. Since the distances between orbitals are fixed and specific for an atom of a particular element, these energy differences are fixed

too. So when electrons move between orbitals, the atom emits or absorbs energy in fixed "packets" called quanta that are characteristic for each atom. The energy is absorbed or emitted as a photon of electromagnetic radiation, such as light (see p.262).

The energy of a quantum (photon) corresponds to a particular wavelength of radiation—the higher the energy of the quantum, the shorter the wavelength. Therefore, atoms of each element are associated with particular wavelengths of radiation. In the visible part of the electromagnetic spectrum, wavelength corresponds to color, so each element is associated with light of particular colors, and its atoms emit light of characteristic colors when excited (given energy, for example, by heating in a flame). These colors are used in spectroscopic analysis, in which an instrument called a spectrometer is used to study and measure a spectrum. By detecting the characteristic colors in the spectra from a flame or even the light from a distant star, scientists can detect the presence of specific elements.

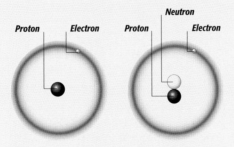

Figure 3: TWO ISOTOPES OF HYDROGEN
A hydrogen atom (left) has just one proton in its nucleus. An atom of its isotope hydrogen-2 (right) has a neutron as well as a proton. Hydrogen-2 is sometimes called deuterium.

Isotopes and radioactivity

Although the number of protons in the atoms of a particular element (the atomic number) is always the same, the number of neutrons can vary between atoms. Atoms of the same element that have different numbers of neutrons are called isotopes (see Figure 3). The mass number (the sum of the number of protons and neutrons in an atom) therefore varies for different isotopes. Isotopes are chemically identical—they react in the same way to form compounds—but their atoms have different masses. The nuclei of some isotopes, called radioisotopes, are unstable—they spontaneously emit radiation (particles or electromagnetic radiation) in order to become more stable. There are three main types of radiation: alpha, beta, and gamma. Alpha particles are the nuclei of helium atoms, consisting of two protons and two neutrons. Beta particles are fast-moving electrons or positrons (identical to electrons but positively charged).

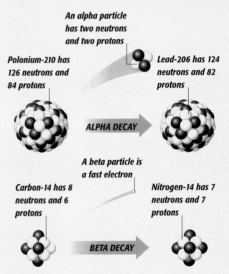

Figure 4: ALPHA AND BETA DECAY
In both of these types of radioactive decay, the new atomic nucleus formed is of a different chemical element from the original.

Gamma rays are a form of electromagnetic radiation similar to X-rays but of even shorter wavelength. When a radioisotope emits alpha or beta radiation, it turns into an isotope of another element (see Figure 4). For example, carbon-14 (with 8 neutrons and 6 protons) is an unstable isotope of carbon. The common carbon isotope, carbon-12, is chemically identical but has only 6 protons. Eventually, one of the neutrons in carbon-14 emits a beta particle and turns into a proton. In this way, the carbon-14 nucleus becomes the nucleus of an atom of nitrogen-14.

Electricity and Magnetism

Matter possesses a property called electrical charge. There are two opposite types of charge, positive and negative. Two particles with unlike charges (one positive, one negative) are attracted to each other. Two particles with similar or like charges (both positive or negative) repel each other. Atoms have equal numbers of negatively charged electrons and positively charged protons, leaving the atom as a whole electrically neutral (uncharged). Charge cannot be created or destroyed.

An atom can lose or gain electrons to become a charged ion (*see p.258*); similarly, a large piece of matter, made of billions of atoms, can have a deficit or surplus of electrons. The resulting positive or negative charge is large enough to measure.

Static charges and insulators

A neutral object can be charged easily. For example, rubbing a balloon with some wool causes electrons to move from the wool to the balloon, giving the balloon an overall negative charge. Two such negatively charged balloons brought close together will repel each other (*see Figure 1*). Another way of charging an object is by induction. If a charged balloon is brought close to an uncharged balloon (without touching it), the negative charge on the surface of the first balloon will repel electrons from the near surface of the second balloon. These electrons remain on the second balloon, because they cannot easily pass through air, so that balloon remains neutral, but one side is positive and the other negative. The facing surfaces of the balloons are now oppositely charged and attract each other.

The charges on the balloons described above are called static charges—they do not pass easily from the balloons through air to other objects, because air is an insulator. An insulator is a material that does not contain charged particles that are free to move; other examples of insulators are plastics and most nonmetals. Static charges can only build up if a charged object is separated from neighboring objects by an insulator. But even the best insulators break down (allow charges to move through them) if the charge difference between neighboring objects is large enough. When charges move through air, some of their energy is turned into the sound and light of an electrical discharge, or spark. Lightning is a massive discharge that occurs when the charge

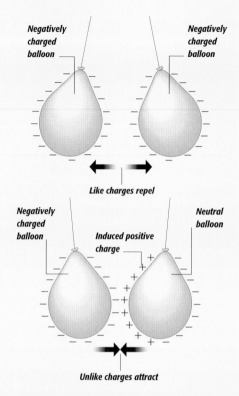

Figure 1: ELECTRICAL CHARGE EFFECTS
Negatively charged balloons repel each other (top). A charged balloon induces an opposite charge in a neutral balloon (bottom), and the two are attracted to each other.

difference between the clouds and the ground is sufficiently large for the insulation of air to break down so that it conducts electricity.

Conductors and electric current

Some materials, including nearly all metals, are good electrical conductors. This means that they contain plenty of charged particles that are free to move. In solid metals, the charge carriers are electrons that move through the crystal structure. If there is a charge separation across a piece of conductor—that is, if one side is relatively positive and the other relatively negative—charged particles will move until the charge is spread as evenly as possible. This movement of charged particles is an electric current. For current to flow continuously, the conductor must form a complete electric circuit, or loop, from an energy source (a cell or generator) and back again.

In metal wires and most electrical appliances, current is carried by negatively charged electrons.

Electrons flow from the negative terminal (cathode) of the cell or generator around the circuit to the positive terminal (anode). However, when electric current was first discovered, electrons had not been identified, and it was agreed that current would mean a flow from positive to negative. Note that the direction of this "conventional" current is always opposite to the direction of electron flow.

Depending on how it is generated, electric current always flows in the same direction through a circuit (direct current, or DC), or it flows alternately in one direction, then reverses (alternating current, or AC). The domestic electricity supply is AC and (in the US) reverses direction 60 times per second. AC has some advantages over DC, one is that it can be transmitted over long distances more efficiently.

Electrochemistry and cells

Atoms of some materials tend to form positive ions by losing electrons, while others tend to acquire electrons to form negative ions. If a pair of such materials is dipped into a substance called an electrolyte, these tendencies can be harnessed to drive an electric current. This is the principle of the galvanic cell, which converts chemical into electrical energy (*see Batteries p.178*). One simple galvanic cell consists of strips of zinc and copper dipped into dilute sulfuric acid. Zinc forms the negative terminal or electrode, and copper the positive electrode; the acid is the electrolyte. If the electrodes are connected to an external circuit, a current flows through the wire and electrolyte. Energy conversion

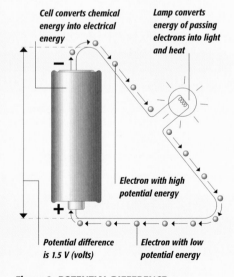

Cell converts chemical energy into electrical energy

Lamp converts energy of passing electrons into light and heat

Electron with high potential energy

Potential difference is 1.5 V (volts)

Electron with low potential energy

Figure 2: POTENTIAL DIFFERENCE
A cell creates a "potential hill," giving each electron energy that it then transfers to components in the circuit.

can also be carried out in the reverse direction: an electric current can be used to drive a chemical change in a type of cell called an electrolytic cell. Cells of this type are used in electroplating and the extraction of aluminum metal from its ore.

Measuring electricity

Electrical charge is measured in coulombs (C). One coulomb is equal to the charge on more than 6 million trillion electrons. Current is the rate of flow of charge, and is measured in amperes or amps (A). A current of 1A is the flow of 1C of charge per second.

An electric "field" permeates the space around a charged particle, and any other charged particle in this field will feel a force. This force is what drives a current around a circuit. Current does not flow on its own—there must be a cell or generator to create and maintain a driving force, which is called the electromotive force (EMF). EMF, also called potential difference or voltage, is measured in volts (V).

The size of current that can flow in a circuit depends on the EMF and also on the resistance that the circuit offers to current flow. This resistance, which can be thought of as "electrical friction," is measured in ohms (Ω). The greater the resistance in a circuit, the smaller the current an EMF can drive.

The power of an electrical device is the rate at which it converts energy to or from the electrical form. Power is measured in watts (W) and depends on both current and EMF. An EMF of 1V driving a current of 1A works at the rate of 1W.

Magnets and electromagnetism

Magnetism and electricity are closely related—they are both aspects of a fundamental force called the electromagnetic force. All matter is magnetic, but

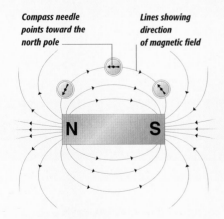

Figure 3: MAGNETIC FIELD OF A MAGNET
The north and south poles of a magnet create a magnetic field between them.

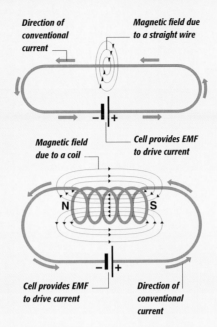

Figure 4: MAGNETIC FIELDS OF WIRE AND COIL
Magnetic fields are induced by currents in straight and coiled conductors.

some materials, such as certain ceramics, iron, steel, cobalt, and nickel, are very powerfully so, and are used to make the familiar bar magnets.

Magnets have two poles—north and south—at opposite ends (see *Figure 3*). As for electrical charges, like poles repel each other while unlike poles attract. If a magnet is suspended so it is free to turn, its north pole always points north. This is because the Earth acts like a giant magnet with its poles near the north and south geographic poles.

A current flowing in a wire creates a magnetic field (a region where a magnetic force is felt) around it (see *Figure 4*). If the wire is wound into a coil, the fields generated by each turn of wire reinforce each other, and it acts like a bar magnet while a current flows in it. Such a coil is known as an electromagnet.

Generators and motors

If an electrical conductor connected to a circuit moves through a magnetic field, an electric current is induced, or caused to flow, in it. This is the principle of the electric generator (*see Power stations, p.170*). If, on the other hand, a wire carrying an electric current passes through a magnetic field, then a force is exerted on the wire. This is the principle of the electric motor (*see Figure 6*). For electromagnetic devices such as generators and motors, the directions of the conventional current flow, the magnetic field (from north to south), and the movement or force felt by a conductor in the

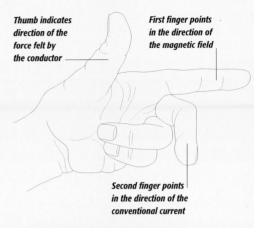

Figure 5: FLEMING'S LEFT-HAND RULE
This rule is applied by pointing the left thumb, first and second fingers at 90° to each other as shown above.

magnetic field are related by rules known as Fleming's rules. For example, Fleming's left-hand rule (*see Figure 5*) defines the direction of the force on a conductor carrying a current in a magnetic field. It is the rule applicable to electric motors.

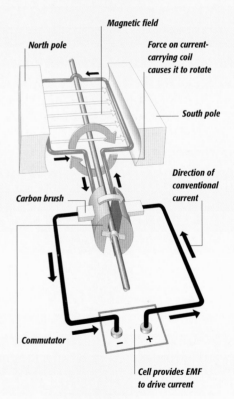

Figure 6: ELECTRIC MOTOR
A motor turns due to the force on a current-carrying coil in a magnetic field. The commutator ensures that the current in the coil reverses direction every half-turn, thus maintaining the direction of the force.

Light

Light is a form of electromagnetic radiation—radiant energy which consists of a combination of oscillating electric and magnetic fields. Other forms of electromagnetic (EM) radiation include radio waves, infrared and ultraviolet radiation, X-rays, and gamma rays. Collectively, these phenomena form the electromagnetic spectrum (see Figure 1). EM radiation requires no physical medium to carry it, and travels through a vacuum at about 186,000 miles per second (300,000,000 meters per second)—the speed of light. Different types of EM radiation pass through different types of matter at various, lower speeds.

Nature of EM radiation

Light and other types of EM radiation behave mainly like waves. The underlying difference between the types is in their wavelength—the distance between successive peaks in their waveforms (see Figure 2). If two waves are superimposed on one another and the peaks of one wave coincide with the troughs of another, the two cancel each other out. If, on the other hand, the peaks of the two waves coincide, they reinforce each other, giving a wave of twice the original amplitude. This effect is called interference.

Wavelength is related to frequency, the rate at which the peaks of a wave pass a fixed point. Frequency is measured in hertz (Hz). One hertz is

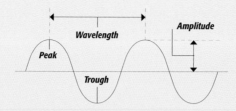

Figure 2: ANATOMY OF A WAVE
A wave is characterized by its wavelength (the distance between successive peaks) and its amplitude, which is a measure of its intensity.

one cycle or vibration per second. The frequency of a wave equals its velocity divided by its wavelength. When EM radiation enters a relatively dense medium, such as glass, it slows and its wavelength becomes shorter, but its frequency remains constant.

EM radiation also displays some particle-like properties. The photoelectric effect—the emission of electrons by certain materials when EM radiation falls on them—is satisfactorily explained only by thinking of the radiation as a stream of tiny particles. These particles, which have no mass and travel at the speed of light, are called photons. A photon's energy depends on the frequency of the radiation—the higher the energy, the higher the frequency.

Visible light

Visible light consists of the portion of the EM spectrum with wavelengths between 0.4 and 0.7 micrometers (millionths of a meter). Light of different wavelengths is perceived as different colors—from red (long wavelengths) through to blue and violet (short wavelengths). What we see as white light is a mixture of all these wavelengths of light (see How color works, p.82).

Figure 1: THE ELECTROMAGNETIC SPECTRUM
The EM spectrum covers a continuous and large variety of radiant energy forms. Their wavelengths range from less than 10^{-12} m (one millionth of a millionth of a meter, or a picometer), which is much smaller than the width of an atom, to tens of kilometers.

Light reflection

When light passes from one medium to another, it may be reflected, transmitted, or absorbed. If the surface of the second medium is smooth and shiny, some or all of the light is reflected. If the second medium is transparent (for example, glass), some of the light is reflected and some is refracted—transmitted through the transparent material, but at a different angle (see Figure 3). If the surface is

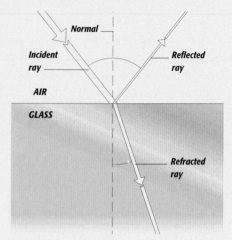

Figure 3: REFLECTION AND REFRACTION
When light strikes the boundary between two transparent media, such as air and glass, some of it is reflected and some is refracted. A little light is also absorbed.

rough and opaque, most of the light is absorbed and turned into heat energy. Only a totally black surface absorbs all the light that hits it.

To understand what happens when light hits a surface, imagine a line (called the normal) at right angles to the surface at that point. Light that strikes the surface at an angle to the normal is reflected at the same angle, but on the other side of the normal.

Mirrors are highly polished surfaces that reflect most of the light that strikes them. The image seen

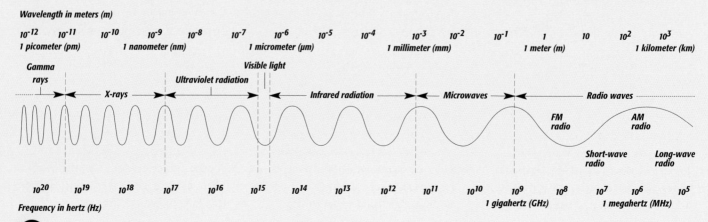

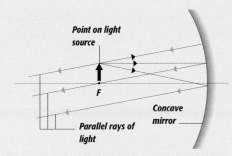

Figure 4: CONCAVE MIRROR (SPOTLIGHT)
A concave mirror can be used to produce a narrow beam of light by placing a light source at the focal point (F) of the mirror. The mirror reflects the rays coming from each point in the light source to produce a set of parallel rays.

in a flat (plane) mirror appears to be behind the mirror, and is called a virtual image—it cannot be projected onto a screen. Curved mirrors may produce "real images," which can be projected.

Curved mirrors

There are two main types of curved mirrors—concave (whose shiny surface curves inward), and convex (which curve outward). These mirrors can produce a variety of images, magnified or diminished (smaller than the object), and real or virtual, depending on the object's distance from the mirror. For example, light rays from a very distant object are nearly parallel, and a concave mirror will reflect them so that they converge at a point in front of the mirror (the focal point) to produce a small real image. This principle is used in reflecting telescopes (see *Optical telescopes, p.228*) and solar power plants (see *Solar energy, p.174*). Concave mirrors are also used in projectors and lamps. In a spotlight, a light source placed at the focal point of a concave mirror produces a beam of parallel rays (see *Figure 4*). Convex mirrors can be used to produce diminished virtual images to give a wide-angle view, as used in automobile mirrors.

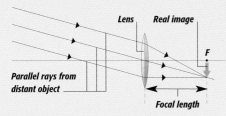

Figure 5: CONVEX LENS (TELESCOPE)
The objective (front) lens of a refracting telescope focuses parallel light rays from a distant object to a single point at its focus (F). There, a real but inverted image is formed.

Light refraction

When a light ray passes from one medium to another of different density, for example, when it passes from glass to air or from warm air to cold air, its speed and wavelength changes and it is refracted (bent). Underwater objects viewed from above the

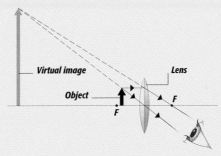

Figure 6: CONVEX LENS (MAGNIFYING GLASS)
When an object between the lens and its focal point (F) is viewed through the lens, an enlarged virtual image is seen on the same side of the lens as the object.

water surface appear closer than they really are because of this phenomenon. Refraction occurs in lenses (curved, polished pieces of glass or other transparent material) and enables light to be bent in a controlled manner. Convex lenses, which curve outward, make light rays that pass through them converge. Concave (inward-curving) lenses make light rays diverge. Convex lenses are used in refracting telescopes to produce a real, diminished image of a distant object (see *Figure 5*), and in magnifying glasses to produce an enlarged, virtual image of an object (see *Figure 6*).

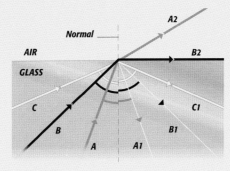

Figure 7: TOTAL INTERNAL REFLECTION
Three rays of light, A, B, and C, pass from glass into air. At the glass/air boundary, ray A is partly reflected (A1) and partly refracted (A2). Ray B is incident at the critical angle—the refracted ray (B2) is parallel to the glass/air boundary. There is also a reflected ray, B1. Ray C is incident at greater than the critical angle and is totally internally reflected (C1).

Total internal reflection

When light strikes a boundary between two media, some of it is refracted, and some is reflected. The proportion of reflected to refracted light and the angle through which it is refracted depend on the angle of incidence at the boundary and on a property of the media called their refractive indices. A ray moving into a more dense medium (for example, from air to glass) is bent toward the normal; a ray traveling in the opposite direction is bent away from the normal. If a light ray passing into a less dense medium strikes the boundary between the media at the "critical angle," it will be refracted along the boundary edge. At incident angles greater than the critical angle, the ray is entirely reflected back into the more dense medium. This phenomenon, called total internal reflection (see *Figure 7*), occurs in optical fibers (see *Cable technology, p.46*), light guides (see *Lighting, p.28*), and in SLR cameras (see *Cameras, p.92*).

Polarization

The electric and magnetic fields that make up light waves normally oscillate in many different directions, all at right angles to the direction of travel. However,

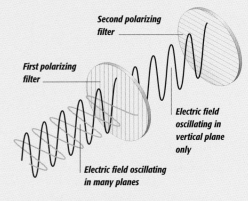

Figure 8: POLARIZATION
When light passes through a polarizing filter, only those waves whose electric fields oscillate in one particular plane are able to pass through. These waves are blocked by a second polarizer set at right angles to the first.

some crystals and other materials transmit only light waves that oscillate in a particular direction—other waves are absorbed (see *Figure 8*). Light that only oscillates in one direction is called polarized light. Filters that polarize light are used in liquid crystal displays (see *Computer output devices, p.68*). Light may also be polarized when it is reflected. Some sunglasses and camera lenses have polarizing filters that absorb this light and reduce glare.

Electronics

Electronics is the branch of science and engineering concerned with varying electric currents that usually represent data signals such as audio and video signals or logical operations in computer circuits.

Semiconductors

Most substances either allow a flow of charged particles to pass through them (they are conductors) or they do not (insulators; see p.260). However, the conductivity of some substances varies significantly with external conditions. Semiconductors such as silicon (see p. 58) fall into this category.

Pure silicon is a poor electrical conductor, unless tiny amounts of certain impurities are added to it in a carefully controlled process known as doping. Two types of impurities may be added. A "donor" material such as phosphorus supplies "free" electrons, turning silicon into a so-called n-type semiconductor ("n-" for negative). An "acceptor" material such as boron has spaces in its structure known as "holes," into which electrons can fit, and makes a p-type semiconductor ("p-" for positive). If some n-type

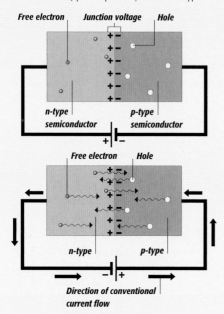

Figure 1: SEMICONDUCTOR DIODE
A diode consists of a crystal of n-type semiconductor (containing free electrons) joined to a crystal of p-type (containing holes). If a voltage is applied as shown at top, no current flows. If the voltage is applied in the opposite direction (bottom), the electrons and holes flow, producing a current.

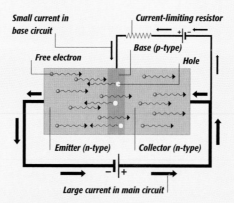

Figure 2: TRANSISTOR
An NPN transistor is a sandwich of p-type semiconductor (the base) between two pieces of n-type (emitter and collector). No current flows in the main collector-emitter circuit unless a small current is allowed to flow into the base as shown.

and p-type silicon are brought together, some free electrons migrate from the n-type to the p-type, and some holes move the other way, creating a voltage across the semiconductor p-n junction. If an external voltage is applied across the semiconductor by a cell, then, depending on its orientation, it either overcomes this junction voltage and makes a current flow, or it acts with the junction voltage so that no current flows. A diode is a device that has such a semiconductor junction and allows current to flow in one direction only (see Figure 1).

Diodes and other semiconductor devices may be made as discrete components for connection by wires into a circuit. They may also be fabricated on a microscopic scale as part of an integrated circuit, by etching and depositing materials on a silicon chip.

Transistors

Transistors consist of a piece of n-type or p-type semiconductor sandwiched between two pieces of the other type. The central piece is called the base, and the outer pieces, the collector and the emitter (see Figure 2). Transistors have two semiconductor junctions, and can be used to amplify or switch currents. A small current flowing into or out of the base (depending on whether it is a p- or n-type) controls a much larger current between the emitter and the collector. Transistors are used as amplifiers in audio amplifiers, and as switches in digital circuits.

Other electronic components

Other semiconductor devices include photodiodes and light-emitting diodes (LEDs). Photodiodes are light-sensitive diodes that allow current to flow when photons hit the p-n junction. Photodiodes convert light into electric signals in CD players (see Sound reproduction, p.80). Image-sensing CCDs in digital still and video cameras usually consist of photodiode arrays (see Cameras, p.92 and Video technology, p.104). In an LED, the reverse process occurs—an applied voltage causes electrons to cross the p-n junction into the p-type semiconductor and fill holes, causing the emission of photons (see p.262).

Other electronic components, such as resistors and capacitors, are not based on semiconductors. Resistors are widely used devices that have a specific amount of resistance to current flow and are used to limit currents in electric and electronic circuits. (All electrical conductors except superconductors resist the flow of current to some extent, converting some electric energy into heat.) Transistors are usually used with a current-limiting resistor in the base circuit.

Capacitors consist of two metal plates separated by an insulator. When a voltage is applied to the plates, a negative electric charge builds up on one plate and a positive charge on the other (see Figure 3). A charged capacitor can be discharged by connecting it to a circuit. Capacitors have a wide range of applications, including charge storage in memory chips (see Computer memory, p.66).

Transistors in logic gates

One of the most important uses of transistors is as high-speed electronic switches in the digital circuits of computers. By connecting these transistorized

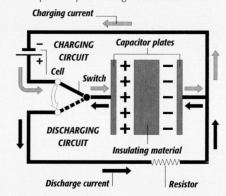

Figure 3: CAPACITOR
When the capacitor is connected to the cell, current flows in the charging circuit to build up charge on the capacitor plates. When the charged capacitor is connected to the resistor, charge flows through the discharging circuit.

switches in arrangements known as logic gates, it is possible to perform simple logical operations. The gate takes electric pulses representing binary numbers (high voltage or "on" for "1," low voltage or "off" for "0") as an input, and produces an

appropriate output pulse representing another binary number (see Figures 4 and 5). Logic gates are based on a branch of math called Boolean algebra. For each type of gate, the output pulse for a particular input is strictly defined and is given by a "truth table." For example, the truth table for an AND gate states that it outputs "0" unless both of its inputs are "1." Millions of such logic gates are connected together in computer microprocessors, enabling them to carry out complex calculations.

Half adders and full adders

By combining three gates, a unit called a half adder can be devised (see Figure 6). This takes two single-bit (binary digit) inputs and outputs the sum of these digits as a two-bit binary number. One step up is a full adder (see Figure 7). This can be made from

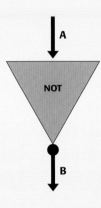

Figure 4: NOT GATE
The simplest type of gate, a NOT gate, contains a transistor that produces an output of 1 if the input is 0 ("NOT 1"), an output of 0 if the input is 1 ("NOT 0"). A NOT gate inverts the input pulse, so it is also called an inverter.

INPUT	OUTPUT
A	**B**
0	1
1	0

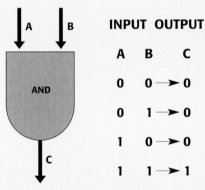

INPUT	OUTPUT	
A	**B**	**C**
0	0 →	0
0	1 →	0
1	0 →	0
1	1 →	1

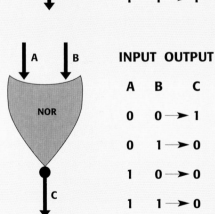

INPUT	OUTPUT	
A	**B**	**C**
0	0 →	1
0	1 →	0
1	0 →	0
1	1 →	0

Figure 5: AND and NOR GATES
These gates have two inputs, A and B. An AND gate outputs a 1 only if inputs A AND B are 1, or else it outputs a 0. A NOR gate outputs a 1 only if neither A NOR B are 1 (that is, both are 0).

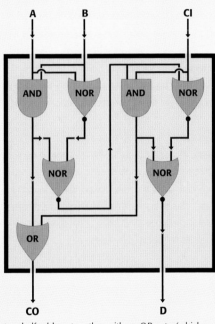

two half adders, together with an OR gate (which outputs a 1 if either one or both of its inputs is 1). A full adder differs from a half adder in that it takes an additional input, called CI (for carry in), combining this with its other two inputs to give a 2-bit output. One part of its output, called CO (for carry out) can be used as the carry in for another adder in a series. In this way a "cascade" of adders can be built up.

Adding two 4-bit numbers

By arranging four full adders in a cascade with a carry in to the first adder and a carry out from the last, it is possible to add together two 4-bit binary numbers and output their sum as a 5-bit binary number (see Figure 8). A larger cascade can be used to add larger numbers.

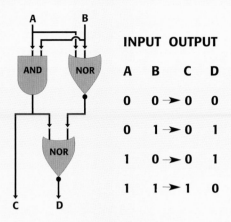

INPUT		OUTPUT	
A	**B**	**C**	**D**
0	0 →	0	0
0	1 →	0	1
1	0 →	0	1
1	1 →	1	0

Figure 6: HALF ADDER
An arrangement of an AND gate and two NOR gates takes two inputs (of 0 or 1) and outputs their sum as a 2-digit number (00, 01, or 10).

INPUT			OUTPUT	
A	**B**	**CI**	**CO**	**D**
0	0	0 →	0	0
0	1	0 →	0	1
1	0	0 →	0	1
1	1	0 →	1	0
0	0	1 →	0	1
0	1	1 →	1	0
1	0	1 →	1	0
1	1	1 →	1	1

Figure 7: FULL ADDER
The combination of two half adders and an extra gate called an OR gate (left) makes a full adder. This has three inputs and two outputs.

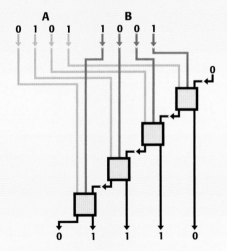

Figure 8: 4-BIT ADDITION (4 FULL ADDERS)
Here the binary numbers 0101 (5 in the decimal system) and 1001 (decimal 9), plus a carry in of 0, are added to output 01110 (decimal 14).

Mechanics

Mechanics is the branch of physics concerned with forces, motion, work, energy, and the principles that underlie the operation of simple machines.

Mass and weight

All physical objects—whether solid, liquid, or gas—have a property called mass; it is a measure of the amount of matter they contain. This mass depends on the number of atoms (see p.258) and the mass of those individual atoms. Mass is measured in grams (g) and kilograms (kg) in the metric system, and in ounces (oz) and pounds (lb) in the Anglo-American system. According to the special theory of relativity (proposed by Albert Einstein), an object's mass varies with its velocity. However, the difference from an object's rest mass is significant only when its velocity approaches the speed of light.

Mass is not the same as weight, which is the force on an object produced by the downward pull of gravity. Physicists measure weight, as they do all forces, in units called newtons (N). An object with a mass of 1 kg experiences a gravitational force of 9.8 newtons (known conventionally as 1 kg weight) at the Earth's surface. This is its weight on Earth. (Equally, the Earth experiences a gravitational force of 9.8 newtons toward the object, because gravity is a force that attracts two objects toward each other.) The same object would experience a gravitational force of only about 1.63 newtons on the Moon (so it would weigh only one-sixth of its weight on Earth), because gravity on the Moon is much weaker. Nevertheless, the object's mass would still be 1 kg.

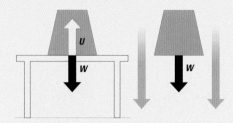

Figure 1: FORCES ON A STATIONARY OBJECT
If an object is stationary, the forces that are acting on it must be in balance. The gravitational force (W) pulling down on this weight (left) is balanced by an equal force (U), called the "normal reaction," exerted upward on the weight by the table. If the table is removed (right), only the gravitational force remains, so the object falls downward.

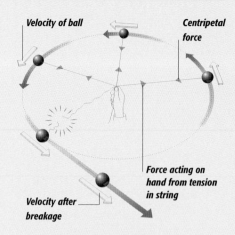

Figure 2: CENTRIPETAL FORCE
A ball swung around in a circle is subject to a centripetal force exerted by the tension in the string. The ball's speed is constant but its velocity (which includes direction as well as speed) continually varies due to this force. If the string breaks, inertia keeps the ball going in a straight line (its velocity is now constant).

Density

Density is the amount of mass in a certain volume (three-dimensional space). If two objects are exactly the same size but one has double the mass of the other, the more massive one has twice the density of the other. Density is measured in kilograms per cubic meter (kg/m^3).

Force, inertia, and movement

A force is a push or a pull. A force cannot be seen, but it can be felt and measured because of its effects. It may act through something attached to or in contact with the object—such as a string, or a table on which the object sits. Or it may act at a distance—that is, through space, without any apparent physical connection—such as the force of gravity and the electromagnetic force.

Unless a force acts on it (or the forces acting on it do not balance), an object stays in the same position or continues moving steadily in a straight line. This property is called its inertia. An object sitting on a table has a downward force exerted on it by the pull of gravity and an equal upward force, called the "normal reaction," exerted on it by the table. The object is said to be in a state of static equilibrium (see Figure 1). Equally, a perfectly round, smooth ball spinning on a perfectly smooth

flat surface continues spinning steadily as long as these conditions persist. Again, gravity and a push upward from the surface balance each other. No forces act in the direction of the ball's motion, but the ball's inertia keeps it moving. The ball is said to be in a state of dynamic (moving) equilibrium.

When an object is swung around on a length of string, its inertia tends to make it move in a straight line; only the pull exerted on it by the string keeps it moving in a circle (see Figure 2). This pull is called the centripetal force. (Many people think that a "centrifugal force" pulls the object outward, but in fact there is no such thing, just the object's inertia.) It is the centripetal force of the Sun's gravity that keeps the Earth and other planets orbiting around it.

Force as a vector

Force is a vector quantity—it has direction as well as size. A force can be depicted as an arrow indicating the direction of a force, with the length of arrow indicating its magnitude. Forces acting through the same point can be added (bearing in mind their size and direction) to give a net or resultant force. Conversely, forces can be resolved (broken down) into vertical and horizontal components (see Figure 3). This can help elucidate the forces acting in various examples of static and dynamic equilibrium.

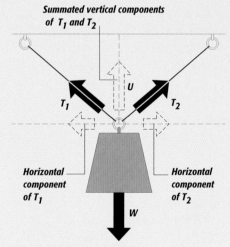

Figure 3: RESOLVING FORCES
This mass is being acted on by three forces—the weight due to gravity (W) and tension forces (T_1 and T_2) in two pieces of rope acting in different directions. The tension forces can be resolved into horizontal components (which cancel each other out) and vertical components, which when added together (U) equal the downward pull of gravity. In consequence the mass remains stationary.

Figure 4: FREE FALL
Ignoring air resistance, an object subject only to the force of gravity accelerates (its velocity increases) downward at a constant rate. Shown right, at intervals of 0.035 seconds, is how a small object drops at the Earth's surface.

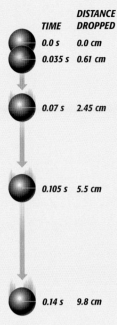

TIME	DISTANCE DROPPED
0.0 s	0.0 cm
0.035 s	0.61 cm
0.07 s	2.45 cm
0.105 s	5.5 cm
0.14 s	9.8 cm

Acceleration

When a net force acts on an object, it progressively changes the object's velocity. Velocity is defined in terms both of an object's speed and its direction of motion, so a change in velocity may involve a change in speed or direction or both. Any such change is called an acceleration and it continues for as long as the force is applied. Acceleration is the rate of change of velocity with respect to time and is measured in meters per second per second—for example, an acceleration of 5 meters per second per second (5 m/s^2) means that an object's speed increases by 5 meters per second every second.

When a tennis racket hits a ball, the acceleration experienced by the ball is brief (only for as long as racket and ball are in contact) but for that brief time reaches a very high magnitude. When you put your foot down on a car's gas pedal, or when a rocket's motors ignite, the acceleration is lower but lasts much longer—until you ease off to a steady speed or the rocket's motors cut out.

An important example of acceleration is that caused by gravity. If an object is dropped at the Earth's surface, it accelerates at a rate of about 32 feet (almost 10 meters) per second per second (see Figure 4). This is the same for objects of any mass (if air resistance, a form of friction, is ignored) because acceleration equals force divided by mass, and since the force of gravity on any object is directly in proportion to its mass, the object's mass has no resultant effect. On the Moon, the acceleration due to lunar gravity is much lower because the gravitational pull on the object is lower, but the

object's mass has not changed. As the astronauts who went there found, if you drop something on the Moon, it takes a surprisingly long time to reach the lunar surface.

Action and reaction

As mentioned above, the forces acting on a stationary object sitting on a surface are equal and opposite. In fact, forces always act in pairs like this, even when causing acceleration. Such opposing forces are called action and reaction.

For example, when a car's wheels push against the road surface (an action), the road pushes back equally in the opposite direction on the tires (the reaction). The action of the exhaust gases shooting out of a rocket's engine causes an opposite reaction on the rocket vehicle, causing it to accelerate forward. A roller-skater pushing on a wall accelerates backward due to the reaction force (see Figure 5).

Friction

In the real world—except in a perfect vacuum, such as outer space, and at extremely low temperatures—any moving object encounters an opposing force called friction (or dynamic friction) that acts to impede and slow down its motion. In fact, there is also friction (called static friction) between two stationary objects. For example, friction prevents an object on a slope from sliding unless the slope is steep enough for the component of the object's weight acting down the slope to overcome friction.

Friction depends on the nature of the two surfaces in contact, particularly on the roughness of the surfaces, which is measured by a number called the coefficient of friction. A rubber tire on the road, for example, is subject to much more friction than a smooth stone sliding on a frozen pond. Friction also depends on the forces acting between the surfaces.

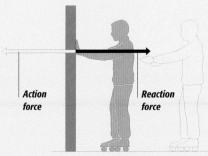

Action force | **Reaction force**

Figure 5: ACTION AND REACTION
A roller skater pushing on a wall produces what is termed an action force. The wall, being immobile, produces an equal force in the opposite direction called a reaction force. This causes the skater to roll away from the wall.

Where the surfaces are horizontal, these forces comprise the object's weight and the opposing normal reaction from the ground.

Friction can be both an advantage and a hindrance. It gives grip—for example between tires and roads, or between the hand and the handle of a tennis racket, or between the loops of a knot. But it also wastes energy in machines, generating useless heat and reducing their efficiency. (A car traveling at a steady speed expends all its fuel on overcoming friction from the road and air and the internal friction in its engine and transmission.)

Friction may be reduced by several methods (see Figure 6). Lubricants such as oil and grease are slippery substances that reduce friction between machine surfaces (such as between a piston and a cylinder; see Piston engines p.110). Bearings are sets of balls or cylinders that reduce friction between moving and stationary parts of a machine—for example, around axles. Bearings roll between surfaces, reducing friction because much less friction is caused by rolling motion than sliding motion.

Rough areas creating friction

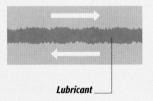

Lubricant

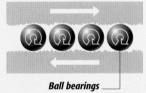

Ball bearings

Figure 6: REDUCING FRICTION
Two objects sliding past each other experience friction (above left), because of rough areas on the surface rubbing together. Two methods of reducing friction are (middle) to introduce a lubricant that separates the surfaces slightly (each now rubs only against slippery lubricant molecules) or (above right) to introduce small, hard-wearing balls or rollers between the surfaces. When the surfaces move, these items roll, with minimal friction. This is the principle of ball and roller bearings.

Mechanics

Energy and work

Every action or change in the physical world involves energy—energy "consumed," "created," or converted from one form to another. In fact, energy is never created nor destroyed. Instead, every process involves converting energy from one form into another (*see Figure 1*). The law of energy conservation states that the total amount of energy in a particular system always remains constant.

Many different energy forms exist, ranging from nuclear energy (bound up in atomic nuclei; *see p.258*) and chemical energy (the energy of chemical molecules) to heat, light, and electrical energy (*see p.260*). In mechanical systems, two important forms of energy are potential energy—energy stored in an object when it is stretched, squeezed, twisted, or raised to a height—and kinetic energy, which is the energy of a moving object (*see Figure 2*).

Energy is closely related to work—in fact, they are two aspects of the same thing and are measured in the same units. The word "work" has a common meaning in everyday language, but it has a special, more precise definition in physics and engineering. It is the transfer of energy to or from an object (including a person or machine) and is equal to the force used multiplied by the distance that the object moves in the direction of the force (*see Figure 3*).

The unit of work and energy is the joule (J). One joule is the work done (or energy expended) when a force of one newton moves through one meter. Another everyday unit of work and energy, used, for example, to measure domestic electricity consumption, is the kilowatt-hour (kWh).

Power

In science and engineering, power is the rate at which work is done or energy converted. The unit of power is the watt (W)—equal to one joule of work done or energy converted per second. Watts are more familiar than joules—a light bulb rated at 60 W transforms 60 joules of electrical energy into light and heat per second, and a one-kilowatt heater transforms 1,000 joules per second. One kilowatt-hour is the amount of energy converted to heat and light by a one-kilowatt appliance in one hour.

Another everyday unit of power and energy is the horsepower (hp)—originally the power of one horse, but now defined more precisely. It is still often used to express the power of engines and motors. One horsepower is equivalent to 745.7 watts.

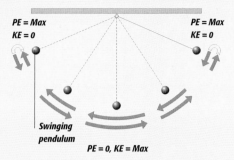

Figure 2: POTENTIAL AND KINETIC ENERGY
As this pendulum swings back and forth, a continuous cycle of conversion occurs between potential energy (PE) and kinetic energy (KE). At the extremes of the swing, the ball is momentarily still, so it has no kinetic energy, but being at its highest point, it has maximum potential energy. At the bottom of each swing, KE is maximum and PE is zero.

Efficiency

When a machine converts energy into work (or energy of one form into energy of another form), not all the energy ends up doing the intended job. Some is wasted—usually as heat. This is, in fact, a fundamental law of nature. The ratio of useful work to work or energy input—usually expressed as a percentage—is the machine's efficiency. As a general rule, machines that convert one kind of mechanical work into another (such as a lever or pulley) are much more efficient—reaching nearly 100 percent—than ones that convert a different form of energy (such as electricity or chemical energy) into mechanical work.

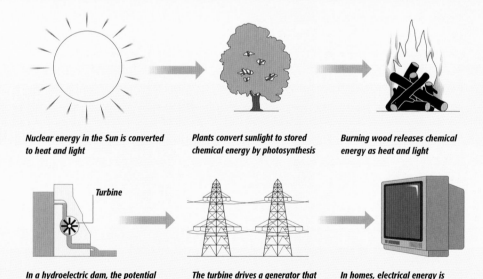

Nuclear energy in the Sun is converted to heat and light

Plants convert sunlight to stored chemical energy by photosynthesis

Burning wood releases chemical energy as heat and light

In a hydroelectric dam, the potential energy of the lake water is converted to the kinetic energy of the turbine

The turbine drives a generator that converts kinetic energy into electrical energy for transmission to homes

In homes, electrical energy is converted to light, sound, and heat by electrical appliances such as TVs

Figure 1: EXAMPLES OF ENERGY CONVERSIONS
Energy is continually being converted from one form into another—some example conversions are shown above. A high proportion of the energy available to us on Earth derives ultimately from radiant solar energy, which itself derives from nuclear fusion processes within the Sun.

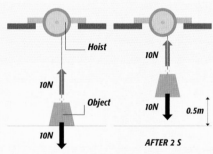

Figure 3: WORK AND POWER
A hoist lifts an object weighing 10 newtons (10N). To do so, it exerts an upward force just in excess of 10 N. After 2 seconds (right), it has raised the object 0.5 meters. The work done is force x distance moved = 10 N x 0.5 m = 5 joules. The power output of the hoist is work divided by time = 5J/2s = 2.5 watts.

Simple machines

In mechanics, a simple machine is a device that changes a force applied to it so that the force acts in a more useful way. For example, it may convert a small force acting over a long distance into a much greater force acting over a shorter distance.

As a result, a force of 20 newtons might lift an object weighing 100 newtons. The ratio of force produced to force applied is called the mechanical advantage or force ratio of the device. In the above example it is 100 (force produced) divided by 20 (force applied), equaling 5. Of course, the law of conservation of energy means that you cannot get something for nothing, so in this example, moving the applied force by 10 centimeters would lift the object by only 2 centimeters—assuming that the machine's energy efficiency is 100 percent, so that the total work input (force x distance) equals the work output (also force x distance).

Many of the everyday things we think of as "machines" have one or more simple machines built into them. The best known ones are pulleys, levers, gears, and wheels and axles. Cranks (which can convert a to-and-fro movement into a rotary one, or vice-versa) and cams (which can only convert rotary motion into a to-and-fro action) are examples of simple machines used in most internal combustion engines (see Piston engines, p.110).

Some other simple machines are difficult to think of as "machines" at all. An inclined plane, or slope, is a simple machine, because it is easier to push a load up it than to raise the same load vertically—it gives a mechanical advantage.

Pulleys

The simplest pulley (consisting of a single wheel) changes the direction of an applied force without giving any mechanical advantage. Depending on the number of pulley wheels used (and the number of times the rope loops over them) a mechanical advantage of two or more can be obtained.

Levers

A lever is a rigid rod hinged around a pivot called the fulcrum. There are three classes of levers (see Figure 4). Class 1 and 2 levers enable a large load to be moved over a short distance by a small effort force that moves over a large distance. Class 3 levers convert a small movement of a large effort force into a larger movement of a small load. In each case, the relative magnitudes of the load and effort forces are in inverse ratio to their distances from the fulcrum.

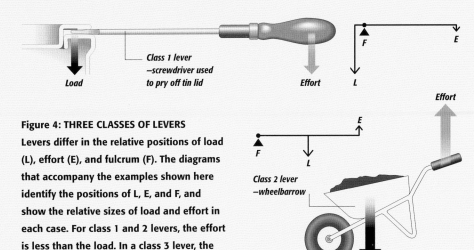

Class 1 lever —screwdriver used to pry off tin lid
Load
Effort
F
E
L

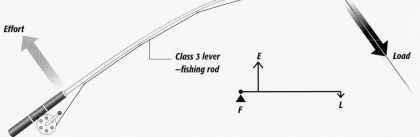

Figure 4: THREE CLASSES OF LEVERS
Levers differ in the relative positions of load (L), effort (E), and fulcrum (F). The diagrams that accompany the examples shown here identify the positions of L, E, and F, and show the relative sizes of load and effort in each case. For class 1 and 2 levers, the effort is less than the load. In a class 3 lever, the effort is greater but the load moves further.

Effort
E
F
L
Class 2 lever —wheelbarrow
Load

Effort
Class 3 lever —fishing rod
Load
E
F
L

Gears

If a gear with 15 teeth turns another with 30 teeth, the larger gear turns at only half the speed of the smaller but the turning force of the axle attached to the larger gear is twice the turning force applied to the smaller gear. The mechanical advantage of the gear pair would thus be 2. In general the mechanical advantage of a gear pair is the number of rotations that the turning gear makes for each rotation of the turned gear, or the ratio of the number of teeth of the turned gear to the turning gear. Some types of gears (such as worm gears), have very high mechanical advantages (see Figure 5).

Springs

A spring is a piece of metal (or sometimes, plastic) that is specially formed and tempered (usually in a helical or corkscrew-shape) so that it will return to its original shape after it is stretched or compressed. A spring is not a machine, but a device for storing potential energy—which is converted back into kinetic energy when the spring is released. The extension or compression of a spring is directly proportional to the force applied to it—if a 1-N force stretches it by 1 cm, then 2 Ns will stretch it by 2 cm.

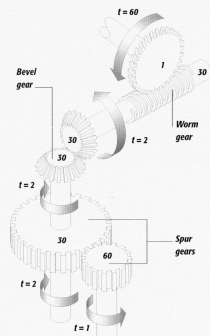

$t = 60$
Bevel gear
1
30
Worm gear
$t = 2$
30
30
$t = 2$
30
Spur gears
60
$t = 2$
$t = 1$

Figure 5: GEAR TRAIN
Several different types of gears are shown here. The numbers on each gear indicate how many rotations each would make for 60 rotations of the small spur gear at the bottom. The "t" numbers for each gear indicate the relative amount of turning force exerted by each.

Mechanics

Liquids and gases

Liquids and gases (including air) are both called fluids, because they flow to fill any space open to them. They share some properties, but they also exhibit some important differences.

The study of how fluids behave is called fluid mechanics. It includes hydrostatics, the study of stationary liquids; hydrodynamics, the study of moving liquids (for example, the flow of water around a moving ship); and aerodynamics, the study of gases in motion (particularly over the surfaces of cars and aircraft) and the forces they exert.

Buoyancy and flotation

As well as filling any empty space, a fluid also flows out of the way of any solid object, which is said to displace the liquid or gas. This displacement of the fluid causes an upward force on the object called the upthrust. The upthrust on an object acts in the opposite direction to its weight and tends to lift it, giving it a property called buoyancy.

The buoyancy of a solid, lightweight object such as a block of wood immersed in water is evident. But every object, however heavy, experiences an upthrust in any fluid, whether liquid or gas. An important physical principle, known as Archimedes' principle after the Ancient Greek who discovered it, states that the upthrust on an object immersed in a fluid is equal to the weight of the fluid the object displaces (see *Figure 1*). If the object is light enough,

so that the upthrust at least equals its weight, then it will float on the surface. There, just enough of it will remain submerged to displace its own weight of the fluid. If it is so heavy that the weight of the fluid displaced can never match its own weight, then it will sink to the bottom.

That is why a steel ship, perhaps weighing 10,000 tons, will float while a solid piece of steel will not—the hollow hull ensures that the ship easily displaces a sufficiently large volume, and thus weight, of water. Pontoons and floats support heavy objects because they are hollow and increase the amount of water displaced. In the same way, when a submarine "blows" its buoyancy tanks, replacing

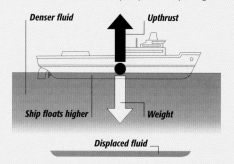
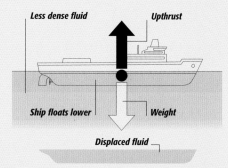

Denser fluid — Upthrust
Ship floats higher — Weight
Displaced fluid

Less dense fluid — Upthrust
Ship floats lower — Weight
Displaced fluid

Figure 2: FLOATING IN FLUIDS OF DIFFERENT DENSITIES
These sister vessels of the ship in Figure 1 are floating in (left) a denser and (right) a less dense fluid. Their weights (W) and the upthrust on them (U) are equal in each case. The weight of fluid displaced is identical, but since the fluids have different densities, the volumes displaced are different. The ship floating in the denser fluid has to displace less fluid for the upthrust to equal the ship's weight, and so floats higher. Conversely, the ship in the less dense fluid floats lower.

water with compressed air, it rises to the surface. Air is much less dense than water, and the submarine as a whole now weighs more less than the water that it displaces. The upthrust on the submarine therefore exceeds its weight.

Exactly the same principle applies to balloons and airships (see p.122). They are filled with a gas such as hot air or helium that is much lighter than the air in the surrounding atmosphere, and their solid structure is lightweight. As a result, the air that they displace weighs more than the whole balloon or airship—so the upthrust overcomes the weight of the balloon or airship and lifts it off the ground.

The weight of fluid that an object displaces depends on the fluid's density (see p.266). Sea water in some parts of the world is saltier and denser than in other places. As a result, ships do not need to displace so much water in order to float,

and they "ride high" (see *Figure 2*). The Dead Sea (an extremely salty body of water in the Middle East, between Israel and Jordan) is an extreme example, where ships (and people) float high on the surface. The least dense water, where ships lie lowest, is warm fresh water.

Pressure

Pressure is the amount of force on a specific area. The force of a 1 ton weight spread over a square foot exerts far less pressure than the same weight spread over a square inch. This is why wide "balloon" tires are fitted to vehicles that travel on soft surfaces such as sand or snow, which cannot exert very large supporting forces. Balloon tires exert less force per unit area (pressure) on the surface than ordinary tires. The surface therefore has to

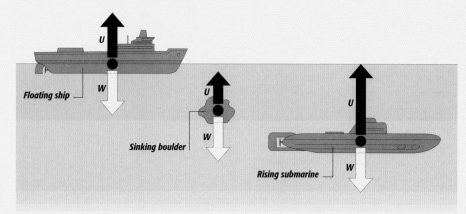

Floating ship
U
W

Sinking boulder
U
W

Rising submarine
U
W

Figure 1: BUOYANCY
The ship, the boulder, and the submarine all weigh the same (W is equal in each case), but the upthrusts on them (U) differ, because they displace different volumes of water. The submarine experiences a larger upthrust than its weight, so it rises. The upthrust on the boulder is less than its weight, so it sinks. The upthrust on the ship equals its weight, so it neither rises nor sinks.

exert less reaction force back on the tire in order to be able to support the vehicle.

Pressure is usually measured in pounds per square inch, or in newtons per square meter, also called pascals (Pa). Other units that are mainly used for fluid pressure include the atmosphere (atm), which equals 101,325 pascals, and the bar (1 bar is 100,000 pascals).

Fluid pressure effects

Pressure is an important consideration in the design of buildings and machines, but pressure effects are particularly significant in liquids and gases.

The first important law is that fluid pressure acts equally in all directions, not just downward. So water will squirt out not just from a hole made in the bottom of a container but also from a hole in the side. The second—called Pascal's law after its French discoverer—is that pressure acts equally throughout a fluid. So if you increase the pressure in one part of a fluid (by using a pump, for example), it increases the pressure by the same amount throughout.

Pascal's law does not mean that pressure is the same throughout a volume of gas or liquid. Earth's gravitational pull causes fluid pressure to increase with depth, due to the weight of liquid or gas above. As a result, air at sea level has an average pressure of 101,325 pascals (1 atmosphere), but at an altitude of 20,000 m (66,000 ft), the pressure is only one quarter as much. In water, pressure increases by 1 atmosphere for about every 10 m (33 ft) increase in depth (see Figure 3).

One important difference between gases and liquids is that gases can be compressed (squeezed) by increasing their pressure—doubling the pressure halves the volume that a particular mass of gas occupies at a given temperature. Liquids cannot be compressed significantly by pressures of the magnitudes encountered on Earth. Liquids and compressed gases have two useful properties—they can transmit forces, and they flow. These properties are exploited in pneumatic and hydraulic machines.

Pneumatics and hydraulics

Compressed air can be stored and piped when needed to a machine containing a piston or a turbine, producing linear (straight line) or turning forces respectively. Pneumatic machines include dentists' drills, which are spun by a tiny turbine, and pneumatic drills, which have a piston.

Hydraulics involves the application of liquid pressure by means of master and slave cylinders

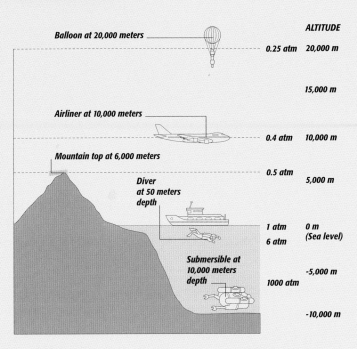

Figure 3: PRESSURE VARIATIONS
Average air pressure at sea level is 1 atmosphere (1 atm). With increasing altitude, pressure drops—to 0.5 atm at 6,000 meters and 0.25 atm at 20,000 meters. Underwater, pressure increases dramatically, by 1 atm for each 10 meter increase in depth, because water is much denser than air.

Balloon at 20,000 meters
Airliner at 10,000 meters
Mountain top at 6,000 meters
Diver at 50 meters depth
Submersible at 10,000 meters depth

ALTITUDE	
0.25 atm	20,000 m
	15,000 m
0.4 atm	10,000 m
0.5 atm	5,000 m
1 atm	0 m (Sea level)
6 atm	
	-5,000 m
1000 atm	
	-10,000 m

linked by pipes. This technology is often used to magnify a small force that moves a long distance into a much larger force that moves a shorter distance (see Figure 4). In other words, it yields a mechanical advantage (see p.269). Hydraulic machines can use long pipes to transmit forces over great distances. Hydraulic systems are used in the braking systems of cars and other wheeled vehicles (see Cars, p.108), and to move the control surfaces, such as the rudder, of large aircraft.

Bernoulli's principle

One of the most important principles of fluid mechanics is Bernoulli's principle. Named for its Swiss discoverer, this states that the pressure in a fast-moving fluid is lower than in a slow-moving one. Bernoulli's principle helps explain how aircraft fly, how tennis balls swerve in the air, and how the

Figure 4: HYDRAULIC SYSTEM
The "master cylinder" has a cross-sectional area of one square meter (m^2). A force of 10 newtons (N) is exerted by the master piston, creating a pressure within the fluid of 10 N per m^2, or 10 pascals (10 Pa). The fluid transmits this pressure through a pipe to the "slave cylinder", of cross-sectional area 10 m^2. The same pressure (10 Pa) acts over the surface of the slave piston, producing a force of 100N (a mechanical advantage of 10). The work done (see p.268) is the same, because the slave piston moves 1 meter for every 10 meters that the master cylinder moves.

read-write head of a computer's hard drive skims very close to the disk's surface.

For example, the curved upper surface of an aircraft wing (see Generating lift, p.125) makes air travel farther and faster over it than underneath. The pressure above the wing is therefore lower than the pressure below it, and the resultant upward force on the wing overcomes the aircraft's weight. Similarly, in a ball that spins as it moves, air passes at a faster relative speed on the side moving in the same direction as the ball's overall direction of movement. This faster air movement creates lower pressure on that side of the ball, causing it to swerve.

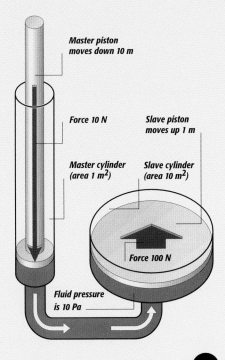

Master piston moves down 10 m

Force 10 N

Slave piston moves up 1 m

Master cylinder (area 1 m^2)

Slave cylinder (area 10 m^2)

Force 100 N

Fluid pressure is 10 Pa

GLOSSARY

Words and phrases in *italics* indicate glossary entries.

absolute zero In theory, the lowest possible *temperature*, at which *atoms* stop all movement: –459.67°F (–273.16°C). In practice, the lowest attainable temperature is a fraction of a degree warmer than this.

acceleration Rate of change of an object's *velocity* (*see p.266*). Deceleration (slowing down) is acceleration in the opposite direction to the object's velocity.

acid Chemical *compound* that forms hydrogen *ions* when it dissolves in water. It reacts with a *base* to form a *salt*.

acoustics Study of *sound* and how it is perceived or heard.

action and reaction Opposing *forces* that act whenever an object is moved. The force causing the movement is the action; the object pushes back with an equal force, called the reaction. *See p.267*.

aerodynamics Study of air and other *gases* in motion, especially their flow over *solid* objects such as moving vehicles and the wings of airplanes.

airfoil or aerofoil Curved surface of an aircraft's wing or similar surface that creates a lifting *force* when air flows over it. *See p.124*.

air (or wind) resistance *Friction* between a moving object and air *molecules*.

alkali *See base*.

alloy *Mixture* of *metals* or, in some cases, of a metal with a nonmetal such as carbon. Alloys are designed to have specific properties (such as hardness).

alpha particle *Subatomic particle*, consisting of two *protons* and two *neutrons*, emitted by some radioactive *isotopes*.

alternating current (AC) *Electric current* whose direction of flow changes. Mains electricity supply alternates 60 times per second.

alternator *Generator* that produces an *alternating current*.

amino acids Simple *organic* chemical *compounds* that can join together to make much more complex *proteins*.

amp (A) Short for ampère, the unit of measurement for *electric current*.

amplitude Maximum height or displacement of a *wave*—related to its *power*, or the *energy* it transmits each *second*.

amplitude modulation (AM) Way of carrying information such as speech or music on a *wave* by altering its *amplitude* in time with the *signal* containing the information.

analog Way of representing or recording *data* by a continuously variable *signal*, as opposed to *digital*.

anode *Positive electrode*, which attracts *electrons* or negatively charged *ions*.

antenna or aerial *Metal* mast, rod, or wire used to transmit or receive radio waves or other *electromagnetic waves*.

armature Central, usually rotating part of an electric machine such as a *motor* or *generator*. It commonly consists of wire coils wound around an iron core.

artificial intelligence (AI) Computer system designed to mimic elements of the human mind, such as the ability to analyze problems and learn from experience.

ASCII Binary code used by most computers to represent alphanumeric (letter and number) data. It stands for American Standard Code for Information Interchange.

atmosphere (atm) Unit of *pressure* equal to the average air pressure at sea level: 101,325 *pascals*, 1,013.25 millibars, or 760 mm mercury.

atom Smallest particle of a chemical *element* that can exist. *See p.258*.

atomic mass; atomic mass number *Mass* of one *atom* of a chemical *element*, expressed relative to that of the most common form of carbon atom, whose atomic mass is 12.

atomic number Number of *protons* in the *nucleus* of an *element*—equal to the number of *electrons* moving around the nucleus.

automation Use of self-controlling *machines* in manufacturing and other processes. Such machines measure their own output and adjust their operation by *feedback*.

background radiation Natural particulate or *electromagnetic radiation* in the environment emitted by radioactive minerals or reaching Earth from space.

bandwidth The range of *frequencies* on either side of the main *carrier wave* of a *signal*.

bar (b) Unit of *pressure* equal to 100,000 *pascals*, or 100,000 *newtons* per square meter—just under one *atmosphere*.

barcode Pattern of black and white lines representing coded *data*. A *scanning* device converts it into a series of *pulses* for interpretation by a computer.

base *Compound* that reacts with an *acid* to form a *salt*. A base that dissolves in water is called an *alkali*.

beta particle High-energy *electron* or *positron* emitted by some *elements* when they undergo radioactive *decay*. *See p.259*.

binary number Number represented only by the digits 0 and 1. *See p.54*.

bionic Describes the use of mechanical and/or electrical devices to mimic or enhance the function of parts of the body.

biotechnology Use of biological processes and living organisms in industry, medicine, etc; also the use of technology to modify organisms such as crop plants.

bit (b) Short for "binary digit," the smallest unit of data used by a computer, represented by a *binary number*. A byte (B) is group of bits (typically 8, 16, or 32) that represents a single character, such as a number or letter.

boiling point *Temperature* at which a *liquid* boils and turns entirely into a vapor or *gas*. It depends on *pressure*. At sea level, water boils at 212°F (100°C), but at the top of a mountain where air pressure is lower, the boiling point is also lower.

buoyancy The tendency of an object to float in a *fluid* (*see p.270*).

byte *See bit*.

calorie (cal) Little-used unit of *energy*. A calorie is the *heat* needed to raise the *temperature* of 1 *gram* of water by 1°C. The kilocalorie (kcal) or Calorie (with capital C), which equals 1,000 cal, is used to measure the energy content of food.

cantilever Beam fixed at one end and dependent on its own stiffness to support the free end. It is used in some bridge designs (*see p.23*).

capacitor Electronic device for storing *electric charge*. *See p.264*.

capillary action Tendency of a *liquid* to flow through a narrow space (e.g. a thin tube or the pores of blotting paper) due to forces between the liquid molecules and the molecules that bound the space.

carrier wave *Electromagnetic wave* that is *modulated* to carry information.

catalyst Substance that speeds a chemical *reaction* but is not used up by the reaction.

cathode *Negative electrode*, which emits *electrons* or negatively charged *ions*.

cathode-ray tube (CRT) Type of *electron tube* in which a beam of *electrons* (also called cathode rays) is scanned across a *phosphor*-coated screen, causing it to glow. CRTs are used in TVs and monitors (*see p.101*).

cell (biological) The basic unit from which living organisms are built.

cell (electrical) Basic electricity-generating unit of a battery that converts chemical energy into electrical energy. A "primary" cell is discarded after using up its chemical *energy*. "Secondary" cells can be recharged using an *electric current* to reverse the chemical reaction that drives the cell.

center of mass/center of gravity The point in an object where its whole *mass* can be considered to be concentrated. It is the point of balance for the body.

centi- (c) A hundredth of a unit. One centimeter (cm) is one-hundredth of a *meter*.

central processing unit (CPU) Main *data*-processing part of a computer.

centrifuge *Machine* for separating things with different *densities* by spinning them rapidly. The denser component is flung toward the outside by *inertia*.

centripetal force *Force* needed to keep an object moving in a circle. There is no such thing as the supposed "centrifugal force" —it is the object's *inertia* that tends to keep it moving in a straight line (*see p.266*).

ceramic Nonmetallic material, usually formed by high-temperature firing. See *p.206*.

chain reaction Chemical or *nuclear reaction* in which the reaction products promote the reaction, tending to accelerate the reaction.

chip *See microchip*.

circuit Series of electrical devices and wires joined together in a continuous, closed path with a source of electric energy so that an *electric current* can flow around it.

coaxial Describes a cable having one conductor, usually a simple wire, inside another (usually in the form of wire mesh) and insulated from each other. See *p.46*.

coherent Describes *light* or other *electromagnetic radiation* in which all the *waves* have the same *wavelength* and are exactly parallel and in step with each other, as produced by a *laser* (*see p.220*).

color The perceived characteristic of *light* that depends on its *wavelength*, ranging from short wavelength (violet and blue) to long wavelength (red).

compact disc (CD) *Digital* recording medium consisting of a 5-in (12-cm) diameter disc in which tiny pits represent *data*. See *p.81*.

compound Chemical substance containing *atoms* of two or more *elements* joined together by chemical bonds (*see p.258*).

compression Making smaller in length or volume. Also, a pushing force. The compression of a *gas* is directly proportional to the *pressure* applied. The term is also used to mean the processing of *digital data* so that it occupies less *memory* than before.

conduction Passage of *heat* or electricity along or through a material, called a "conductor." *Metals* are the best conductors of both *heat* and *electric current*.

convection Transfer of *heat* due to the fact that a hot *fluid* (*liquid* or *gas*) is less dense, or more buoyant, than a cold one; the heated fluid rises, carrying the heat away, and is replaced by cold fluid.

coulomb (C) The unit of *electric charge*, equal to an *electric current* of one *amp* flowing for one *second*.

counterweight Balancing weight.

crystal *Solid* whose *atoms* or *molecules* are arranged in a regular pattern called a lattice. A crystal usually also has a regular external shape, such as a cube.

cyberspace A computer-generated virtual world in which humans may interact. See *p.74*.

cylinder A three-dimensional shape with a circular cross-section of uniform size. In engineering, it is found in *piston* engines (*see p.110*) and *hydraulic* systems (*see p.271*).

damper Device for absorbing *vibrations* and smoothing out sudden movements.

data Facts, or their numerical representation in a computer system.

decay Spontaneous change of the unstable *nucleus* of an *atom* into the nucleus of a different *element*, accompanied by radioactive emission. See *p.259*.

density *Mass* of an object or material per unit volume, measured in *pounds* per cubic *foot* or *kilograms* per cubic *meter*. See *p.266*.

dichroic mirror *Mirror* that reflects *light* of a certain *wavelength* (equivalent to *color*) but allows other colors to pass through.

diffraction Bending of *light* or other *electromagnetic waves* passing around an object or through a narrow opening. A diffraction grating is a plate of glass ruled with narrow lines that acts much like a *prism*, splitting light into a *spectrum*.

diffusion Gradual mixing of substances such as two different *gases*, miscible (mixable) *liquids*, or dissolved *solids* as their *molecules* mingle. Also the slow passage of a gas through a porous solid.

digital A way of representing or recording *data* by a code of discrete *binary numbers*, as opposed to a continuous *analog* wave.

diode Electronic device, usually made from a *semiconductor*, that allows an *electric current* to pass in one direction only.

direct current (DC) *Electric current* that flows constantly in one direction, in contrast to *alternating current*. The direction of flow is said by convention to be from *positive* to *negative*; however, current is generally carried by negatively charged *electrons* flowing from negative to positive.

discharge Flow of electricity in the form of *electrons* or charged *ions* through a *gas* (including air) or a vacuum. Electric arcs, sparks, lightning, and the corona in neon tubes are examples of discharges.

distillation Separation of mixed *liquids* (or a *liquid* from a dissolved *solid*) by boiling, which leaves behind dissolved substances, and then condensing the pure vapor back to a liquid. See *p.182*.

DNA Short for deoxyribonucleic acid, the chemical basis for *genes*.

doppler effect Shift in perceived *frequency* of *sound* or *electromagnetic waves* as the source moves relative to an observer.

drag Frictional *force* that opposes the movement of an object through a *fluid* (*liquid* or *gas*); includes *air resistance*.

efficiency Ratio of *work* or *energy* output of a *machine* or system (such as a *motor* or *generator*) to its work or energy input.

effort *Force* applied to a simple *machine*, such as a *lever* or *pulley*, in order to overcome another force, the *load*.

elasticity Ability of many *solid* objects to return to their original shape after being stretched, compressed, or deformed.

electric charge Amount of electrical imbalance in an object. It is described as *negative* if there is a surplus of *electrons*, *positive* if there is a deficit, or neutral if there is neither surplus nor deficit. More generally, it is the electrical property of certain *subatomic particles* that creates an *electric field*.

electric current A flow of electrically charged particles along wires or other conductors.

electric field Invisible field of *force* caused by an *electric charge*. As a result of their fields, opposite charges (*positive* and *negative*) attract each other whereas like charges repel.

electrochemistry Branch of science and technology dealing with the relationship of electricity and chemical *reactions*. It includes electricity generation by batteries, and chemical processing by *electrolysis*.

electrode Usually metallic wire or rod used to conduct electricity to or from a medium such as an *electrolyte* or the vacuum in an *electron tube*. A positively charged *electrode* is called an *anode*; a negatively charged one is called a *cathode*.

electrolysis Causing chemical change by passing electricity through an *electrolyte* (e.g. in *metal* extraction; see p.187).

electrolyte Molten or dissolved chemical substance that conducts electricity.

electromagnet *Magnet* made from a coil of wire, usually with a soft iron core. It becomes magnetic only when an *electric current* flows through the coil.

electromagnetic force One of the fundamental forces of nature, acting between electrically charged *subatomic particles*. It is responsible for *magnetism*, for electrostatic attraction and repulsion (see *electric field*), and partly for holding *atoms* together.

electromagnetic radiation Collective term for *electromagnetic waves* of all *wavelengths*.

electromagnetic wave *Wave* consisting of combined magnetic and *electric fields* that travels at the *velocity of light*. *Light* consists of electromagnetic waves, as do *radio waves*, *microwaves*, *X-rays*, *infrared*, *ultraviolet*, and *gamma radiation*. See p.262.

electromotive force See *voltage*.

electron *Subatomic particle* with a *negative electric charge* and a *mass* that is 1/1,836 that of a hydrogen *atom*. It is a basic

component of all atoms (*see p.258*) and is the cause of all electrical phenomena.

electron tube or valve Electronic device consisting of two or more *electrodes* in an evacuated enclosure, usually made of *glass*. Like modern *transistors*, it can act as a *rectifier*, switch, or amplifier. See p.80.

electron volt (eV) Unit of *energy* used in studying *subatomic particles*. It is equal to the energy gained by an *electron* accelerated by a potential difference of one *volt*.

element Substance that cannot be broken down into anything simpler by chemical means. See p.258.

elementary particle See *subatomic particle*.

energy Capacity to do *work*.

energy level One of the fixed, permitted amounts (*quanta*) of energy in an *atom* or *molecule*—for example, the amount of energy possessed by an *electron* by virtue of the *orbital* it occupies in an atom (*see p.221*).

equilibrium State of balance in a physical or chemical system. In static equilibrium the system is stationary. Dynamic equilibrium involves change, but with no net change to the system—for example, a body moving at constant velocity.

evaporation Change from a *liquid* to a *vapor* or *gas* at a *temperature* below the liquid's *boiling point*.

expansion Enlargement (in length, area, or volume), generally caused by heating or, in the case of *gases*, by a reduction in *pressure*.

external combustion engine Engine in which fuel is burned outside the mechanism producing movement. Steam engines and steam *turbines* are examples.

farad Unit of capacitance—the amount of electric charge a *capacitor* can store.

feedback Mechanism by which a *machine's* output is measured and *signals* are transmitted back to the input.

fiber optics Use of *optical fibers*.

fission Splitting, in particular of an *atom's nucleus*, often with the release of much *energy*. See also *fusion*.

fluid Any material that can flow to fill a container—a *gas* or a *liquid*.

fluorescence Emission of *light* by some substances when bombarded with *electrons*

or *ultraviolet radiation*. Phosphorescence is similar to fluorescence but may persist for a short time after the bombardment stops.

focus Point to which a convex *lens* or concave *mirror* concentrates parallel *rays* of *light*; or point from which a concave lens or convex mirror appears to disperse such parallel rays. The distance from the lens or mirror to the focus is called its focal length. *See pp.226, 228, 263.*

foot (ft) Basic unit of length in the Anglo-American system, equal to 12 *inches*.

force Any push or pull acting on an object. Unless it is exactly balanced by an opposing force, it causes *acceleration*.

fossil fuel Any fuel created in the distant past by the partial decomposition of plant or animal remains, including coal, oil, and gas. Burning such fuels increases the level of carbon dioxide in the atmosphere, contributing to the *greenhouse effect*.

freezing point Also called melting point. *Temperature* at which a *liquid* turns into a *solid*. It varies with *pressure*, but less so than the *boiling point*.

frequency Number of *oscillations* or *vibrations* of a *wave* or other phenomenon per unit of time—usually per *second*.

frequency modulation (FM) Method of superimposing a *signal* on a *carrier wave* by varying the carrier wave's *frequency*.

friction *Force* that opposes the movement of an object that is in contact with another. It is caused by roughness of the objects' surfaces and can be reduced by use of a lubricant or bearings (see p.267).

fuel cell Device for converting chemical *energy* in a fuel (such as hydrogen) directly into electricity. See p.179.

fulcrum Pivot on which a *lever* operates.

fusion Joining together, particularly of two atomic nuclei to form a heavier *nucleus* with the release of much *energy*. See also *fission*.

gamma radiation Energetic, extremely short*wavelength*, *electromagnetic radiation* emitted by many radioactive *isotopes* and in *nuclear reactions*.

gas State of matter with no fixed volume or shape, whose *molecules* or *atoms* have much more energy than those of a liquid, and which move to fill up available space.

gas turbine *Internal combustion engine* in which the main moving part is a *turbine* turned by hot *gases* produced by burning fuel in a combustion chamber. *See p.128.*

gear Toothed wheel on a shaft. The teeth mesh with those of another gear to transmit and change a rotary motion. *See p.269.*

gene Unit of heredity, responsible for a discrete characteristic of a living organism. Physically, a gene consists of a section of *DNA*. *See p.214.*

generator *Machine* for converting mechanical *energy*—usually a turning motion—into electricity. *See p.170.*

geothermal energy *Energy* extracted from below the Earth's surface and used to generate electricity. It derives from the *compression* of rocks when the Earth was formed and from *radioactivity* deep in the Earth's interior. *See p.176.*

giga- (G) Billion (thousand million) times a unit. One gigabyte (GB) is 1,000,000,000 *bytes*, or 1,000 megabytes.

glass Usually transparent *ceramic* material made by melting together silica, limestone, and often other additives.

GMO Short for "genetically modified organism." *See p.212.*

gram (g) Metric unit of *mass*. However, in the SI (international system) of units, the basic unit of mass is the kilogram (kg), equal to 1,000 grams.

gravity Fundamental *force* of nature that attracts any two bodies to each other. It is proportional to the bodies' *masses* but is very weak compared to other basic forces; it is usually noticeable only when at least one of the bodies is extremely massive—e.g. a planet or star. Gravity decreases with the square of distance—at double the distance, the force is one-quarter.

greenhouse effect Warming of the Earth's climate because of carbon dioxide and other so-called greenhouse gases in the air, trapping heat from the Sun.

gyroscope Device containing a heavy spinning wheel that keeps its orientation, however, the frame it is mounted in moves. As a result, it can be used as an accurate compass. *See p.135.*

half-life Time it takes for half of the atomic *nuclei* in a sample of radioactive material to *decay*. After one half-life, it takes another half-life for half the remaining nuclei to decay, and so on.

heat Form of *energy* associated with rapid *vibration* of *atoms* and *molecules* and with *infrared* radiation. Reducing or increasing the amount of heat in an object lowers or raises its *temperature*.

hertz (Hz) Unit of *frequency*. One hertz equals one cycle or *oscillation* per second.

horsepower (hp) Unit of *power* originally based on the ability of a horse to raise a *load*, now used only to rate the power of engines and *motors*. It equals 745.7 *watts*.

hydraulics Study and use of *fluids*—especially *liquids*—under *pressure* or when moved by a pump or other device.

hydrocarbon Any chemical substance whose *molecules* contain only hydrogen and carbon *atoms*. The best known are gasoline and other petroleum products.

image Representation of an object by an optical device such as a *mirror* or *lens*. A "real" image can be projected on a screen; a "virtual" image only appears to be at a certain point in space and cannot be shown on a screen. *See p.263.*

inch (in) Unit of length equal to one-twelfth of a *foot*. It equals about 25.4 millimeters.

inclined plane A slope. In mechanics, it is considered a simple *machine* because it gives a *mechanical advantage* when raising an object against Earth's *gravity*.

induction Process of producing an *electric charge* or *electric current* in an object, or of making it magnetic.

inertia Tendency of an object to stay still or to continue moving at constant *velocity*. An external *force* must be applied in order to overcome inertia.

infrared Type of *electromagnetic radiation* whose *wavelength* is longer than that of visible *light* but shorter than *microwaves* (*see p.262*). It is also known as "radiant heat" because it is given off by hot objects and causes heating when it is absorbed.

inorganic In chemistry, related to any substances other than those found in or made by living things and their related carbon-based substances (*see organic*).

insulator Object or material that is a poor conductor of *heat* and/or electricity.

integrated circuit *Microchip*, usually of silicon, with many electronic components such as *transistors* and *capacitors*, together with fine wires connecting them, formed on its surface to make a complete *circuit*.

intensity Strength of an electromagnetic or other *field*, measured in terms of *power* per unit area—e.g. *watts* per square meter.

interference Interaction of two or more *waves* so they totally or partly reinforce or cancel each other out. The result may be a pattern of intense and lesser disturbance called an "interference pattern."

interferometer Measuring instrument that works by detecting the *interference* pattern created by two *light* or other *waves*.

interlacing Formation of an *image* on a *cathode-ray tube* by scanning even- and odd-numbered lines alternately, so displaying half the image at a time. Used to reduce flicker.

internal combustion engine Engine in which fuel is burned within the mechanism producing movement. *Piston* engines and *gas turbines* are examples.

Internet Worldwide communications network linking computers. *See p.70.*

ion Electrically charged *atom* or *molecule*. It is formed by removing or adding one or more *electrons*—a process called ionization.

isotopes Forms of an *element* with different *atomic masses*, having different numbers of *neutrons* in their *nuclei*.

jet engine *See gas turbine.*

joule (J) Main unit of *energy* in the metric system, produced when a *force* of 1 *newton* moves through 1 *meter*, or a 1-amp current flows for 1 *second* through a potential difference of 1 *volt*.

kilo- (k) Thousand times a unit. One kilogram (kg) is 1,000 *grams*.

kinetic energy *Energy* possessed by an object because of its movement. It equals mv^2, where m is the object's *mass* and v is the object's *velocity*.

laser *See p.220.*

latent heat *Heat energy* absorbed when a *solid* melts or a *liquid* boils—or given up during freezing or condensation.

lens Curved piece of *glass* or other transparent material that refracts *light rays* (alters their direction of travel) to form an *image*. A "convex" lens makes parallel rays converge (come together), a "concave" one makes them diverge (spread).

lever Rod or bar that turns on a *fulcrum* or pivot. Used to transmit or change a force. *See p.269.*

lift Upward *force* produced by the wings of an airplane, rotor of a helicopter, or foils of a hydrofoil.

light Visible form of *electromagnetic radiation*. It consists of a continuous *spectrum* of radiation, with *wavelengths* from about 400 nm (blue) to 800 nm (red).

light-emitting diode *Semiconductor* device that emits light of a particular color when current crosses its p-n junction (*see p.264*).

light-year Distance travelled by a *ray* of *light* in one year, equal to about 5.9 trillion miles (9.5 trillion km).

linear motor Electric *motor* with no rotating parts, producing movement directly, by electromagnetism (*see p.120*).

liquid State of matter in which *molecules* are bound less strongly than in solids, but more strongly than in gases. Liquids have fixed volume, but no fixed shape.

liquid crystal *Liquid* that behaves in some ways like a *crystal*, with its *molecules* lined up regularly. As a result it can act as a polarizer. *Heat* or an *electric field* changes this alignment, making liquid crystals combined with a source of *polarized light* useful as display devices (*see p.69*).

liter (l) Metric unit of volume used mainly for *fluids*. It is equal to 1,000 cm³.

load In mechanics, the *force* pulling down or resisting movement.

logic gate Basic component of computers used for processing *signals*. Each type of logic gate gives a defined output for any given pattern of input. *See p.265*.

loudspeaker *Transducer* for turning electric *signals* into *sound* (*see p.81*).

luminescence *Fluorescence* or phosphorescence; also emission of *light* of particular *wavelengths* caused by heating, used for dating objects (*see p.217*).

machine Any mechanical contrivance. In mechanics, a "simple machine" is any device that changes a *force* applied to it—e.g. *gears*, *levers*, *pulleys*, and *inclined plane(s)*.

magnet Any object capable of producing the force of *magnetism*.

magnetism Invisible *force* produced by *magnets* and *electromagnets* that attracts other magnets and objects made of iron, steel, cobalt, nickel, and certain other materials. The area of influence of such a force is called a "magnetic field." *See p.261*.

mass Amount of matter in an object, measured in *grams* or kilograms in the metric system, or *pounds* or *ounces* in the Anglo-American system. See also *weight*.

mass-energy The *energy* equivalent of *mass*, or vice-versa. According to the theory of relativity, mass and energy are interconvertible aspects of the same phenomenon.

mass number *See atomic mass*.

mechanical advantage Ratio of *load* to *effort* in a simple *machine*; also called "force ratio."

mega- (M) Million times a unit. One megawatt (MW) is 1,000,000 *watts*.

memory In computers, means of storing *data* either temporarily (in *RAM*) or longer-term (e.g. on a hard disk).

metal *Element* or *alloy* that is strong, hard, shiny, a good conductor of heat and electricity, and usually malleable (capable of being formed into shapes by hammering).

meter (m) Principal unit of length in the metric system. It is defined in terms of a particular *wavelength* of light given off by *atoms* of the *element* krypton. One meter equals about 39.4 *inches*.

micro- (µ) One-millionth of a unit. One microvolt (µV) is one-millionth of a *volt*.

microchip An *integrated circuit*.

microcomputer Personal computer using a *microprocessor*.

micrograph Highly enlarged printed *image* produced by a *microscope*.

microphone *Transducer* for converting *sound* into electric *signals* (*see p.76*).

microprocessor *Integrated circuit* that contains all the basic components of a computer. Also called a CPU (central processing unit).

microscope Instrument for magnifying and observing small objects, which may be invisible to the naked eye, using *light* or *electrons* (*see p.226*).

microwave *Electromagnetic radiation* with a *wavelength* between about 1 mm and 30 cm —longer than *infrared* but shorter than *radio waves*. Microwaves are used for *radar*, communications, and heating food.

milli- (m) One-thousandth of a unit. One millimeter (mm) is one-thousandth of a *meter*.

mirror Polished surface that reflects *light* or other *electromagnetic radiation*. *See p.262*.

mixture Material that contains two or more substances blended together without them being chemically combined.

modem Device for converting computer *data* into a form that can be transmitted along telephone wires, and vice-versa. Short for modulator-demodulator.

modulation Superimposing a *signal* (representing *sound*, video, etc) on a *carrier wave* prior to transmission. Demodulation is the reverse: extracting the signal from the carrier wave. *See pp.48–49*.

molecule Smallest particle of a chemical *compound* that can exist. *See p.258*.

momentum *Mass* of a moving object multiplied by its *velocity*.

monomer Chemical "building block" (the repeating unit) of a *polymer*.

motor Any *machine* that converts *energy* (such as electricity or *heat*) into useful *work* —usually in the form of movement. Examples include *external* and *internal combustion engines*, electric *motors*, and *rocket engines*.

multiplexing Technique of carrying more than one *signal* on a single *carrier wave* or along a single cable. *See p.47*.

nano- (n) One-billionth (thousand-millionth) of a unit. One nanometer (nm) is one-billionth of a *meter*.

nanotechnology Very small-scale engineering (*see p.59*), used to make micromachines.

negative In math, less than zero; a "minus" number. In electricity, describes the sense of the *electric charge* associated with the *electron*.

neutral Having no net *electric charge*. In chemistry, neither *acid* nor *alkaline*.

neutron *Subatomic particle* with no *electric charge* and a *mass* similar to that of the *proton*. It is found as a component in all atomic *nuclei* except the lightest *isotope* of hydrogen. Bombardment with neutrons may cause nuclei to split or undergo other *nuclear reactions*.

newton (N) Metric unit of *force*: the force needed to accelerate a *mass* of 1 kilogram by 1 *meter* per *second* per second. *See also weight*.

noise In electronics, communications, etc,

random unwanted *sounds* or *signals* that may obscure the desired signal.

nuclear radiation *Radiation* (including *alpha particles*, *beta particles*, *gamma radiation*, *neutrons*, and *X-rays*) emitted by atomic *nuclei*, either due to *radioactive decay* or as a result of a *nuclear reaction*.

nuclear reaction Change in nature of an atomic *nucleus* due to *fission*, *fusion*, *radioactivity*, or bombardment with *subatomic particles*.

nucleus Central core of an *atom*, containing almost all of the atom's *mass* and consisting of *protons* and *neutrons*. *See also fission; fusion; nuclear reaction*.

ohm (Ω) Unit of electrical *resistance*. A *voltage* of 1 *volt* drives an *electric current* of 1 *amp* through a resistance of 1 ohm.

operating system Set of *programs* that control the basic operation of a computer.

optical fiber Very fine flexible *glass* or plastic thread used to guide light (*see p.47*).

orbit Path of a planet, moon, or artificial *satellite* around its parent body, to which it is attracted by *gravity*.

orbital Region of an *atom* within which an *electron* moves. According to *quantum* mechanics, electrons can occupy only certain "permitted" orbitals.

ore Natural mineral from which a *metal* can be extracted. *See p.186*.

organic Originally, a substance found in or derived from living organisms. Now includes nearly all carbon-containing substances.

oscillation *Vibration*, especially electrical rather than mechanical.

oscillator Electric *circuit* that generates an electrical *wave*, widely used in radio and related communications (*see p.48*).

osmosis Passage of *liquid* from a weak *solution* to a stronger one through a so-called semipermeable membrane—a barrier whose pores are big enough to allow solvent (liquid) *molecules* to pass, but too small for molecules of the solute (dissolved substance).

ounce (oz) Unit of *mass* equal to one-sixteenth of a *pound*.

oxide *Compound* of an *element* with oxygen.

particle *See subatomic particle*.

pascal (Pa) Metric unit of *pressure*, equal to a *force* of 1 *newton* per square *meter*.

pendulum Mass hanging from a string or rod and allowed to swing from side to side. For small swings, a pendulum's period (the time taken for one complete swing) depends only on the length of the string or rod. *See p.41*.

periodic table Chart of chemical *elements* listed in order of *atomic number* and arranged so that "families" of elements with similar chemical properties fall in columns.

petrochemicals Chemical *compounds* derived or made from petroleum oil.

pH Measure of how *acidic* or *alkaline* a *solution* is. Pure water has a pH value of 7 and is neutral; a figure below 7 is acid, above 7 is alkaline.

phase Particular point in an *oscillation* or *wave*. If two waves' crests and troughs coincide, they are said to be "in phase."

phosphor Substance used on the screen of a *cathode-ray tube* that emits light when struck by a beam of *electrons* (*see p.101*).

photoconductivity Phenomenon shown by certain *semiconductor* materials whereby they allow an *electric current* to flow when *light* falls on them.

photodiode *Diode* that allows *electric current* to flow only when *light* falls on it.

photoelectric effect Dislodging of *electrons*—producing an *electric current*—by *light* or *ultraviolet radiation* falling on the surface of a *metal* or certain other materials.

photon *Quantum* of *light* or other *electromagnetic radiation*.

photovoltaic cell *See solar cell*.

pico- (p) One-trillionth (million-millionth) of a unit. One picofarad (pF) is one-trillionth of a *farad*.

piezoelectric effect Property of some *crystals* that produce an *electric voltage* if they are physically distorted—or that change shape if a voltage is applied to them.

piston Close-fitting part of a hydraulic system or piston engine (*see p.110*) that moves within a *cylinder*.

pixel Smallest element of a *digital image*, consisting of a spot of monochrome *light* or a trio of red, green, and blue spots of light. Also, a light-gathering or light-emitting cell that corresponds to such an element.

plasma *Gas*-like form of matter consisting of charged *ions* and separate *electrons* rather than neutral *atoms*. It occurs at very high

temperatures and in electrical discharges.

plastic Literally, easily shaped, especially when hot. More generally, a *polymer*, most of which can be so shaped (*see p.184*).

polarized light *Light waves* whose associated electric (or magnetic) fields all oscillate in the same direction. (In ordinary light, the oscillations occur in many directions.)

polymer *Molecule* consisting of very long chains of much simpler molecules called *monomers* bonded together, with or without complex cross-linkages between the chains. *Plastics* are polymers. *See p.184*.

positive In math, a number greater than zero. In physics, the electrical opposite of *negative*, associated with the *proton* and with a deficit of *electrons*.

positron *Subatomic particle* similar to an *electron* but with a *positive electric charge*.

potential, electric *Potential energy* of an *electric charge*, equivalent to its *voltage*. The voltage between two points in a *circuit* is also known as the "potential difference."

potential energy *Energy* of an object due to its position (such as its height above ground) or electrical or chemical state.

pound (lb) Principal unit of *mass* in the Anglo-American system. One pound (lb) equals about 454 *grams*.

power Rate of doing *work* or changing *energy* from one form to another. The main units are the *watt* in the metric system and *horsepower* in the Anglo-American system.

pressure Measure of the *force* acting per unit surface area, measured in *pascals*, *bars*, *atmospheres*, or in the Anglo-American system in *pounds* per square *foot* or per square *inch*.

prism Angled block of *glass* or other transparent material. It is used to split white *light* into its constituent *colors*, or to deflect the path of light *rays* in optical instruments such as cameras (*see p.92*).

program Set of instructions in *digital* form for performing specific tasks in a computer.

protein One of many types of large, complex *molecules*, formed from *amino acid* units, that are vital to all living organisms. Enzymes (biological *catalysts*) are proteins, as is much of the structural material of humans and animals.

proton *Subatomic particle* with a *positive electric charge* that is found in all atomic *nuclei.* Its *atomic mass* is about 1.

pulley Wheel mounted in a frame over which a rope passes, forming a simple *machine*.

pulse Short burst of electromagnetic or electric *energy*, often representing a *bit*.

quantum "Packet" or particle of *electromagnetic radiation*.

radar Stands for "radio detection and ranging"—the use of short-wavelength *radio waves* to detect objects such as ships and aircraft over long distances.

radiation Generally, any form of "rays"—either *electromagnetic waves* or *subatomic particles*; particularly, high-energy forms such as *gamma radiation*, *X-rays*, or high-speed particles that harm living organisms.

radioactivity Property of emitting *radiation* found in unstable *elements* and *isotopes*.

radio wave *Electromagnetic wave* with a *wavelength* longer than 11¾ in (30 cm) (sometimes also including shorter *microwaves*), produced by an oscillator and used mainly for communications. *See p.48.*

RAM Short for "random-access memory," the type of *memory* used in computers to store current *data* and *programs*.

ray General term for any narrow beam of *electromagnetic radiation*, including *light*.

reaction In chemistry and nuclear physics, any process in which the end products are different from the starting materials. The results are different chemical *molecules* or different *nuclei* respectively. In mechanics, *see action and reaction*.

reciprocating motion To-and-fro motion, like that of the *piston* of an engine.

rectifier Electronic device that converts *alternating current* into *direct current*.

red shift Apparent change of *wavelength* of *light* from distant stars or galaxies toward the red end of the *spectrum*, due to their rapid movement away from us. The shift is caused by the *doppler effect*.

reflection "Bouncing" of *light* or other type of *electromagnetic radiation* from a surface.

refraction Change in the direction in which *light* or other *electromagnetic radiation* is propagated when it changes speed on passing from one transparent medium into another (such as from air to *glass*).

refractive index Measure of how much a

particular transparent material retards the transmission of *light* compared with its speed in a vacuum. (In a material with refractive index 2, light travels at half its speed in a vacuum).

resin Plastic-like *polymer*, particularly one used as an adhesive or as the matrix in composite materials (*see p.144*).

resistance Property of a conductor to resist the flow of an *electric current*. Resistance converts electric energy into heat.

resistor Electronic component with more or less high *resistance*, used to control the flow of *electric current* or reduce *voltage* across another component in a *circuit*.

resolution In optics, ability to pick out fine detail in an *image*.

resonance Strong *vibration* caused when an object is made to *oscillate* at the same *frequency* as its natural frequency of oscillation.

rocket engine Type of *motor* in which thrust is created by hot *gases* (produced by burning fuel) shooting from a nozzle. Rocket engines can be used in space because they carry their own oxygen. *See p.238.*

ROM Short for "read-only memory," type of computer *memory* whose *data* cannot be changed. It is used in computers and other electronic devices to hold the basic operating instructions.

salt Substance made in a chemical *reaction* between an *acid* and a *base*. Common salt—sodium chloride—is an example.

sampling Technique of measuring the *amplitude* of a wave many times per *second* to turn it into a *digital signal*.

satellite Any object, natural or artificial, that orbits a star or other celestial body.

scanning Viewing an object, or creating an *image*, in many narrow strips.

second (s) Basic unit of time in both the metric and Anglo-American systems. It is defined in terms of a particular property of of cesium *atoms* (*see p.224*).

semiconductor *See p.264.*

sensor Electronic device for detecting a physical phenomenon (e.g. *temperature*, *light*, or *radiation*).

signal Anything (such as a *wave*) representing information such as *sound*, video, or computer *data*.

simple machine *See machine* and *p.269*.

smart A system that has a degree of *artificial intelligence*.

software Computer *programs* and *data*, as opposed to hardware (computers themselves).

solar cell Device for generating electricity from sunlight, using the *photoelectric effect*.

solenoid *Electromagnet* formed by a coil of wire, usually containing a movable iron core. Used as a switch or similar device. An *electric current* through the coil creates a magnetic field, which attracts the iron core into the coil to operate a switch mechanism.

solid State of matter in which the *atoms* or *molecules* are in fixed positions, giving the solid a fixed shape and volume.

solution Uniform *mixture* in which one component (the "solute") is uniformly dissolved—dispersed as *molecules* or *ions*—in the other (the "solvent").

sonar Stands for "sound navigation and ranging"—the use of *sound waves* rather like *radar* to detect objects at long distance, especially in water.

sound Mechanical *pressure* waves—in which the carrying medium (a *gas*, *liquid*, or *solid*) is alternately *compressed* and rarified between about 20 and 20,000 times a *second*—that are detectable by the human ear. Sound *waves* cannot travel through a vacuum. Their speed depends on the medium and *temperature*—about 1,085 ft (330 m) per second in air at 0°C; 1,128 ft (344 m) per second in air at 20°C; and 4,865 ft (1,483 m) per second in water.

spectrometer Instrument for studying and measuring the *spectrum* of *light* from stars and other sources.

spectrum Bands of *color* formed when *light* is split into its component wavelengths by a *prism* or *diffraction* grating. Bright or dark lines indicate the presence of particular chemical *elements*.

speed *See velocity.*

static electricity Electrical phenomena due to stationary *electric charges*.

stereo(phonic) Describes *sound signals* fed through at least two channels to the left and right ears, to give an impression of space.

stereoscopic Describes *images* fed separately to the left and right eyes, to give the appearance of depth.

strain In mechanics, a measure of deformation caused by a *stress*.

stress In mechanics, a *force* applied to a material that causes a *strain*.

subatomic particle Any particle that occurs inside *atoms* or takes part in *nuclear reactions*. Such particles include *protons* and *neutrons* (both of which contain even smaller particles called quarks), and also so-called "elementary particles," which cannot be subdivided. The latter include *quarks*, *electrons*, and *positrons*.

superconductivity Phenomenon of almost zero electrical *resistance* shown by many materials at *temperatures* close to *absolute zero*. Some *ceramics* (*see p.206*) have been discovered that are superconducting at considerably higher temperatures, around -220°F (-140°C).

supercooling Abnormal cooling of a *liquid* below its normal *freezing point*. Any disturbance of a supercooled liquid (such as striking its surface) can cause abrupt freezing.

supersonic Faster than the speed of *sound*.

surface tension An effect observable at the surface of a *liquid* (such as water) due to the natural attraction of *molecules* of the liquid for each other. The attractive *forces* make the surface behave like a stretched skin.

telescope Instrument for producing an enlarged *image* of a distant object. *See pp.228, 230.*

temperature Property of an object—its hotness or coldness— that governs the direction that *heat energy* flows. Heat always flows from high temperature to low.

tension Pulling *force*.

tera- (T) Trillion (million million) times a unit. One terabit (1Tb) is 1,000,000,000,000 *bits*.

terminal End-station. In computing, usually refers to a "dumb" *data*-entry machine attached to a network that does not have any processing power of its own.

thermodynamics Study of *heat energy*, its transfer from one point to another, and its ability to do *work*. It explains the working(s) of *machines* such as *external* and *internal combustion engines*, steam *turbines*, etc.

thrust Any pushing *force*, especially that generated by a *rocket* or *jet engine*.

ton; tonne (t) Units of *mass*. Ton is an Anglo-American unit equal to 2,000 *pounds* (short

ton) or 2,240 pounds (long ton). Tonne is a metric unit equal to 1,000 kilograms (about 2,204 pounds).

torque Turning *force*. Also called a moment.

torsion Twisting. A torsion bar is a type of spring consisting of a metal rod that is twisted.

total internal reflection *See p.263.*

transducer Device for converting one form of *signal* into another—e.g. *sound* to or from electrical *waves*.

transformer Electromagnetic device for stepping up or down the *voltage* of an alternating current.

transistor *See p.264.*

turbine Fan- or propeller-like structure that is turned by a fast moving *gas* (e.g. steam or engine exhaust gases) or *liquid* (e.g. water).

ultrasound *Sound waves* at frequencies above the range of normal human hearing, which is about 20 kHz (20,000 hertz).

ultraviolet radiation *Electromagnetic radiation* with a *wavelength* shorter than that of the shortest wavelength of visible *light*, but not as short as that of *X-rays*.

upthrust Upward *force* on any *object* immersed in a *fluid*. It equals the *weight* of fluid displaced (*see p.270*).

velocity Rate and direction of movement—in other words, speed in a particular direction.

velocity of light *Velocity* of *electromagnetic radiation*, including *light*, in a vacuum—about 186,000 miles (300,000 kilometers) per *second*. According to the theory of relativity, no object can attain this speed, but *subatomic particles* in huge accelerators (*see p.232*) almost reach it.

vibration Rapid to-and-fro movement.

viscosity The resistance of a *liquid* to flow.

volt (V) Unit of electrical *potential* difference or voltage.

voltage Difference in the electrical *potential* between two points in an electric *circuit*.

watt (W) Metric unit of *power*—rate of doing *work* or expending *energy*—equal to one *joule* per *second*.

wave *Oscillation* of particles (as in *sound* waves) or of a field (as with *electromagnetic waves*) that moves through matter or space,

transferring *energy* from one point to another.

wavelength Distance between two adjacent crests or troughs of a *wave*.

wave-particle duality Exhibiting characteristics of both *waves* and particles—a property of *light*, other forms of *electromagnetic radiation*, and *subatomic particles*.

weight *Force* of *gravity* on an object. The weight of one kilogram at the Earth's surface is about 9.8 *newtons*.

work Transfer of *energy* when a *force* moves its point of application. Energy and work are two aspects of the same phenomenon—work is the activity resulting from the use of energy—and have the same units.

X-rays Energetic *electromagnetic waves* with a *wavelength* shorter than that of *ultraviolet radiation* but longer than *gamma radiation*.

yard (yd) Anglo-American unit of length equal to three feet, or about 91.5 cm.

Common abbreviations for units	
A	amp
atm	atmosphere
b	bar; bit
B	byte
c	*centi-
C	coulomb
cal	calorie
eV	electron volt
F	farad
ft	foot; feet
g	gram
G	*giga-
hp	horsepower
Hz	hertz
in	inch
J	joule
k	*kilo-
l	liter
lb	pound
m	meter; *milli-
M	*mega-
μ	*micro-
n	*nano-
N	newton
Ω	ohm
oz	ounce
p	*pico-
Pa	pascal
s	second
t	ton; tonne
T	*tera-
V	volt
W	watt
yd	yard
	*prefixes

Internet addresses

General Science and Technology Sites

home.about.com/industry
home.about.com/science
howthingswork.virginia.edu
ideafinder.com/home.htm
www.asap.unimelb.edu.au/hstm
    /hstm_fields.htm
www.bbc.co.uk/science
www.britannica.com
www.Colorado.EDU/physics/2000
www.enc.org/classroom/dd
www.eurekalert.org
www.exosci.com/main/news
www.explorations.org
www.explorescience.com
www.howstuffworks.com
www.links999.net
www.matweb.com
www.members.tripod.com/~IgorIvanov
    /physics
www.nationalgeographic.com
www.nature.com
www.newscientist.com
www.particleadventure.org
www.pbs.org/saf/
www.pbs.org/neighborhoods/science/
www.physlink.com
www.sciam.com
www.sciam.com/askexpert/
www.sciamarchive.com
www.scicentral.com
www.sciquest.com
www.shef.ac.uk/chemistry/web-elements
www.studyweb.com
www.world-science.com

Urban and Domestic

home.wxs.nl/~grijns/subway/metro.html
www.branta.connectfree.co.uk
    /bridges.htm
www.ci.nyc.ny.us/html/dep
www.dot.gov/technology.htm
www.gallawa.com/microtech/mwfaq.html
www.horology.com
www.misty.com/~don/light.html
www.nwl.ac.uk/ih
www.reed.edu/~reyn/transport.html
www.skyscrapers.com
www.time.gov/exhibits.html
www.vacweb.com/vacuums
www.waterweb.org

Communications

cybercollege.com/tvp001.htm
recordingeq.com/GlosPubAE.htm
www.adamwilt.com/DV.html#1394
www.amtechdisc.com/file/amtech
    /CDPAPER.HTML
www.bjphoto.co.uk
www.cybercomm.net/~chuck/phones.html
www.digitalcentury.com
www.dvddemystified.com/dvdfaq.html
www.ee.surrey.ac.uk/Contrib
    /Entertainment/homeent.html
www.ee.washington.edu/conselec
    /CE/ConsElectHome.html
www.karbosguide.com
www.minidisc.org
www.mit.bme.hu/~bako/zaozeng
    /chapter1.htm
www.netlingo.com
www.pbs.org/transistor
www.pctechguide.com
www.rfwilmut.clara.net/repro78
    /repro.html
www.rps.org
www.swcfla.com/swcfla-defs.htm
www.teleport.com/~samc/hdtv
www.WebOPaedia.com
www.whatis.com

Transportation

faculty.washington.edu/%7Ejbs/itrans
nationalacademies.org/trb
www.airbus.com
www.boeing.com
www.conceptcar.co.uk
www.evworld.com
www.fia.com
www.futurecarcongress.org
www.ott.doe.gov/oaat
www.trimble.com/gps
www.vwc.edu/library_tech/wwwpages
    /gnoe/avd.htm
www4.nationalacademies.org/trb/
    homepage.nsf/web/links

Crime and Security

dmoz.org/Science/Science_in_Society
    /Forensic_Science
web.idirect.com/~mjp/mjpwww.html
www.afb.org.uk/public/glossuk1.html
www.biometricgroup.com
www.gwu.edu/~fors/maafs/imaglink.htm
www.netsurf.com/nsf/v01/03
    /nsf.01.03.html
www.pbs.org/wgbh/nova/moolah

Power and Industry

ee.unsw.edu.au/~p2139851/alten-au.html
electrochem.cwru.edu/estir
socrates.berkeley.edu/~rael
solstice.crest.org/renewables/index.shtml
www.doe.gov
www.epa.gov
www.eren.doe.gov
www.eurorex.com/technolo.asp
www.nal.usda.gov/bic
www.nrel.gov
www.offshore-technology.com
www.usda.gov/agencies/biotech

Medicine and Research

dmoz.org/Computers/Robotics
members.aol.com/Bossdoll
www.apl.washington.edu/Biennial
    /crum.html
www.cdc.gov/health/diseases.htm
www.fnal.gov
www.graylab.ac.uk/omd
www.holoworld.com
www.jb.man.ac.uk
www.laserfaq.org/laserfaq.htm#faqwil
www.mayohealth.org/mayo/9908/htm
    /imaging.htm
www.naic.edu
www.ncbi.nlm.nih.gov
www.nhgri.nih.gov/HGP
www.npl.co.uk
www.pesgb.org.uk
www.time.gov/exhibits.html
www.unl.edu/CMRAcfem/em.htm
www.users.aol.com/murrk/index.htm
www.vec.bgsm.edu
www.worldmall.com/erf/autopsy.htm
www1.cern.ch

Space

www.education.nasa.gov
www.es.rice.edu/ES/humsoc/Galileo
www.hubble.stsci.edu/
www.ksc.nasa.gov
www.nasa.gov
www.today@nasa.gov
www.science.ksc.nasa.gov/shuttle
www.spaceflight.nasa.gov
www.spacescience.nasa.gov
www.usgs.gov

Index

Entries in *italics* indicate images.
Entries in **bold** indicate titles of double-page features.

Acknowledgments

Marshall Editions would like to thank the following:

John Wright, Technical Specialist at 3M UK plc for illustration reference and technical information about heliostats/light pipes; BBC, London, U.K.; Dr. Mike Barnes, Lecturer in Power Electronics, UMIST, Manchester, U.K.; Roger Gollop, Microscopy & Thermal Services, Product Quality & Services, Blue Circle Technical Centre, Blue Circle Industries Plc, for SEM image of cement crystals; BP Educational Service, Dorset, UK; Ping Wong at Canon (UK) Ltd. for illustration reference for inkjet printers; Deep Ocean Engineering, San Leandro, CA., U.S.; Dyson Appliances Ltd. (Dyson Duel Cyclone™); Epson (UK) Ltd.; Equator Corporation, Houston, Texas, U.S.; Dr. Martin Evison, Dept of Forensic Pathology, University of Sheffield, U.K., for supplying text for article on Facial Reconstruction and help with obtaining images; Russell Lewis F1 Graphics, Hampshire U.K.; Norman Foster Associates, London, U.K.; Framestore, London, U.K.; Professor Roger Goodall, Professor of Control Systems Engineering, Loughborough University, U.K;. Honshu-Shikoku Bridge Authority, Japan; Jodrell Bank Observatory, Cheshire, U.K.; NASA; Pii Pipetronix GmbH, Germany; Shell Education Service, Berkshire, U.K.; SNCF, Paris, France; Sony U.K;. TAGMcLaren Marketing Services, U.K.; Rex Beckett, European Technical Manager, Digital Imaging at Texas Instruments Inc. (Digital Light Processing Cinema™ and Digital Micromirror Device ™ are trademarks of Texas Instruments Inc.).

Indexer: David Tyler
Editorial assistant: Ben Horslen
Additional editorial assistance: Felicia Bromfield, Antonia Cunningham, Cathy Meus, Philip Morgan, Connie Novis, Andy Oppenheimer, Susan Watt
Additional text: John Barrett, John Farndon, Mike Flynn, Brian Johnson, Dr. Claire Seymour, Colin Uttley, Philip Wilkinson
Special thanks to Marion Dent for her commitment and good humour, and to Bob Mecoy and Diane McGarvey for their speed and patience.

Picture credits

Page 5 l. Getty One Stone/Deborah Davis, r. Getty One Stone/Peter Poulides; **6** l. Image Bank/Mark Loiseau, c. John Downer/www.osf.uk.com, r. Getty One Stone/ Donavan Reese; **7** l. SPL/Andrew Syred, c. SPL/NASA, r. SPL/ David Parker; **10** Getty One Stone/Deborah Davis; **12** Corbis/Kit Kittle; **14** Courtesy Pipetronix; **17** Corbis/ART on FILE; **19** Jubilee Line Press Office; **21** QA Photos; **22** Honshu-Shikoku Bridge Authority; **22** The Stock Market; **24** Blue Circle Technical Centre/Roger Gollop; **25** t. SPL/John Mead, b. Arcaid/Ezra Stoller/ESTO; **27** SPL/Sittler Jerrican; **28** SPL/Alex Bartel; **29** Courtesy Heliobus Systems; **30** Corbis/Ted Spiegel; **39** SPL/Eye of Science; **42** Getty One Stone/Peter Poulides; **46** Global Marine Systems; **51** SPL/US Geological Survey; **52** t. SPL/Los Alamos National Laboratory, b. Science & Society Picture Library; **53** t. SPL, b. Kaizo; **57** m. SPL/John Mead, b. Corbis/Henry Diltz; **58** SPL; **59** Courtesy Sandia National Laboratories Intelligent Micromachine Initiative; **60** t. Courtesy Intel, b. Apple Bite Communications; **62** Andrew Sydenham; **65** The Stock Market; **68** Alva B.V. Thea van der Heuvel; **74** t. Sony, b. SPL/James King-Holmes; **75** tl. SPL/Hank Morgan, tc. SPL/Hank Morgan, b. SPL/Philippe Plailly/Eurelios; **78** t. SPL/Philippe Plailly, b. SPL/Andrew Syred; **80** Courtesy Glasplatz; **91** Ilford; **98** BBC Worldwide; C**99** t. Crawley Creatures; **101** SPL/Francoise Sauze; **102** NHPA/Gerard Lacz; **106** Image Bank/Mark Loiseau; **114** c. Courtesy Mercedes/Daimler/Chrysler, bl. Australian Picture Library/Peter Menzel, br.SPL/Martin Bond; **115** SPL/G.Brad Lewis; **119** SPL/Martin Bond; **121** Corbis/Bill Ross; **122** The Stock Market; **126** Courtesy Boeing Commercial Airplane Group; **129** aviationpictures.com; **131** aviationpictures.com; **134** aviationpictures.com; **136** SPL; **140** t. Getty One Stone/Kim Westerskov, b.Getty One Stone/Jim Corwin; **141** Courtesy Stena Line; **142** Corbis/Amos Nachoum; **143** Corbis/Ralph White; **144** Corbis/Sergio Carmona; **145** t. Getty One Stone / Robert Aschenbrenner, b. Getty One Stone/James Andrew Bareham; **146** John Downer/www.osf.uk.com; **152** The Bank of England; **153** t. The Department of the Treasury Bureau of Engraving & Printing, bl. The Bank of England; **154** Mirror Syndication; **154** Science & Society Picture Library; **158** Courtesy Recognition Systems, copyright 2000 Ingersoll-Rand Company; **159** t. SPL/David Parker, c+b. IriScan; **162** Dr Martin Evison; **163** Dr Martin Evison, cr.SPL/Alfred Pasieka; **165** SPL/Geoff Williams; **166** tr. The Stock Market, cl. Science & Society Picture Library, cb Science & Society Picture Library; **167** t. The Stock Market, b. Science & Society Picture Library; **168** Getty One Stone/Donavan Reese; **170** The Stock Market; **173** SPL/Arthus Bertrand; **174** t. SPL/Hank Morgan, b. SPL/Sheila Terry; **175** SPL/Tony Craddock; **176** t. SPL/Simon Fraser; **181** tl. The Stock Market, tr. SPL/Hattie Young; **183** Getty One Stone / Mark A Leman; **184** t. Corbis/Bettmann; b.SPL; **185** t.SPL/James Holmes/ZEDCOR, b. Andrew Sydenham; **186** Corbis/Kevin Fleming; **188** t. SPL/Jim Gipe, b. Getty One Stone/Peter Dean; **189** t. Getty One Stone/Andy Sacks, b. SPL/Hank Morgan; **190** SPL/Dr Jeremy Burgess; **192** SPL/Andrew Syred; **195** SPL/Bernard Benoit; **196** SPL; **197** SPL/Clinique Ste. Catherine/CNRI; **198** SPL/Alfred Pasieka; **199** SPL/Dr John Mazziotta et al/Neurology; **206** t. SPL/NASA, b. SPL/David Parker; **207** SPL/Chris Knapton; **208** Courtesy Otto Bock; **209**SPL/Volker Steger; **211** Sony UK; **217** SPL/Sheila Terry; **218** SPL/ Precision Visuals/Teleco Drill Tech. Inc; **220** SPL/John Greim; **222** SPL/Philippe Plailly; **225** SPL/Alexander Tsiaras; **226** SPL/A.B. Dowsett; **227** t. SPL/Andrew Syred, b.SPL/Philippe Plailly; **228** NASA; **229** NASA; **230** t. SPL/NRAO, b. Corbis/Jim Sugar Photography; **232** SPL/CERN; **234** SPL/NASA; **236** l. Hulton Getty, r. SPL/NASA; **237** t. SPL/NASA, b. SPL/Marshall Space Flight Center/NASA; **238** aviationpictures.com; **240** SPL/NASA; 241 SPL; **243** SPL/NASA; **246** t. SPL/European Space Agency, b. SPL/NASA; **247** NASA, 249 NASA; **251** NASA; **253** NASA; **255** r. Digital Vision, b. Anglo-Australian Observatory; **256** SPL/David Parker; **Endpapers** SPL/CERN